Fundamentals of Economics for Engineering Technologists and Engineers

S. KANT VAJPAYEE

University of Southern Mississippi

Upper Saddle River, New Jersey
Columbus, Ohio

Library of Congress Cataloging-in-Publication Data

Vajpayee, S. Kant.
Fundamentals of economics for engineering technologists and engineers / S. Kant Vajpayee.
p. cm.
Includes bibliographical references and index.
ISBN 0–13–674383–8
1. Engineering–Economic aspects. I. Title.

TA177.4.V35 2001
658.15–dc21

00-063701

Vice President and Publisher: Dave Garza
Editor in Chief: Stephen Helba
Executive Editor: Debbie Yarnell
Production Editor: Louise N. Sette
Project Supervision: York Production Services
Design Coordinator: Robin G. Chukes
Cover Designer: Becky Kulka
Cover Art: © Becky Kulka
Production Manager: Brian Fox
Marketing Manager: Jimmy Stephens

This book was set in Clearface by York Graphic Services, Inc. It was printed and bound by R. R. Donnelley & Sons Company. The cover was printed by The Lehigh Press, Inc.

10 9 8 7 6 5 4 3 2 1
ISBN: 0-13-674383-8

To my first teachers:

My late father, who taught me to love learning
and my mother, who taught me to love work

PREFACE

I have been teaching a course in engineering economics since the 1980s. In spite of my best efforts to make it enjoyable and interesting, I found many of my students "sweat" in this course. Some of them might have blamed me for my shortcomings as an instructor, but all along I felt that the real problem has been the textbook. I considered other texts too, and found them equally unsatisfactory, primarily due to the overemphasis on engineering science. So I decided to write a text myself. This is it!

Fundamentals of Economics for Engineering Technologists and Engineers has been developed as a one-semester introductory text. It focuses on the basics, with minimal mathematics and theory. It is aimed at majors in engineering technology, engineering, and industrial technology. It should prove useful also to majors in business and in the physical sciences such as physics, chemistry, mathematics, and computers. It should be equally helpful to practicing engineers and technologists. Enough example problems have been provided in each chapter to illustrate the fundamentals. End-of-chapter exercises comprise discussion and multiple-choice questions along with numerical problems.

The main features of the text are as follows:

1. The writing style is novel. I have tried to *talk* to the reader, as if delivering a lecture.
2. Most of the problems relate to engineering projects and are as close to the *real world* as the treatment of the material would allow. To enliven the discussions, a few personal finance problems have also been included.
3. The treatment has been kept simple and straightforward, focusing on the learning of concepts. Material likely to be difficult for the average student has been thoroughly explained.
4. The time constraint of a one-semester, three-credit-hour course (three 50-minute meetings each week over 16 weeks) has been kept in mind in determining the extent of coverage.
5. The book is targeted at readers interested in "how to apply" the economic principles. Theoretical derivations and equations have been kept to the bare minimum. Concepts and first principles have been emphasized instead.

The text is supported by an instructor's manual containing solutions to the chapter exercises. The manual also offers a suggested course outline and hints on how to render engineering economics interesting. It also contains my email address and telephone and fax numbers for instructors adopting the book. Feedback from them and other readers, especially students, is welcome for improving future editions.

I am thankful to the University of Southern Mississippi for the sabbatical leave during which most of the manuscript was developed. I would like to thank the reviewers of this text: M. M. Bakr, University of Arkansas at Little Rock; Dennis E. Kroll, Bradley University; Kenneth G. Merkel, University of Nebraska–Lincoln; Charles E. Mobley, University of North Carolina at Charlotte; and Philip W. Patton, Montana Tech of the University of Minnesota. I also appreciate the support of Prentice Hall staff, especially Steve Helba.

Finally, I wish to thank my family and friends who suffered with, or perhaps without, me during the writing of the manuscript. I feel guilty for not being in the nascent worlds of play and fun of my grandchildren Sapna, Shyam, Kunal, and Ram as much as we all would have liked.

S. Kant Vajpayee

Kant.Vajpayee@usm.edu

BRIEF CONTENTS

CONTENTS

PART III REALISM 293

SYMBOLS[1]/ABBREVIATIONS

ΔBCR	Incremental benefit–cost ratio
ΔROR	Incremental rate of return
/	÷ (division)
μ	Mean
σ	Standard deviation
π	22/7 = 3.1416
A	End-of-period cashflows in a uniform series
A_1	The first cashflow in a geometric series
AB	Annual benefit
ACRS	Accelerated cost recovery system
AGV	Automated guided vehicle
AM	Annual maintenance cost
APR	Annual percentage rate
APY	Annual percentage yield
AW	Annual worth
AWs	Annual worths, plural of AW
BCD	Benefit–cost difference
BCR	Benefit–cost ratio
BS	Bachelor of Science
CAM	Computer-aided machining or manufacturing
CD	Certificate of deposit
CNC	Computer numerical control

[1]Source: The *Manual of Standard Notation for Engineering Economy Parameters and Interest Factors* by the American Society for Engineering Education, Engineering Economy Division, Washington, D.C.

CPA	Certified public accountant
CPI	Consumer price index
DDB	Double-declining balance
EPA	Environmental Protection Agency
EUAB	Equivalent uniform annual benefit
EUAC	Equivalent uniform annual cost
e	Base of natural logarithm = 2.171828
F	Future sum of money, future payment, final payment
f	Inflation rate, function of
FW	Future worth
FWs	Future worths, plural of FW
G	Arithmetic gradient (fixed increase in arithmetic series)
g	Geometric gradient (rate of increase in geometric series)
GIGO	Garbage in, garbage out
I	Total interest for the period
i	Interest rate for the compounding period (r/m)
i_{eff}	Effective annual interest rate
i_f	"Real" interest rate that accounts for inflation
IRA	Individual retirement account
IROR	Internal rate of return
IRS	Internal Revenue Service
ISO	International Organization for Standardization
k	Prefix for kilo, meaning 1,000
LCM	Least common multiple
m	Number of compounding periods per year
MACRS	Modified accelerated cost recovery system
MAPI	Machinery and Allied Products Institute
MARR	Minimum attractive or acceptable rate of return
MBA	Master of Business Administration
N	Useful life
n	Number of compounding periods, useful life
NASA	National Aeronautics and Space Administration
NPW	Net present worth
OSHA	Occupational Safety and Health Administration
P	Present sum of money, initial investment, principal
PC	Personal computer

PW	Present worth
PWs	Present worths, plural of PW
r	Nominal interest rate (per year)
ROR	Rate of return
SI	System International
SL	Straight-line (depreciation method or rate)
SOYD	Sum-of-years-digits
x	Independent variable
$y(x)$	Dependent variable y as a function of x

PART I

Fundamentals

The first part of the text is its *foundation*. In its four chapters we discuss the basic principles essential for comprehending and solving engineering economics problems. This part prepares the reader to follow the rest of the coverage. As mentioned in the preface, mitigating the "fear" and "anxiety" of engineering economics as a course is a major objective of this text. Part I contributes immensely to the attainment of this objective.

Chapter 1 introduces engineering economics. It emphasizes that engineering economics is basically decision making. The concepts of cashflow tables and diagrams are presented in Chapter 2. Chapters 3 and 4 discuss the time-valued equivalency of cashflows, which is the core of engineering economics.

CHAPTER ONE

Introduction

IN THIS CHAPTER YOU WILL LEARN ABOUT

- What engineering economics is
- The decision making process
- Interest rates
- Simple interest
- Compounding of interest
- The time value of money
- A six-step solution procedure

At the dawn of the 21st century, as the world shrinks to what is being called the *global village,* more and more goods and services cross geographic boundaries, with the result that their production, distribution, and consumption are becoming truly international. The globalization and fierce competition of the marketplace demand that economic decisions be both precise and accurate. Recent management trends of re-engineering, downsizing, restructuring, total quality management (TQM), and continuous improvement testify to the fact that the competition for the market of goods and services has really swung into the top gear. Economic factors play a much more crucial role in industry today than at any time in the past.

As a distinct group of professionals, engineers and technologists continually strive to enhance the productivity and quality of the products and services they are involved with. They are often asked to do more with less (resources). It is becoming increasingly important that these professionals be thoroughly skilled in engineering economics so that they can contribute rationally to capital investment and other cost-related decisions. To this end, an undergraduate course in engineering economics is almost mandatory for all majors of engineering and engineering technology.

This text discusses the basic principles of engineering economics and illustrates their applications to industrial projects involving costs and benefits. Though addressed primarily to engineers and engineering technologists, it will also be useful to engineering managers, industrial technologists, business managers, and applied scientists.

1.1 ENGINEERING ECONOMICS

Economics is concerned with decision making relating to design, production, distribution, and consumption of goods and services. *Engineering economics* is a specialty of economics that focuses on engineering projects. It deals with the economic aspects of product, equipment, service, or technical support. Its knowledge is a "must" for engineers and engineering technologists—in fact for all technical and management professionals.

If you have $20,000 to invest and are deciding which company shares to buy, then the decision involves *business economics,* or simply economics. If, on the other hand, you decide to use this sum as capital to make and market a better mousetrap, then the associated sets of decisions fall in the arena of *engineering economics*. Engineering economics presumes some technical knowledge on the part of the decision maker, while (business) economics does not. Since both economics and engineering economics share the same fundamental principles, a person skilled in one *may be able* to venture into the other.

1.2 ECONOMIC DECISION MAKING

We frequently make economic decisions for a variety of problems. But only the well-thought-out, rational decisions lead to successful conclusions. If the problem is simple, we may be able to think through it in our heads. Suppose you need to buy a pencil. The local bookstore sells pencils for 20 cents each, or a pack of five for 60 cents. You immediately ask yourself, Should I buy a pack? You mentally determine that the pack works out cheaper (12 cents each, instead of 20 cents if bought separately) and, let us say, decide to buy a pack. In so deciding you probably also considered the following:

1. Should I spend 60 cents in buying the pack, rather than 20 cents for one?
2. Can I spare the 60 cents now? Do I have another immediate need for this sum—for example, a can of coke?
3. What shall I do with the other four pencils if I buy the pack? Shall I be able to use them later? Oh yes, my daughter may need them, since her school begins next week.

Decision-making problems as simple as this are obviously too trivial for a textbook. At the other extreme, we sometimes face enormously complex problems. Besides the economic considerations, such problems also involve noneconomic factors that may be political, social, or ethical in nature. Most of these factors are not quantifiable in monetary terms. Consider for example the task of preparing the national budget. Besides its size, some relevant questions are how to finance the expenditures, how much to borrow, whom to tax, and how the budget will affect people living below the poverty line. Such questions raise issues that are much more than economic. In such problems political and social considerations may overtake the economic ones, rendering decision making really difficult. Moreover, of the several alternatives or solutions, none may be satisfactory to the decision maker and/or the people affected by the decision. Such enormously complex problems are also beyond the scope of this text.

In between the *trivial* and the *enormously complex,* is an array of moderately difficult problems that are primarily economic in nature. Such engineering-related problems fall within the realm of engineering economics and are the subject matter of this text. Their solutions require that we

1. State and comprehend the problem,
2. Collect and analyze the associated data,
3. Carry out the calculations, and finally
4. Decide as to which alternative offers the best solution.

Engineering economics is much more than carrying out the calculations for a problem. It involves the above four tasks. That is why it is said that *engineering economics* is basically *decision making*. Consider for example Linda, an engineering supervisor, who has been facing frequent breakdowns of the plant's only stamping machine. At the annual budget time, she has to decide whether to replace the machine this year or to keep it for another year. If she decides to replace it, which of the five makes of machine available in the market should she buy and thus budget for? Such problems fall within the scope of this text and are discussed later in Chapter 14.

Economic decision making is a rational process. In general, it involves the following nine steps.

1. Recognition of the problem
2. Goal(s) of the solution
3. Data collection/gathering
4. Research for feasible alternatives
5. Criterion for selecting the best alternative
6. Mathematical modeling and associated calculations
7. *What-if* analyses with the model
8. Selection of the best alternative
9. Postimplementation follow-up

Some of these steps may be absent in engineering projects, not because they are not required but because they have already been completed by someone. For example, the recognition of the problem might have been done by the plant manager. From a business viewpoint, step 1 may sometimes represent an opportunity, rather than being a problem, to invest capital for realizing profits. The goal in step 2 may be cost savings, increased throughput, higher profit, better quality, and so on.

In this text, we focus primarily on steps 5, 6, and 8 through discussions, examples, and chapter-end exercises. The other six steps are usually learned on the job. The last step helps the engineer to learn from the decision so that better decisions can be made in the future. Step 7 is performed mostly with commercial software. The textual problem statements have been structured to contain sufficient relevant data that in industry are normally generated at steps 1 through 4. In the real world, data may not be readily available; their collection (step 3) consumes significant time and effort.

1.3 INTEREST RATE

Interest is the cost of using someone else's money. In that sense, it may be thought of as rent. The money you borrow from the bank belongs to savers whom the bank pays interest. In turn, the bank charges you interest. What you pay to the bank as interest is more than what the bank pays to the saver. The bank is thus the "middle man," pocketing the difference in interests.

Interest rate is a measure of the interest for a given loan during the loan period. Suppose you borrow \$200 for one year from the local bank at an interest rate of 5% per year. For the year the interest will be 5% of \$200, that is, \$10; your payback will thus be \$200 principal plus \$10 interest—a total of \$210. The term *payback* is used here in a literal sense to mean the total of what is paid back to the bank. In Chapter 5, it is used in its criterion sense as practiced by engineering economists.

The prevailing interest rate is a function of both time and place. The rate may be different today from what it was last month or year. It may be different in the United States from what it is in India or Kenya. It may even differ from bank to bank in the same town. The interest rate also depends on the creditworthiness of the borrower (individual or company), that is, on the risk involved in lending as judged by the lender. The overall economic climate and government policies also play critical roles in the prevailing range of interest rates.

Interest rate is usually denoted by i and expressed in percentage (%). An annual interest rate of 8% is stated as $i = 8\%$ per year. Let us say that you borrow a sum of \$350 for one year. This sum is called the *principal*. The relevant calculations for the year-end payback (principal plus interest) are these:

$$\begin{aligned}
\text{Interest rate per year} &= 8\% \\
&= 8 \text{ percent}^{1} \\
&= 8 \text{ per hundred} \\
&= \$8 \text{ per hundred dollars} \\
\text{Number of hundreds in the \$350 principal} &= 3.5 \\
\text{Therefore, interest for one year} &= \$8 \times 3.5 \\
&= \$28 \\
\text{Year-end total payback} &= \text{principal} + \text{interest} \\
&= \$350 + \$28 \\
&= \$378
\end{aligned}$$

In these calculations, we used the number of hundreds in the loan (\$350) because the interest rate was expressed in percent. If the 8% rate had been expressed in fractional form as $8/100 = 0.08$, that is, \$0.08 for every dollar of the principal, we would have multiplied 0.08 with \$350, getting the same result of \$28 for the interest:

[1]Here, *cent* means "hundred."

$$
\begin{aligned}
\text{Interest rate per year} &= 8 \text{ percent} \\
&= 8 \text{ per hundred} \\
&= 8/100 \text{ per unit} \\
&= 0.08 \text{ per unit} \\
&= \$0.08 \text{ per unit dollar}
\end{aligned}
$$

$$
\text{Number of units in the \$350 principal} = 350 \text{ dollars}
$$

$$
\begin{aligned}
\text{Interest for one year} &= (\$0.08/\text{dollar}) \times 350 \text{ dollars} \\
&= \$28
\end{aligned}
$$

To summarize,

> *The interest rate is usually denoted by* i *and expressed in percent (%) per year. To calculate the amount of interest earned for a given principal, first convert* i *into its fractional value by dividing by 100. Then multiply the fractional value of* i *with the principal to determine the interest for the year.*

1.4 SIMPLE INTEREST

Interest is calculated for each period of the agreed loan duration. The period may be a year, a month, a quarter of a year, or any other time window. The loan duration comprises one or more periods. Interest charged for a loan is called *simple* if the interest earned for a period does not become a part of the principal for the next period. In other words, the interest earned for a period *does not* earn interest in the subsequent period. Instead, it is paid to the lender. If it is not paid out every period, it simply accumulates without earning any interest. At the end of the loan, the principal and the accumulated interest are paid together. Thus, the interest earned during each period leaves the principal unaffected; as a result, the principal does not increase during the loan duration.

If you borrow $300 for 3 years at 7% annual interest rate, then under *simple interest* your total payback at the end of the loan will be $363. Of this, $300 is the principal and $63 the accumulated interest[2] for three years.

In equation form, total simple interest I is expressed as

$$I = Pin \tag{1.1}$$

where P = principal
i = interest rate (in fraction) per period
n = number of periods

---

[2]
$$
\begin{aligned}
\text{Interest for one year} &= \text{Interest rate} \times \text{Units of dollar} \\
&= 0.07 \times \$300 \\
&= \$21
\end{aligned}
$$
$$
\text{Interest for three years} = \$21 \times 3
$$

In the preceding illustrative example, where period was expressed in years,

$$P = \$300$$
$$i = 0.07 \text{ (fractional value of 7\%) per year}$$
$$n = 3 \text{ years}$$

Therefore, from Equation (1.1)

$$I = \$300 \times 0.07 \times 3$$
$$= \$63$$

Equation (1.1) involves four variables: I, P, i, and n. Given the values of any three, the fourth can be evaluated.

Equation (1.1) can be extended to determine the final payback F by adding the total interest I to the principal P. Thus,

$$\begin{aligned} F &= P + I \\ &= P + Pin \\ &= P(1 + in) \end{aligned} \tag{1.2}$$

Equation (1.2) also involves four variables: F, P, i, and n. Given the values of any three, the fourth can be evaluated.

Equations (1.1) and (1.2) are handy and should normally be used to work out problems involving simple interest, but it is often relatively easier to follow the first principles[3] rather than use the equation; simply trace the procedure that was used to derive the equation.

Interest rate is usually quoted for the year (*annual*). Sometimes, it may be quoted for the month or other time periods such as quarter (one fourth of the year, i.e., three months). Whatever the time period, *i and n must be expressed in compatible units*. For example, if $i = 1\%$ per month, then the value of n for a three-year investment is 36 months (not 3 years). For this rate and duration, the total interest for an investment of \$300 will, from Equation (1.1), be

$$\begin{aligned} I &= Pin \\ &= \$300 \times 0.01 \times 36 \\ &= \$108 \end{aligned}$$

---

[3]What has been discussed as an illustration in the second paragraph of Section 1.4 and its footnote is a good example of first principles. By first principles we mean a solution approach based on fundamentals, not on equation(s). For example, to determine the year-end interest, we multiply the principal with the fractional value of the annual interest rate. The alternative approach is to use Equation (1.1) with $n = 1$. You may wonder what the difference between first principles and the equation actually is. Not much. In fact, an equation is derived following the first principles. Thus, by first principles we mean the procedure that yields the equation. Of the two, use the one that renders the solution easier.

Simple interest rates are rare in the financial marketplace. Hardly anyone transacts money at simple interest. Relatives and close friends may at times display their generosity by loaning money at simple interest. In the business world, *simple interest* is nonexistent.

EXAMPLE 1.1

After earning a BS (bachelor of science) degree in construction engineering, John decided to start his own business. His rich, generous uncle has agreed to loan him $80,000 at an annual rate of 4% simple interest. John is to pay his uncle yearly interest on the loan anniversary and return the principal on the fifth anniversary. What is John's loan payment schedule?

Solution

By payment schedule we mean the scheme by which interest and the principal are paid back. In other words, how much will be paid when. In this case, the interest for the year is paid on the loan anniversary, while the principal is paid back on the fifth anniversary along with the interest for the fifth year.

We need to determine the interest for the year, since it is due on each anniversary. Since no portion of the principal is paid until the fifth anniversary, the principal remains $80,000 during the entire loan duration.

We can use Equation (1.1) to determine the annual ($n = 1$) interest. The given data are

$$P = \$80{,}000 \qquad i = 4\% \qquad n = 1$$

Remember to use the fractional value of the interest rate (i.e., $i = 4/100 = 0.04$).

Therefore,

$$\begin{aligned} I\,(\text{interest for the year}) &= Pin \\ &= \$80{,}000 \times 0.04 \times 1 \\ &= \$3{,}200 \end{aligned}$$

Thus, John will pay his uncle $3,200 as interest on each of the first four loan anniversaries. The payment on the fifth anniversary will be

$$\begin{aligned} \text{Principal} + \text{Interest for the fifth year} &= \$80{,}000 + \$3{,}200 \\ &= \$83{,}200 \end{aligned}$$

Thus, John's payment schedule is $3,200 on each of the first four loan anniversaries and $83,200 on the fifth anniversary.

In the preceding example, a table such as the one that follows might have been appropriate to summarize the loan payment schedule. Note that the table columns have been given proper headings.

Loan Anniversary	Payment Due
1	$3,200
2	3,200
3	3,200
4	3,200
5	83,200

Engineers and technologists usually express themselves through charts, graphs, tables, or modern displays such as animation. These modes of expression summarize the data and highlight the important pieces of information. Even when the problem (project) does not specifically ask for a table, graph, or chart, it is desirable to use them to enhance the presentation of results.

1.5 COMPOUNDING

In most cases, if the interest is not withdrawn, or paid out when due, it is allowed to augment the principal at the end of the (interest accounting) period. The total of the principal and the interest for the period becomes the new principal, which earns interest during the following period. Consequently, *the interest earns interest.* This augmenting effect of interest on the principal is called *compounding*.

The dictionary meaning of the word *compounding* is putting together or combining. In here, it means combining together the interest earned for a period with the *beginning* principal for the period. The total of the two yields the *ending* principal for the period, which becomes the beginning principal for the subsequent period. The following discussion illustrates the concept of compounding.

Consider that you borrow $200 for two years at 20% yearly interest with annual compounding. Thus, the beginning principal P is $200, and the value of i is 0.20. Since compounding is annual, the interest period is year. The interest for the first period or year, payable at its end = $200 × 0.20 = $40. This interest is combined with the beginning (original) principal $200. Thus, the principal at the end of first year = $200 + $40 = $240. This $240 now becomes the principal at the beginning of the second period. Therefore, the interest for the second year = $240 × 0.20 = $48. Thus, the final (total) amount F payable at the end of second year is

Principal at the beginning of second year + Interest for the second year
= $240 + $48
= $288

Had the above $200 loan been at simple interest (not compounded), the value[4] of F, payable at the end of the second year, would have been

---

[4] One can use Equation (1.2), from which $F = P(1 + in) = \$200\,(1 + 0.20 \times 2) = \$280$.

$$
\begin{aligned}
&\$200 + \text{Total simple interest earned over two years} \\
&= \$200 + (\$200 \times 0.20) \times 2 \\
&= \$280
\end{aligned}
$$

The final payment of $280 under simple interest is less than when compounding was done ($288). The interest earned by the lender under compounding is $8 more, which you paid as a borrower. Thus, compounding is beneficial to the lender and expensive to the borrower—by exactly the same amount.

All borrowing and lending transactions in today's marketplace use compound interest. That is why all the discussions from now on are based on compounding. *Assume compounding of interest everywhere unless mentioned otherwise.*

EXAMPLE 1.2

Maya owns and operates a business selling used robots. From the profit realized in the current fiscal year, she decides to save a sum of $30,000 for use later to expand her business. The sum is invested in a three-year certificate of deposit (CD) with the local credit union at an annually compounded interest rate of 6% per year. How much will this CD mature to?

Solution

The CD investment is for a term of three years at a yearly interest rate of 6%, compounded annually. We solve this problem following the first principles.

$$
\begin{aligned}
\text{Beginning capital (principal)} &= \$30{,}000 \\
\text{Interest for the first year} &= \$30{,}000 \times 0.06 \\
&= \$1{,}800 \\
\text{Principal at the end of first year} &= \text{Beginning principal} + \text{Interest for the year} \\
&= \$30{,}000 + \$1{,}800 \\
&= \$31{,}800
\end{aligned}
$$

This becomes the principal at the beginning of the second year.

$$
\begin{aligned}
\text{Interest for the second year} &= \$31{,}800 \times 0.06 \\
&= \$1{,}908 \\
\text{Principal at the end of second year} &= \$31{,}800 + \$1{,}908 = \$33{,}708
\end{aligned}
$$

This becomes the principal at the beginning of the third year.

$$
\begin{aligned}
\text{Interest for the third year} &= \$33{,}708 \times 0.06 \\
&= \$2{,}022.48
\end{aligned}
$$

$$\begin{aligned}\text{Total at the end of third year} &= \$33{,}708 + \$2{,}022.48\\ &= \$35{,}730.48\end{aligned}$$

Thus, Maya's $30,000 CD will mature to $35,730.48.

Example 1.2 has illustrated how problems involving compounding of interest can be worked out following the first principles approach. However, this approach becomes lengthy and cumbersome for compounding involving several periods. In such cases, equations are easier to use. We derive such equations in Chapter 3 and illustrate their use there and beyond.

EXAMPLE 1.3

In Example 1.2, how much more did Maya's CD earn in comparison to that under simple interest?

Solution
Total interest earned in Example 1.2 is

$$\begin{aligned}F - P &= \$35{,}730.48 - \$30{,}000\\ &= \$5{,}730.48\end{aligned}$$

Had Maya's CD earned simple interest, the total interest I at the end of three years, from Equation (1.1), would have been

$$\begin{aligned}I &= Pin\\ &= 30{,}000 \times 0.06 \times 3\\ &= \$5{,}400\end{aligned}$$

Thus, compounding yielded[5] an additional earning of

$$\$5{,}730.48 - \$5{,}400 = \$330.48$$

1.6 TIME VALUE OF MONEY

As seen in the preceding three sections, a sum of money has the potential to "grow" due to the interest it can earn. This fact is stated by saying that *money has time value*. The current sum is the principal (or *present* sum) P, which increases due to interest

---

[5]Alternatively, you can work out the difference between the final sum F under compounding and that under simple interest. Since F with compounding is already known to be $35,730.48, use Equation (1.2) to determine F under simple interest. Thus,

$$\begin{aligned}\text{Additional earning} &= \$35{,}730.48 - P(1 + in)\\ &= \$35{,}730.48 - \$30{,}000(1 + 0.06 \times 3)\\ &= \$35{,}730.48 - \$35{,}400\\ &= \$330.48\end{aligned}$$

to a final (or future) sum F. The increase in P is $F - P$, which equals the total interest I. The value of I depends on the principal P, period n, and interest rate i, and whether or not the periodic interest is compounded. In Chapter 3, we learn more about the time value of money and its effect on cashflows.

1.7 THE SIX-STEP PROCEDURE

Engineering economics problems are primarily numerical in nature—the values of certain parameters are given and the value of the unknown is determined. Over the years, I have noticed three common weaknesses in students enrolled in engineering economics course:

1. Difficulty in comprehending the problem statement,
2. Confusion with the mathematics[6] involved, and
3. Hesitancy on how to solve the problem (i.e., the sequence of solution steps).

Most numerical problems in an engineering economics course are stated at a language level expected of college students. The language level of the problem statement in Example 1.2 is typical. Occasionally, students find a problem statement incomprehensible. Other likely difficulties are not knowing how to proceed, which data to use, or which equation(s) to use; worry[7] about not using all the given data; inability to manipulate the equation; and so on.

Here is a six-step procedure[8] that should alleviate some of the difficulties you may face in solving engineering economics problems.

Step 1: ***Comprehend the Problem.***
Read and reread the problem statement until you fully understand what is being asked. If it is long, mentally divide it into logical, comprehensible modules.

Step 2: ***Summarize the Given Data.***
Write down the given data against their customary symbols in the format: *Symbol* = *Data*. For example, $P = \$300$, $n = 5$ years. Problem statements may at times contain more data than you actually need to solve the problem. In real-world project statements, the desired data may be hidden in or entangled with the nonessential data, so be careful while gathering and summarizing them.

Step 3: ***What Is the Unknown?.***
Identify the unknown, and write down its symbol, if any. This is the parameter needing evaluation for the given set of data.

Step 4: ***Search for the Process.***
Ponder on how to proceed to solve the problem. Am I better off following the first principles? Is there an equation that relates the unknown to the knowns (the given data)? Look for a form

$$\text{Unknown} = f(\text{knowns})$$

where f stands for "function of."

---

[6]The mathematics involved in engineering economics is mostly algebra.

[7]Problem statements may contain redundant data.

[8]The procedure can in fact prove useful in solving any numerical problem.

The unknown–knowns relationship may not be obvious. Search for indirect relationships, if necessary. Logically trace the relationship(s) among the given parameters to discover *how to reach the "unknown" from the "knowns."* If needed, rearrange the parameters of the equation to get all the knowns on the right side of the equation.

Step 5: Solve for the Unknown.
Manipulate the knowns with the tools available, such as scientific calculators, compound interest tables, or software.

Step 6: Confirm the Answer, if Possible.
This final step is essential to achieve confidence in the result. Use your "gut feeling," subjective judgment, mental arithmetic, and/or scribbles to check whether the answer seems right. For example, interest rate i cannot be negative, period n must be positive, and F cannot be less than P. If the answer looks suspicious, check the given data, and go through the solution once more.

You can remember the above six steps through the acronym CD-UP-SC, whose letters stand for

C	Comprehend	The problem statement
D	Data	Gather the given data
U	Unknown	What needs to be evaluated
P	Process	Equation or first principles
S	Solve	Solve to get the answer
C	Confirm	Check the answer

Example 1.4, a modified version of Example 1.1, illustrates how to apply the six-step procedure.

EXAMPLE 1.4

After finishing a BS in computer engineering technology, Joshua has set up a computer repair business. He needs to buy a diagnostic system that will expedite repair. The system costs $9,500 and is likely to generate $3,750 per year. Since his current loans are excessive, no financial institution is interested in lending him any more money. He discussed the difficulty of raising the capital with his rich uncle who happens to be generous. His uncle agrees to lend him $9,500 at 4% simple interest provided Joshua pays him the interest annually on time. How much will Joshua earn annually by investing in the system after paying the interest to his uncle?

Solution

Let us go through each of the six steps.

Step 1: Comprehend
Note that the loan earns simple interest. It is not clear when Joshua is to pay back his uncle the $9,500 loan; the problem statement is vague about it. But do we really need this

information to solve the problem? Not really. Note that the interest for the year is paid annually on the loan anniversary. This payment must be subtracted from the annual income of $3,750.

Step 2: *Data*
The given data are

$$P = \$9{,}500 \qquad i = 4\% \qquad n = 1$$

Remember to use the fractional value of the interest rate, that is, $i = 4/100 = 0.04$.

Step 3: *Unknown*
The unknown here is the annual earning after paying for the interest.

Step 4: *Process*
Since the annual income from the use of the system is given, we need to determine the annual interest to evaluate the unknown. Shall we follow the first principles? Or is there an equation that relates the earned interest I with P, i, and n? Yes indeed, there is one—Equation (1.1), discussed earlier. Which of the two—first principles or Equation (1.1)—yields an easier solution? In here, it does not matter, both are equally effective. So let us use Equation (1.1).

Step 5: *Solve*
The interest for the year can be found from Equation (1.1), with $n = 1$, as

$$\begin{aligned} I &= Pin \\ &= \$9{,}500 \times 0.04 \times 1 \\ &= \$380 \end{aligned}$$

Thus, Joshua will pay his uncle $380 each year on the loan anniversary. Since the diagnostic system generates $3,750 per year, his net annual earning will be

$$\$3{,}750 - \$380 = \$3{,}370$$

Step 6: *Confirm*
Let us check if the answer seems correct. As it is a simple problem, we can do it in our "heads." The principal is $9,500. If the interest rate were 1%, the interest for the year would be $95. This $95 can be rounded to $100 to aid the mental arithmetic. So for an interest rate of 4% (four times as high), the approximate interest for the year should be four times $100, that is, $400. Thus, the value of I at step 5 as $380 seems alright[9], since it is close to the mentally-worked-out approximate value of $400.

The usefulness of the six-step procedure might not have been quite obvious in this illustration. Try applying it to other problems in the text. Frequent use of the procedure will hone your skills and reinforce its usefulness.

---

[9]We could have approximated the principal itself to $10,000 to aid mental arithmetic. At a 4% rate, this would have given $400 as annual interest.

Do we have to follow the six-step procedure in a very formal way, as was illustrated? Not really. But keeping the procedure in mind while solving problems is likely to be helpful.

SUMMARY

Engineering economics plays an increasingly critical role in industry as competition for the production and distribution of goods and services intensifies under the impact of globalization. It is basically a decision-making process, comprising nine steps. In engineering economics one learns to solve moderately difficult engineering problems involving costs and benefits. Interest can be thought of as the rent for using someone's money; interest rate is a measure of the cost of this use. Interest rates are of two types: simple or compounded. Under simple interest only the original principal earns interest. Simple interest is nonexistent in today's financial marketplace. Under compounding, the interest earned during a period augments the principal; thus interest earns interest. Compounding of interest is beneficial to the lender. Due to its capacity to earn interest, money has time value. The time value of money is important in making decisions pertaining to engineering projects. A six-step procedure that is helpful in solving engineering economics problems has been presented in Chapter 1.

EXERCISES

Discussion Questions

1.1 Why is the study of economics by engineers and technologists more important today than in the past?

1.2 State in not more than one hundred words an engineering economics problem of your choice.

1.3 Your local government is considering building a new road. Discuss whether or not this is an engineering economics problem.

1.4 Is the six-step procedure discussed in Section 1.7 for solving engineering economics problems helpful? Justify your answer.

1.5 Why do financial institutions never offer loans at simple interest?

Multiple-Choice Questions (Encircle the *best* answer.)

1.6 The decision to go on a two-week vacation represents

- a. an engineering problem.
- b. an economics problem.
- c. an engineering economics problem.
- d. none of the above

1.7 Nasim is overjoyed on winning $500,000 in a lottery. Of this windfall, he decides to spend $20,000 on his marriage next month, buy a motel for $300,000 as an investment, and use the remainder to start a business of making plastic toothpicks. This decision falls under
   a. business economics.
   b. engineering economics.
   c. home economics.
   d. none of the above

1.8 Which of the following is more like an engineering economics problem?
   a. Earning a BS degree
   b. Buying a car for business as well as personal use
   c. Getting married
   d. Preparation of the U.S. federal budget

1.9 Problems most suitable for engineering economics analyses
   1. are sufficiently important.
   2. can't be worked out in our heads.
   3. focus primarily on economic factors.
   4. cost a lot to solve.
   a. 1 and 3
   b. 2 and 4
   c. 1, 2, and 3
   d. 1, 2, 3, and 4

1.10 Simple interest means that the interest
   a. rate is quoted in round figures, not in decimals.
   b. must be paid at the end of each interest period.
   c. does not augment the principal.
   d. is paid whenever convenient to the borrower.

1.11 Time value of money means that
   a. time is money.
   b. time not used sensibly translates into lost money.
   c. the principal is capable of earning interest.
   d. it is hard to enjoy "the good life" without money.

1.12 Engineering economics decision making involves nine steps. Which of the following is *not* one of them?
   a. Recognition of the problem
   b. Determination of the feasible alternatives
   c. Financing of the project
   d. Selection of the best alternative

Numerical Problems

1.13 A young engineer has just started her small construction company. Due to paucity of funds, she is not able to buy a truck the company desperately needs. Her grandpa has an old truck and is willing to help her. The truck's market value is

$3,500, but he sells it to her on credit for $3,000. She agrees to pay him each year 3% simple annual interest for the next five years. On the fifth anniversary of the loan she will also pay him $3,000. Prepare the payment schedule.

1.14 To help his young engineer son, Kashi lends him $50,000 for seven years at simple interest of 3.5% per year. How much will Kashi receive from his son when the loan will be due? Assume that the interest is paid together with the principal as a lump sum.

1.15 Your father deposits $10,000 in a savings account and earns simple interest. If he is paid $205 quarterly as interest, what is the annual interest rate?

1.16 How long will it take $3,000 invested at simple annual interest of 5% to become $5,000?

1.17 How many years will it take a sum to double if it earns simple interest at 9% per year?

1.18 Determine the annual rate of simple interest if $360 is earned as interest at the end of 15 months for an investment of $3,500.

1.19 Jose borrows $40,000 to buy a new machine at an annually compounded interest rate of 9% per year. The loan is to be paid in full on its fourth anniversary. How much will the payment be?

1.20 Mumtaz invested $2,000 in a five-year CD at an annual interest rate of 5%, compounded yearly. Due to an unforeseen situation, she had to withdraw all the money on the third anniversary of the investment. How much did she get if there was a $25 penalty for early withdrawal?

1.21 A small tool and die company takes a loan at a 12% annual interest rate to buy an injection molding machine. The company pays back the loan (capital and interest) on the first loan anniversary in the sum of $62,350. If compounding is annual, how much was the loan?

1.22 Engineering Unlimited takes a loan at a 16% yearly interest rate, compounded quarterly, for purchasing a robotic system. The company pays back the loan (capital and interest) through two payments: $50,000 at the end of six months, and $65,000 on the first loan anniversary. How much was the loan?

1.23 You need to borrow $1,000 to complete your education, intending to pay it back along with the interest next year. Credit union A charges 14.5% interest compounded annually, while B charges 14% compounded quarterly. Which credit union should you borrow from?

CHAPTER TWO

Cashflows

IN THIS CHAPTER YOU WILL LEARN ABOUT

- The concept of cashflows
- Cashflow tables
- Sign convention in a cashflow table
- Auxiliary tables
- How to prepare a cashflow table
- The end-of-period assumption
- Cashflow diagrams
- How to sketch a cashflow diagram

Engineering projects involve transactions of money. Money spent on the project is called *cost, disbursement, expenditure,* or *cash outflow,* and the money earned (or saved[1]) from the project is antonymously called *benefit, receipt, revenue,* or *cash inflow*. These transactions or cashflows occur at different times during the project life. The terms *cash outflow* and *cash inflow* denote the directions in which cash "flows" with reference to the project account. These are simply accountants' terms for *cost* and *benefit* respectively. Engineers and technologists are more conversant with the terms *cost* and *benefit* than with *cash outflow* and *cash inflow*.

An engineering project with its *estimated* costs and *expected* benefits is an engineering economics problem whose solution requires a clear understanding of the associated cashflows. The problem is easier to comprehend and solve if the cashflows and their timings are properly tracked. This is done by

a. Tabulating the cashflows and their timings, and/or
b. Diagraming the cashflows.

The result of (a) is a *cashflow table,* whereas that of (b) is a *cashflow diagram*. In this chapter, we learn how to summarize the cashflows of a project in a table or diagram.

[1]If a project saves money through higher productivity, better product quality, or other ways, the saving is in fact an earning.

Cashflow tables and diagrams are tools. They help us visualize and understand the problem better. They facilitate comprehension of the problem. They may not be essential for solving it. Experienced engineering economists can mentally visualize the cashflows and their timings, and proceed directly to solve the problem without needing the table or the diagram. But even for them mental comprehension becomes unmanageable for complex problems, necessitating the use of tables and/or diagrams. For beginners, the use of tables and diagrams is highly recommended. It is essential that sufficient skills be mastered in developing cashflow tables and diagrams.

2.1 CASHFLOW TABLES

A cashflow table is basically a two-column table. Its first column shows the timing of the cashflows, while the second column shows the amounts of cashflow. The column headings of a typical cashflow table are *year* and *amount* (or *cashflow*).

Suppose you borrow $2,000 today and pay back this loan over the next three years as annual payments of $800 each on the loan anniversaries. The cashflow table will be

Year	Cashflow
0	$2,000
1	−800
2	−800
3	−800

Let us discuss the entries in this table. In the first column, year 0 represents "now" or "today," when the project begins, which in this case is your borrowing (receiving) $2,000 from the lending institution. Year 1 represents the time period from now to this time next year, that is, the end of the first year (loan anniversary). Similarly, the entries of 2 and 3 in the first column mark the ends of year 2 and year 3.

Let us now look at the second column. The first entry of $2,000 is your loan. The entry of 800 for year 1 is the payment on the first loan anniversary. Note that no cash flows *during* the year, but only *on* the loan anniversary when you pay back $800. The same thing happens on the second and third anniversaries. Note that the currency symbol $ has been prefixed only to the first cashflow. The other cashflows are assumed to be in the same currency. This avoids currency symbols swamping the tabulated data.

An important point to note in the second column is the convention of algebraic sign for the cashflows. Each data in a cashflow table must be preceded by either a negative (−) sign or no sign (understood to be positive). Amounts received (inflows) are given no signs and those paid out (outflows) are given − signs. The best way to avoid any confusion with the sign convention is to remember that money received (inflows) increases the balance in the receiver's account, that is, has a positive effect on the account. The reverse is true when transactions are paybacks or costs (outflows).

The cashflow of $2,000 in the table on page 20 actually has a positive sign, but, since positive numbers are not signed, we did not print a + sign in front of $2,000. The signs in front of the other three cashflows (800) are negative. This follows the convention that a negative number must be preceded by a − sign. Each payment of $800 to the lender reduced your account balance by that much, that is, had a negative effect on your account. That is why these paybacks have been preceded by a − sign.

2.1.1 Whose Cashflow Table?

An important consideration while preparing a cashflow table is to ask the question, Whose cashflow table is it? The preceding table was yours (the borrower's). How will the cashflow table of the lender look? From the lender's viewpoint, say the bank's, the cashflow table will be

Year	Amount
0	−$2,000
1	800
2	800
3	800

Note the change in cashflow signs as compared to the previous table. The logic of the sign convention, however, remains the same and can explain the signs in this table too. With reference to the bank's account, the $2,000 loaned to you is an outflow that reduced the bank's balance. That is why it is prefixed with a − sign. Your loan payments of $800 increase (positive effect) the bank's account balance and are therefore labeled positive (no sign).

In general, cashflow signs in the borrower's (one party) cashflow table are just the opposite of those in the lender's (the other party).

To summarize:

Cash inflows/receipts/benefits/revenues	No sign
Cash outflows/disbursements/costs/expenditures	Negative sign

2.1.2 Development of a Cashflow Table

When the cashflows of an engineering project (economics problem) are complex, it may be desirable to develop and use auxiliary tables as an aid to comprehension. Such tables are later summarized as a (final two-column) cashflow table. The important question while developing auxiliary tables is, How many columns to have and for what purpose. The obvious answer is, As many as necessary to develop the cashflow table with minimum confusion. The decision on number and size of the auxiliary tables is one of the creative steps in the solution process. Example 2.1 illustrates the discussions.

EXAMPLE 2.1

Jack, an engineering freshman at college, borrows $5,000 from the local bank at 8% annual interest to buy a used car. He will use his scholarship money to pay back the loan during his four years at college. He plans to pay $1,000 of the principal on each of the first three loan anniversaries along with the interest for the year. On the fourth anniversary (his graduation year), he plans to pay back the remainder of the loan along with the interest. Develop his cashflow table.

Solution
To help us prepare Jack's cashflow table, we will use an auxiliary table containing four columns. Why four columns? The transactions have two components: one principal and the other interest. We need two columns for these, and another (third) column to enter their total. There has to be a column for the time period. Thus a four-columned auxiliary table will suffice.

To keep the auxiliary table development easier, we will exclude year 0 for now and add it later. The skeleton of the auxiliary table will look like this:

Year	Principal	Interest	Total
1			
2			
3			
4			

The first column is always for the period, year in this case. In the second and third columns we will enter the principal paid and the interest for the year, and in the last column the total paid. Note that each column has an appropriate heading.

Let us now do the calculations and post the data in the table as we proceed along.

Year 1: We enter $1,000 in column 2 (principal) as part payment of the principal for the first year. Interest for the first year = $5,000 × 0.08 = $400. Thus, $400 is posted in the interest column. The total of these two ($1,400) is entered in the last column. With these entries the table looks like this:

Year	Principal	Interest	Total
1	$1,000	$400	$1,400
2			
3			
4			

Year 2: Again, we enter $1,000 in column 2. The principal for which interest is earned during the second year = $5,000 − $1,000 = $4,000 (since $1,000 of the principal was paid on the first loan anniversary). Therefore, the interest for the second year

= $4,000 × 0.08 = $320. Thus, $320 is posted in the interest column. The total of these two ($1,320) is entered in the last column.

In a similar way, the data for the other two years can be calculated and posted. With all the data posted in, the auxiliary (or payback) table becomes

Year	Principal	Interest	Total
1	$1,000	$400	$1,400
2	1,000	320	1,320
3	1,000	240	1,240
4	2,000	160	2,160

Note the second-column entry of 2,000 for year 4. This is the final payment of the principal ($5,000 − $3,000), since only $3,000 had been paid by year 3.

This auxiliary table showing the payback schedule can now be summarized to yield Jack's cashflow table. In general, only that column of the auxiliary table which shows the total is retained in the (final) cashflow table.

To complete the table, a new row showing the loan against year 0 is inserted now. With appropriate cashflow signs, the resulting cashflow table is

Year	Cashflow
0	$5,000
1	−1,400
2	−1,320
3	−1,240
4	−2,160

Some engineering economists prefer to keep the calculations intact by retaining all the columns of the auxiliary table in the final cashflow table. Many prefer to see only two columns in a cashflow table and therefore exclude the other columns of the auxiliary table(s) and the calculations therein, as was done here. If the auxiliary table(s) and the associated data are likely to impede the visualization of net cashflows, it is preferred to have only two columns in the (*final*) cashflow table. A two-column cashflow table offers definite advantages when several auxiliary tables are involved, as in complex problems.

There is no standard way of preparing an auxiliary table. Engineering economists use different columns in the auxiliary table(s) for the same problem, but *the resulting two-column cashflow table for a problem must come out to be the same.* Consider, for example, the following auxiliary table developed by another economist, based on his own creative thinking, for solving the same problem (Example 2.1). Note the different headings, one more column, and the inclusion of year 0 in the beginning, as compared to the earlier solution.

Year	Paybacks			Principal Remaining
	P	I	$P + I$	
0				$5,000
1	$1,000	$400	$1,400	4,000
2	1,000	320	1,320	3,000
3	1,000	240	1,240	2,000
4	2,000	160	2,160	0

The purpose of the data in the last column of this table is to enable the tracking of remaining principal so that interest for the following year can be calculated easily. When summarized carefully as a two-column cashflow table, this auxiliary table also yields the same result.

From the discussions here and in Example 2.1, we can see that

1. One or more auxiliary tables may be used to facilitate calculations and minimize confusion.
2. Once the cashflow data have been posted in the auxiliary table(s), they should be summarized as a cashflow table showing the timing and amount of each transaction.

Sometimes the total of the project cashflows is also shown in the table as *net cashflow,* as done here for Example 2.1.

Year		Amount
0		$5,000
1		−1,400
2		−1,320
3		−1,240
4		−2,160
	Net	−1,120

The net flow of −$1,120 in Jack's cashflow table represents the total interest Jack *paid* (since the sign is negative) to the bank for the loan. In the lender's cashflow table, the net flow will be $1,120 (+ sign, opposite of that in Jack's table), representing the interest *earned* by the bank.

The development of the cashflow table in Example 2.1 has been simple. Example 2.2 considers an engineering project whose cashflow table is relatively complex to develop.

EXAMPLE 2.2

The international environmental management standard ISO 14000 requires that Jenny, the plant supervisor, must install suitable pollution control equipment. She is considering two of them, one based on the principle of neutralization and the other on precipitation. They cost $750,000 and $450,000 respectively; both are expected to last five years and have salvage values of $150,000 and $105,000 respectively. The annual maintenance costs are $500 and $750 respectively. The chemical needed for the neutralization-based equipment costs $50,000 per year, the cost for the other is $95,000 per year. Develop a cashflow table to help Jenny visually compare the data pertaining to the two alternatives.

Solution

Since several costs are involved, the use of one or more auxiliary tables is desirable in this case. The question is, How many, and what columns they should have. Let us begin with the first auxiliary table. Others will be prepared as and when necessary. As discussed earlier, the auxiliary tables will be summarized into a final two-column cashflow table.

The first column of the auxiliary table will of course be for the period, year in this case. Since two types of equipment are under consideration, there should at least be two additional columns, one for each. Let us code neutralization equipment as A and precipitation equipment as B. This will keep the table headings short. The skeleton of the table then becomes

Year	A	B
1		
2		
3		
4		
5		

The initial costs of A and B may be kept excluded from the auxiliary table; they will be added to the cashflow table later against year 0, as was done in Example 2.1. Some economists prefer to include these costs in the beginning itself by inserting a row for year 0. In that case, the table gets modified to

Year	A	B
0	−$750,000	−$450,000
1		
2		
3		
4		
5		

The signs for these cost entries and the currency sign $ should be obvious based on the discussions in Section 2.1 and Example 2.1.

If we decide not to include the row for year 0 at this point of the development, our auxiliary table remains unchanged as

Year	A	B
1		
2		
3		
4		
5		

Now we need to decide how many columns there should be under A and B. For both A and B, the cashflow for each year comprises the following two elements:

1. Cost of chemical, and
2. Maintenance cost

Further, there is a third element at the end of the useful life (fifth year), namely salvage value as cash inflow (benefit).

Thus, we should have at least three columns, one for each element. There should be another column for the total of these elements. The four columns may be given proper headings, for example, *Chem.* for element 1 (chemical), *Main.* for element 2 (maintenance), *Sal.* for element 3 (salvage), and *Total* for the last column. The auxiliary table skeleton thus looks like this:

Year	A				B			
	Chem.	Main.	Sal.	Total	Chem.	Main.	Sal.	Total
1								
2								
3								
4								
5								

Note that the table had to be widened to accommodate all the columns. The width need of the auxiliary table was not known in the beginning. Thus, as you proceed, you may have to keep adjusting the table width.

We are now ready to post the data in the appropriate columns. Be careful while doing it; any mistake may lead to an erroneous cashflow table, and therefore to a wrong solution. With the data posted in, the auxiliary table looks like this:

Year	A				B			
	Chem.	Main.	Sal.	Total	Chem.	Main.	Sal.	Total
1	−50,000	−500			−95,000	−750		
2	−50,000	−500			−95,000	−750		
3	−50,000	−500			−95,000	−750		
4	−50,000	−500			−95,000	−750		
5	−50,000	−500	150,000		−95,000	−750	105,000	

Note that the costs have been prefixed with a − sign, and benefits with no sign (i.e., + sign), as discussed earlier. We follow these sign conventions throughout. Also note that the salvage values are entered at the fifth year, when they are realized as lump sum revenues.

Looking at the previous table, we see that the space for data under the Total column is tight. This can be circumvented by using a wider sheet of paper, or by making two auxiliary tables, one for A and the other for B. Let us adopt the latter. In that case, the two auxiliary tables will be

Year	A			
	Chem.	Main.	Sal.	Total
1	−50,000	−500		
2	−50,000	−500		
3	−50,000	−500		
4	−50,000	−500		
5	−50,000	−500	150,000	

Year	B			
	Chem.	Main.	Sal.	Total
1	−95,000	−750		
2	−95,000	−750		
3	−95,000	−750		
4	−95,000	−750		
5	−95,000	−750	105,000	

Each of these auxiliary tables needs to be completed for the net cashflows in the Total column. The usefulness of signed cashflows becomes evident while totaling the various elements. Following the algebraic addition rule, the two auxiliary tables complete in every respect become

Year	A			
	Chem.	Main.	Sal.	Total
1	−50,000	−500		−50,500
2	−50,000	−500		−50,500
3	−50,000	−500		−50,500
4	−50,000	−500		−50,500
5	−50,000	−500	150,000	99,500

Year			B	
	Chem.	Main.	Sal.	Total
1	−95,000	−750		−95,750
2	−95,000	−750		−95,750
3	−95,000	−750		−95,750
4	−95,000	−750		−95,750
5	−95,000	−750	105,000	9,250

These auxiliary tables can now be summarized as a two-column cashflow table. Inserting the row for year 0 to show the initial equipment costs, and decoding the equipment type, the final cashflow table with proper headings becomes

Year	Neutralization Equipment	Precipitation Equipment
0	−$750,000	−$450,000
1	−50,500	−95,750
2	−50,500	−95,750
3	−50,500	−95,750
4	−50,500	−95,750
5	99,500	9,250

Example 2.2 illustrated the use of auxiliary tables, the need for additional care in posting the relevant data, and the creative tasks involved in developing a cashflow table. If you can think of other ways (perhaps better!) to develop the cashflow table for Example 2.2, try them.

Cashflow tables, such as those in Examples 2.1 and 2.2, are useful by themselves, since they help the analyst visualize the problem and the associated data. At times they are used to sketch cashflow diagrams, discussed later in Section 2.2, though the diagrams may be developed directly from the problem statement. Both the table and the diagram contain and display data that are used for further analysis of the problem, as illustrated in the other chapters.

2.1.3 The End-of-Period Assumption

One basic practice in engineering economic analyses is that the cashflows are posted at the end of the period (year, quarter, month, or whatever it is for the given problem). Even when transactions take place at other times within the period, we assume the total cashflow to have occurred only at the end of the period.

Consider, for example, the annual maintenance costs of the two pieces of equipment in Example 2.2 earlier. These have been posted in the table as a cost at the end of the year, say on 31 December. We know that equipment can break down any time during the year (say in June), and that it may be repaired immediately (next week) so that it can be put back to use without delay. Thus, the money earmarked for maintenance is used as and when the equipment breaks down, but, by convention, we post all the maintenance costs for the period together as one lump sum at period end.

Isn't this practice wrong? In a sense, it is, but if we were to consider the cashflows exactly when they occur, their posting might become cumbersome due to numerous timings. Then too, how narrow a time window do we consider? Do we consider the cost every week, or every day? That is why we make the assumption, and in doing so sacrifice some accuracy to achieve practicality, that during-the-period cashflows be treated as if they occurred *at the end of the period.*

Another reason for the period-end assumption is that the funds earmarked for use during the period are kept "liquid" for easy access, usually in checking accounts that earn no interest, so any consideration of the actual timing of a cashflow within the period does not matter. Moreover, the period considered is usually pegged to the accounting and reporting cycles for tax and other business purposes.

2.2 CASHFLOW DIAGRAMS

A cashflow diagram is a graphical sketch of the cashflows and their timings. Since the cashflows are contained in their table, a cashflow diagram can be sketched using the cashflow table. The cashflow table and the corresponding diagram are two displays of the same data, like the two sides of the same coin.

A cashflow diagram depicts the magnitude and direction of the cashflows as vertical lines at specific points (markers) along a timeline. The timeline is horizontal and shows the time periods at which cashflows occur. It is appropriately divided and marked. The left end of the timeline usually represents 0, meaning *now* or *today*. At each of the relevant time markers a vertical line, also called a vector, proportional to the cashflow amount is drawn. The vector is upward, above the timeline, for cashflows that are benefits (inflows), and downward for costs (outflows). The vectors are arrow-headed away from the timeline. Figure 2.1 shows a typical cashflow diagram.

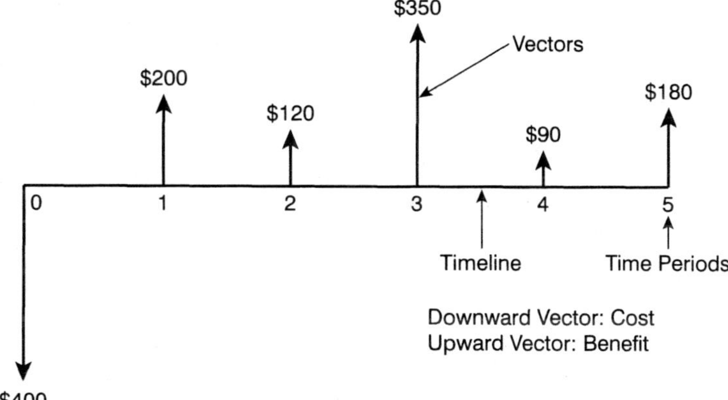

Figure 2.1 A Typical Cashflow Diagram

2.2.1 Development of a Cashflow Diagram

The drawing or sketching of a cashflow diagram involves (i) laying out the timeline and marking it, and (ii) drawing the vertical cashflow vectors. The use of the term *vector,* rather than *line,* is more appropriate, since it indicates the direction (of the cashflow) too. Altogether four steps are involved in developing a cashflow diagram. The steps will be clear in the illustrative example that follows.

Step 1: ***Set out (sketch) a horizontal line.***
The length of the line should be sufficient to accommodate the time periods pertinent to the problem. The line may be 4 to 6 inches long. Since the lengths along the line represent time, we call it a timeline.

Step 2: ***Divide the timeline equally beginning at the left end,***
so that all the periods can be accommodated. Erase any excess length at the right end, and mark the divisions to represent the periods. The left end point is usually marked 0 to represent *now* (or some reference time period).

Step 3: ***Work out a suitable scale[2] to size the cashflows.***
Use the largest cashflow to determine an appropriate scale so that the other vectors will fit within the space available.

Step 4: ***Sketch the vectors to represent the cashflows.***
Keep their lengths in proportion to the cashflow amounts. Take care of the direction. Benefits (inflows) are drawn upward above the timeline, costs (outflows) downward. You can remember this convention easily by recalling that because benefits *increase* the project account balance their vectors are *up,* and because costs *decrease* the balance their vectors are *down.* End each vector with an arrow pointing away from the timeline. Finally, write down the cashflow amounts by the arrows.

Regarding the convention for vector direction, always remember:

Cash inflows/receipts/benefits/revenues	Vector up
Cash outflows/disbursements/costs/expenditures	Vector down

Let us consider an example to illustrate this four-step procedure. Assume that you have borrowed $4,000 today and plan to pay back this loan with interest over the next five years as annual payments of $1,000 each, on the loan anniversaries. Based on what we have learned in Section 2.1, your cashflow table will be

Year	Amount
0	$4,000
1	−1,000
2	−1,000
3	−1,000
4	−1,000
5	−1,000

[2]Though scaling is not essential, especially in a sketch, it leads to a proportionate diagram that facilitates comprehension.

Let us go through the four steps to develop the cashflow diagram.

Step 1: *Timeline*
Set out a horizontal line of reasonable length, as shown in Fig. 2.2(a).

Step 2: *Time Markers*
Beginning at the left end, mark equal lengths on this line to represent all the time periods, five in this case, as in Fig. 2.2(b). Erase the excess length on the right, as in Fig. 2.2(c). Write down the time periods at the markers, as in Fig. 2.2(d). Writing them slightly off to the right frees space for the vectors.

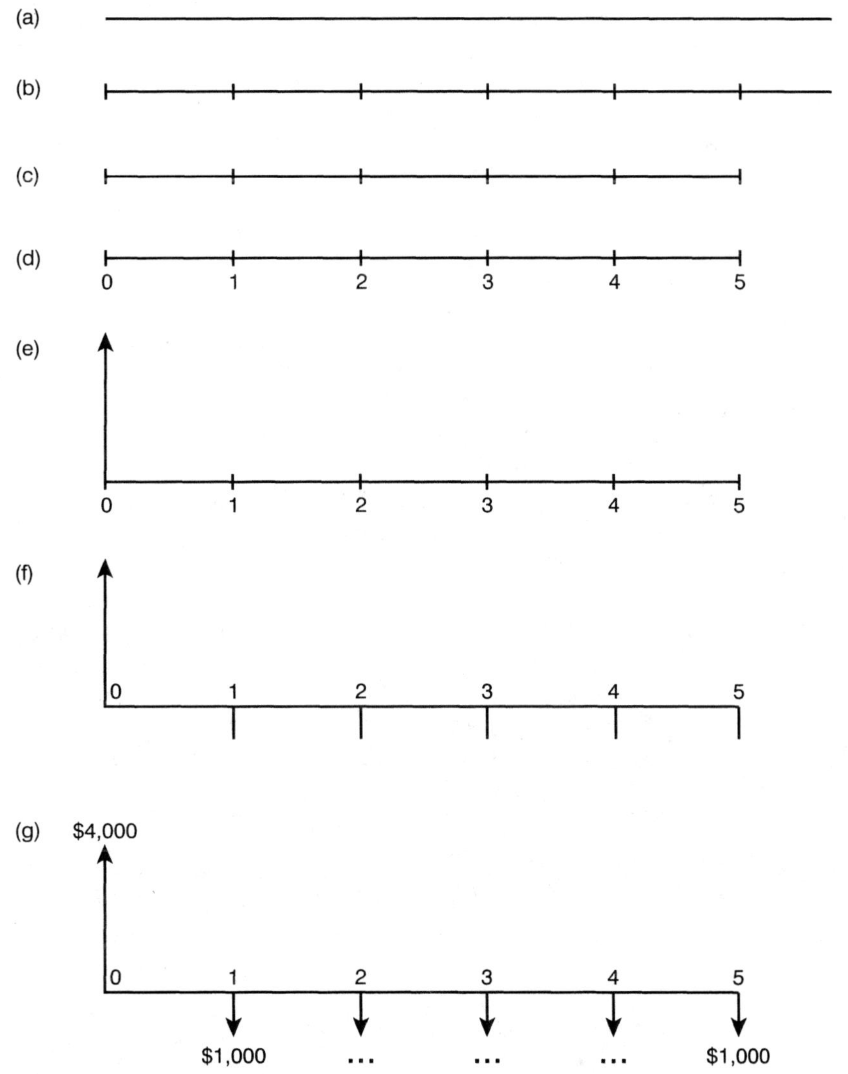

Figure 2.2 The Four Steps in Sketching a Cashflow Diagram

Step 3: *Cashflow Scale*
The largest cashflow in the table is $4,000. What length of the vector should represent it? Let us use a 2-cm length for this purpose, so 1 cm = $2,000. A round figure for the scale yields simpler scaling of the other vectors.

Step 4: *Draw the Vectors*
We now draw the cashflow vectors. Beginning at the left, set out at marker 0 a 2-cm vertically upward line to represent your borrowing of $4,000, as in Fig. 2.2(e). This vector is up, since $4,000 is an inflow. Other cashflow vectors are 0.5-cm long, representing $1,000 each. Since they are all outflows (− sign in the table), these vectors are vertically downward at the appropriate time markers, as in Fig. 2.2(f). Arrow the vector ends and write down the cashflow amounts nearby, as in Fig. 2.2(g), which is the cashflow diagram.

Note that Fig. 2.2 has been drawn step by step to explain the procedure. Each sketch is not necessary; it is only the final, composite diagram—Fig. 2.2(g)—we are interested in. Once you have practiced the four steps enough, the process will become routine and you will be able to sketch the (final composite) cashflow diagram with ease.

Let us now consider the example from the lender's viewpoint (the bank). As discussed in Subsection 2.1.1, the signs of the data in the lender's cashflow table are just the opposite of those in the borrower's. This fact reverses the vector directions in the lender's cashflow diagram as shown in Fig. 2.3; note the reversal in comparison to Fig. 2.2(g). In all other respects the two diagrams (borrower's and lender's) are alike.

As discussed in Subsection 2.1.2 and illustrated in Examples 2.1 and 2.2, we often use auxiliary table(s) to comprehend a given problem, summarizing the data finally in a two-column cashflow table. In the same way, one can draw auxiliary cashflow diagram(s) that may have more than one vector at a time marker corresponding to the various cashflows. Such auxiliary diagrams too should be summarized in a (final) cashflow diagram showing at each time marker only one vector representing the *net* cashflow. The vector may altogether be absent at certain time markers if no cash flows there, or if the net cashflow is zero.

Do we need to develop the cashflow table before sketching the corresponding cashflow diagram? The answer is both yes and no. If it helps, do it; otherwise don't. Preparing the cashflow table first is usually helpful, especially to the beginners. After you have mastered enough skills in sketching the cashflow diagram directly from the problem statement, you can skip the development of the cashflow table, unless it is specifically asked for. In Example 2.3, a direct-to-the-diagram approach is illustrated, while in Example 2.4 the cashflow table has been asked for.

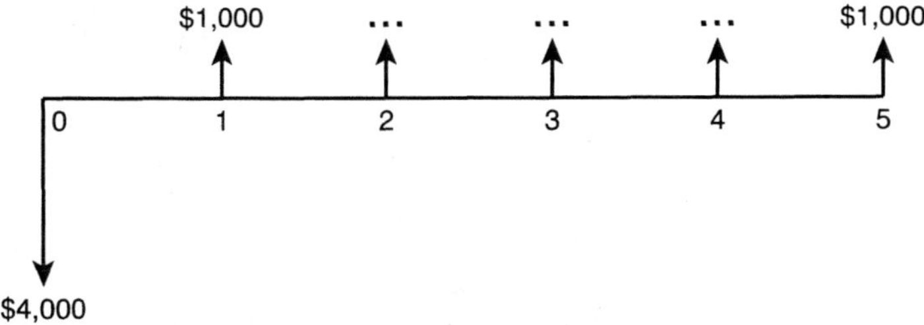

Figure 2.3 Lender Cashflow Diagram—Compare with Figure 2.2(g)

EXAMPLE 2.3

Monica begins her four-year program toward a degree in electronics engineering technology. Her parents are interested in helping her start on graduation an electronics repair business in her home town. Their research shows that she will need start-up capital of $50,000. They plan to save $300 per month, beginning next month, which will at the end of four years be used as the down payment for the bank loan she will need. However, her parents will withdraw when due the interest earned on the monthly savings for their own use. Draw her parents' cashflow diagram, excluding any consideration of the interest earned.

Solution

The question does not ask specifically for the cashflow table. We have two choices: (a) prepare the cashflow table as an aid to the development of the cashflow diagram, or (b) proceed directly to sketch the diagram. What dictates the choice? Basically it is the level of confidence the engineering economist has for (b). Experienced economists choose (b), bypassing the table. The decision also depends, to a large extent, on the complexity of the problem.

This example problem does not seem too complex; let us attempt to sketch the diagram directly, following the four-step procedure outlined in Subsection 2.2.1. Begin with the comprehension of the problem statement. Read it once again (more if you need to). Monica will need $50,000 as start-up capital on graduation four years from now. Her parents' monthly savings at maturity will be used as the down payment toward the bank loan. If the savings accumulate to F, then the loan will be $50{,}000 - F$. Also recall that her parents withdraw the interest earned by their savings. In other words, only the monthly saving of $300 accumulates. Since the savings are monthly over four years, there are $4 \times 12 = 48$ time periods.

With this understanding, we are ready to sketch the cashflow diagram. Let us go through the four steps; keep glancing at Fig. 2.4 as you read.

Step 1: *Timeline*
Set out a horizontal line of reasonable length, as in Fig. 2.4.

Step 2: *Time Markers*
Beginning at the left end, mark equal lengths on this line to represent the early time periods (months). Since the number of periods is large (48), they cannot all be displayed, so show discontinuation using dotted lines and mark the last two time periods (see Fig. 2.4). Write down the time periods at the markers, beginning with 0 at the left end and terminating with 48 at the right end.

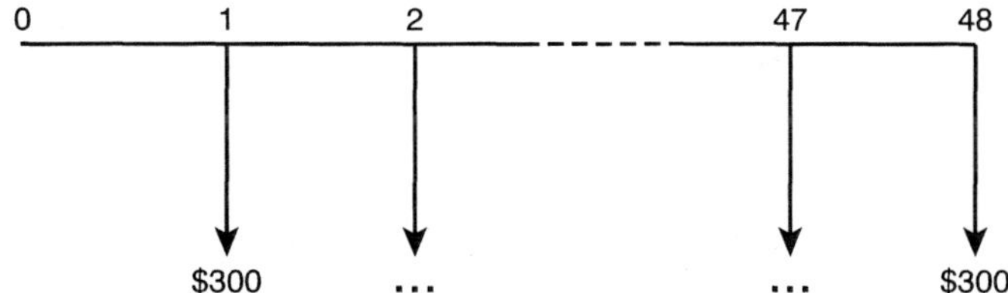

Figure 2.4 Cashflow Diagram for Example 2.3

Step 3: Cashflow Scale
All cashflows in this problem are the same ($300). We can choose any reasonable vector length, since no further scaling is required.

Step 4: Draw the Vectors
Now, sketch the arrowed cashflow vectors. As per the problem statement, no cash flows now; that is, there is no vector at marker 0. The cashflow begins at period 1, where we draw a vector[3] to show the savings of the first month. But which way? Up or down? The answer lies in whether the cashflow is negative or positive. For Monica's parents, the savings are outflows (costs), so the vector should be downward, below the timeline. A smart way to judge whether the vector should be downward or upward is to base it on this: If the cashflow brings the account balance *down,* the vector should be *downward*; if it makes the balance go *up,* the vector should be *upward.* In this case, because the cashflow brings down Monica's parents' account balance, the vector is downward. At the other markers we similarly sketch downward vectors.

The resulting cashflow diagram is shown[4] in Fig. 2.4.

In problems where the timeline has to be long to accommodate all the time periods, as in the preceding example, it is shown broken. Also, some of the cashflows may be disproportionately large in comparison to the others. Again, the taller cashflow vectors may be shown broken. Fig. 2.5 shows a cashflow diagram in which both the timeline and the tallest vector (5,000 at period 0) are shown broken. Breaks having similar meaning are commonly practiced in engineering drawings. Even when the timeline has a break, its

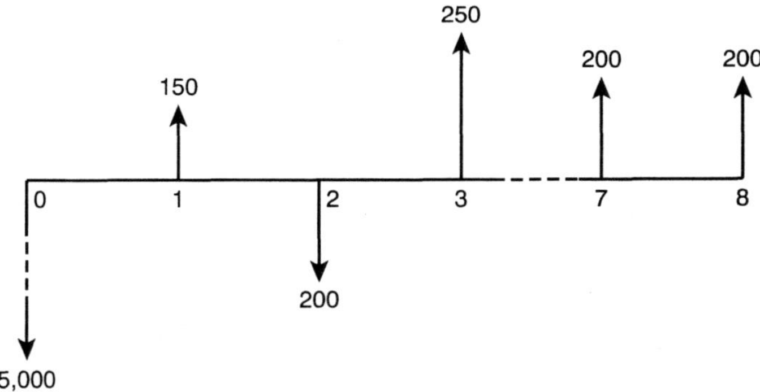

Figure 2.5 Broken Lines Showing Continuation

---

[3]Since it is a sketch, the vectors do not have to be exactly equal; draw freehand, ensuring, however, that they look equal.

[4]We considered month as the time period in this example, as per the given data. Since the interest is withdrawn when due, the monthly savings simply add. We could therefore have used year as the time period. In that case, four cashflows of $3,600 (= $300 × 12) each would have been drawn at four time markers 1 through 4. This would have reduced the number of time periods from 48 to 4 and avoided the need to show a broken timeline. Note, however, that if the interests earned were not withdrawn when due, and were allowed to compound, then replacing the forty-eight $300 monthly cashflows into four $3,600 annual cashflows would have been erroneous.

two ends should be shown complete with appropriate point markers. For the vector too, the break should be in the middle rather than at the ends. Another enhancement to cashflow diagrams may be the absence of a currency symbol to aid visual clarity, as in Fig. 2.5.

EXAMPLE 2.4

Anjali, a plant supervisor, has been informed that the company has made a better than expected profit this year and that she can budget funds to replace the robot. She feels that the current robot can be used for another five years. The replacement robot to be bought then will cost $50,000 and last for five years. The maintenance costs of the new robot will be $1,200 in the first year, increasing thereafter annually by $200 for the remaining four years. Considering the time value of money, she budgets this year for $35,750, which is deposited in a five-year CD to mature to $50,000. Prepare both the cashflow table and the diagram for the project.

Solution

We begin with the problem statement. The time period in this example is year. The current robot is expected to last five more years; the new robot will be purchased then (five years from now).

Anjali takes out $35,750 from this year's company profit and deposits this in a five-year CD. No data is given about the interest rate the CD will earn, but she seems to have worked out that $35,750 will mature to $50,000 at the end of five years. This $50,000 will be used to buy the new robot, which will incur maintenance costs during its five-year life. Thus, this problem involves ten time periods—five years of the current robot's life and five years of the new one's.

With this understanding, let us first prepare the cashflow table, which seems simple. We can do without an auxiliary table and proceed directly to prepare the two-column cashflow table, whose skeleton will look like this:

Year	Amount
0	
1	
2	
3	
4	
5	
6	
7	
8	
9	
10	

We need to gather the relevant data from the problem statement and post them in the table. For row 0, the cashflow is $35,750, the money Anjali sets aside now (this year). Since this is a cost (paid out) to the company[5], it will have a − sign. For the next five years, during which the current robot will continue to be used, there are no cashflows. Though the current robot may need maintenance, the associated costs are not given and hence are not a part of the current project. Such costs might have been taken into account while purchasing the current robot. Thus the cashflows during years 1 through 5 are zero.

At the end of the fifth year, $35,750 would have increased to $50,000. Depending on how this $50,000 is posted, the cashflow table may have two types of entries for year 5.

a. The simplest way to look at the situation is to consider that this $50,000 is paid directly from the CD account to the supplier of the new robot. In that case there is no new cashflow from Anjali's company to the supplier. Looked at this way, there should be no entry for year 5.
b. The approach in (a), which looks simpler, will be objected to by professional accountants. What they will insist on is that the accumulated sum $50,000 be shown as an inflow into company books, and a concurrent payment (outflow) of $50,000 be made out to the supplier. Thus, there will be two entries in the cashflow table corresponding to year 5: one as $50,000 and the other as −$50,000. The former is the inflow (+ sign not shown); the latter is the outflow (payment to the robot supplier). The two entries result in a *net* cashflow of zero for year 5, the same as in (a).

The new robot is acquired at the end of year 5 (the same as the beginning of year 6) and its use begun. The maintenance cost during the first year of the new robot's use is $1,200. This year is the sixth year of the whole project. Thus, for year 6, the maintenance cost of $1,200 is entered with a negative sign. Since the maintenance costs for subsequent years increase annually by $200, the cashflows for the subsequent years will be 1,400, 1,600, 1,800, and 2,000, all with negative signs.

Thus, the cashflow table following approach (a) will be

Year	Amount
0	−$35,750
1	0
2	0
3	0
4	0
5	0
6	−1,200
7	−1,400
8	−1,600
9	−1,800
10	−2,000

[5]Note that we are preparing the company's cashflow table.

If we follow the strict accounting practice explained in (b), the cashflow table will be

Year	Amount
0	−$35,750
1	0
2	0
3	0
4	0
5	50,000
	−50,000
6	−1,200
7	−1,400
8	−1,600
9	−1,800
10	−2,000

We can now proceed to draw the cashflow diagram. Since the cashflow table already exists, we do not need to analyze the problem and its data. Instead, we use the cashflow table to develop the diagram. Follow the earlier discussions of Subsection 2.2.1 and the illustrative Example 2.3 on how to develop a diagram.

For (a)-type entry, the cashflow diagram[6] is Fig. 2.6(a), while for (b)-type entry, it is Fig. 2.6(b). Note a few "cosmetic" enhancements in these diagrams. For example, the time periods have been labeled above the timeline to add clarity. Their usual labeling—below the timeline and slightly off to the right of the time markers—would have been congested. Also note that we did not label each time period, skipping the odd ones 1, 3, . . . ; when space is limited, this is acceptable. Some analysts do not label the time periods at all, especially if there are few. Further, the cashflow amounts have been normalized into thousands, reducing the number of digits required to label them. A qualifying note to this effect must be made at a prime location in the diagram, as in Fig. 2.6. Some analysts use the notation k (always lowercase), which stands for a thousand. In that case, the cashflow data in Fig. 2.6 might have been labeled as 35.75k, 1.2k, 1.4k, . . ., without the qualifying note *Amount in Thousands of Dollars.* Cosmetic enhancements are meant to add clarity to a cashflow diagram. A clear and unambiguous diagram is likely to lead to an error-free solution. Use your imagination to achieve clarity.

[6]Some engineering economists prefer to show all the vectors at their time markers, for example the two vectors at year 5 in Fig. 2.6(b). Others prefer to show only the net cashflow, which being zero at period 5 does not appear in Fig. 2.6(a).

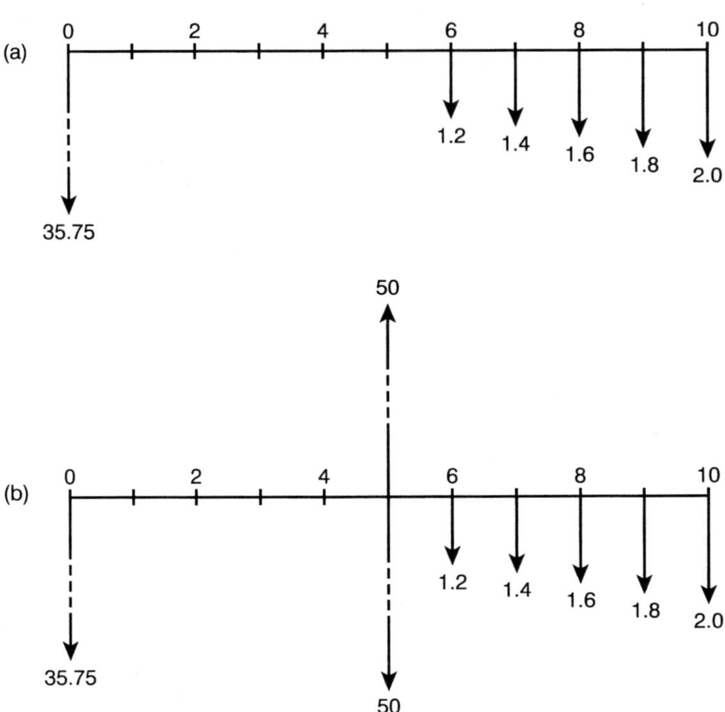

Figure 2.6 Cashflow Diagram for Example 2.4

SUMMARY

Cashflow tables and diagrams are effective tools that aid comprehension of engineering economics problem/project statements. They summarize the given data for further analysis. Their concepts and developments have been discussed in Chapter 2. Engineering economists use them quite often.

A cashflow table is basically a two-column table in which the first column lists the time periods and the second column lists the corresponding cashflows. Cash outflows (costs, disbursements, or expenditures) are preceded by a negative sign. Cash inflows (benefits, receipts, or revenues) have no sign, meaning that they are positive.

A cashflow diagram is a representation of the cashflows and their timings. It is drawn, in fact sketched, by displaying the cashflows as scaled vertical vectors along a timeline that shows the timing of their occurrences. Cash outflows (costs, disbursements, or expenditures) are drawn as downward vectors. Cash inflows (benefits, receipts, or revenues) are drawn as upward vectors. In other words, cashflows with negative signs in the cashflow table are downward vectors, and those with no sign are upward vectors.

The cashflow table or diagram, or sometimes both, may be helpful in solving a problem. They may be developed on their own, using the given data, independent of each other. Alternatively, one may be developed easily from the other.

EXERCISES

Discussion Questions

2.1 Why does a cashflow table usually have only two columns?
2.2 Given the lender's cashflow table, how do you prepare the borrower's cashflow table?
2.3 Explain the role of auxiliary table(s) in the development of a cashflow table.
2.4 Of the two, cashflow table and cashflow diagram, which one offers more visual impact and why?
2.5 Why don't the cashflow data displayed in a cashflow diagram show their signs—whether negative or positive?
2.6 Discuss the four steps involved in developing a cashflow diagram. Which step is most likely to introduce error in the diagram and why?
2.7 Discuss the pros and cons of developing a cashflow diagram directly from the problem statement, without first preparing the cashflow table.

Multiple-Choice Questions (Circle the *best* answer.)

2.8 A cashflow with no sign represents
  a. receipt of money.
  b. disbursement of money.
  c. maintenance cost.
  d. none of the above
2.9 In a cashflow table, a + sign represents
  a. cash inflow.
  b. disbursement.
  c. cash outflow.
  d. none of the above
2.10 *Cash outflow* is another term for
  a. disbursement.
  b. receipt.
  c. revenue.
  d. benefit.
2.11 An auxiliary table is _______ essential in developing a cashflow table.
  a. always
  b. never
  c. sometimes
  d. none of the above
2.12 An auxiliary table must have ______ columns.
  a. 2
  b. 3
  c. 4
  d. none of the above

2.13 Cashflow vectors are drawn upward when the cashflows are
  a. receipts.
  b. disbursements.
  c. costs.
  d. none of the above

Numerical Problems

2.14 Prepare the cashflow table for Example 2.3.

2.15 Perfect Manufacturers decided to finance the company's new car. It paid $5,000 at the time of purchase and agreed to pay $5,000 on each of the next four anniversaries of the purchase. The first year's maintenance is free. The company budgeted for sums of $500, $1,000, $1,500, and $2,000 for the subsequent four years' maintenance. It expects to sell the car at the end of the fifth year for an estimated price of $2,500. Prepare the cashflow table for this project.

2.16 Five years ago, a firm purchased a used personal computer (PC) system for $2,500. The PC did not require any repair during the first two years. For the subsequent three years, the repair costs were $85, $130, and $195 respectively. The computer was sold at the end of the fifth year for $500. Develop the cashflow table and state any assumptions made.

2.17 Engineering Unlimited borrows $100,000 for purchasing a robotic system. It pays back the loan (capital and interest) through payments of $50,000 at the end of six months and of $65,000 on the first loan anniversary. Sketch the cashflow diagram.

2.18 You have borrowed $1,000 from a bank, which is to be paid back in five equal yearly installments of $235 each, beginning on the first loan anniversary. Prepare your cashflow table and sketch the cashflow diagram.

2.19 Prepare the bank's cashflow table and diagram for problem 2.18.

2.20 Sketch the cashflow diagram for Example 2.1.

2.21 Sketch the two cashflow diagrams separately for Example 2.2.

2.22 Cutting-Edge Services plans to invest $300,000 in an international information service based on the Internet. It expects to earn $50,000 in the first year, which is likely to increase annually by $4,000 during the subsequent nine years. After that it will decline annually by $3,000 during the next five years. Draw the cashflow diagram for this project.

2.23 Environmentally Enlightened Farmers is a small cooperative venture based on organic farming. Its cost for gasoline use has been high. It uses 10,000 gallons of gasoline per year. To economize on the fuel cost, it is considering investing in a fuel storage system with an initial cost of $2,500. The system's maintenance cost during the first year is expected to be $50, which will increase annually by $25 during the subsequent nine years. The system will save 5 cents per gallon. Draw the cashflow diagram for the investment by assuming the system's salvage value to be $450 at the end of its useful life of ten years.

2.24 An investment of $22,000 in a hydraulic system returns a benefit of $8,250 in the very first year. The benefits increase annually by 10% during the subsequent four years. At the end of the fifth year, the system is sold for $3,000. Sketch the cashflow diagram after preparing the cashflow table.

CHAPTER THREE

Single Payment

IN THIS CHAPTER YOU WILL LEARN ABOUT

- The basic equation of compounding
- The law of compound interest
- Interest earned under compounding
- The exponential nature of compounding
- Equations versus first principles
- How to solve single-payment problems
- Manipulation of the basic equation's variables
- Functional notations and their use
- How to interpolate
- The use of compound interest tables
- Nominal and effective interest rates

In Chapter 1 we discussed the concept of interest compounding following the first principles. We saw there in Section 1.5 how compounding enables a principal to accumulate faster than simple interest does. In this chapter, we extend those discussions to learn how the first-principles approach becomes cumbersome when n, the number of time periods, is large. Following the first principles we derive equations that are used to solve problems involving compound interest. We develop in this chapter the most *basic equation* of engineering economics—sometimes called the *law of compound interest*. The pertaining discussions emphasize the manipulation of the basic equation and its variations.

This chapter covers single-payment problems that involve two cashflows. All business transactions comprise at least two cashflows—one by the first party and the other by the second party. When you borrow money from a bank, you as a borrower are one party and the bank as a lender is the other party. Some typical transacting parties are individual investors or lenders, financial institutions such as banks or credit unions, insurance companies, and the federal government. The concept of two parties in a transaction is helpful in comprehending a problem.

You may borrow, and pay back, different amounts at different times. In this chapter, however, we consider the simple case where you borrow a lump sum once and pay it back at the end of the loan along with the interest. So, though the bank and you make a single payment each, there are two[1] transactions, that is, a pair of cashflows.

Consider that as a supervisor you spend \$1,000 to improve an existing machine in your plant. Here, you representing your company are the first party and the machine (or its vendor) is the second party. The machine is expected to pay back through higher revenues resulting from increased production and/or better product quality.

In problems involving a pair of cashflows, an initial sum P increases under compounding to become the future sum F. The pairing of cashflows arises from the lending or borrowing of P and its payback later as F. The future sum F is greater than P due to the time value of money, their difference $F - P$ being the total interest earned. Note that symbols other than P and F may also be used to denote the present and future sums.

3.1 THE BASIC EQUATION

Refer to Example 1.2, where interest was compounded following the first principles. For small values of n, as in that example, the first-principles approach is acceptable in terms of solution effort, especially since no equations are involved. One can, in theory at least, follow the first-principles approach even when n is large. But this approach becomes cumbersome, especially lengthy, as the value of n increases. When n is large, it is preferable to use equations, for a more concise solution, than to follow the first principles.

We derive in this section the most fundamental equation of engineering economics. Consider an initial (or present) sum P invested for n periods under compounding. We use the following notations:

P	Present[2] sum (initial principal)
i	Interest rate (in decimal fraction, not in percentage) per period
n	Number of interest compounding periods
F	Final sum (payable at the end of n periods)

$$\text{Interest earned during the first period} = \text{Initial principal} \times \text{Interest rate} = Pi$$

[1]There are situations in everyday life where only one party makes a single payment—and that's all; for example, your giving \$50 to a charity or receiving a cash gift from grandma on your birthday. The other parties in these examples (the charitable organization in the first and you in the second) do not pay back, at least not *materially*.

[2]P is known by different names, such as *present sum, principal, initial sum*, or *investment*; but they all mean the same thing. Note that P is a lump-sum single payment, and so is F.

Due to compounding, this interest is added to the initial principal P. Thus,

Principal at the *end of the first period*
= Initial principal + Interest during the first period
$= P + Pi$
$= P(1 + i)$

This $P(1 + i)$ becomes the principal at the *beginning of the second period.*

Interest earned during the second period
= Principal for the second period × Interest rate
$= P(1 + i)i$

Thus,

Principal at the *end of second period*
= Principal at the beginning of the second period + Interest for the second period
$= P(1 + i) + P(1 + i)i$
$= P(1 + i) + iP(1 + i)$

Taking $P(1 + i)$ out from both the terms gives

Principal at the end of second period $= P(1 + i)(1 + i)$
$= P(1 + i)^2$

Repeating the above procedure,

Principal at the *end of the third period* $= P(1 + i)^3$

⋮

Principal at the *end of the nth period* $= P(1 + i)^n$

Thus, the (initial) principal P increased under compounding to become $P(1 + i)^n$ at the end of the nth period. Noting that this is the final sum F, we can say

$$F = P(1 + i)^n \tag{3.1}$$

This is the *most basic*[3] *equation* in engineering economics. All problems of engineering economics rely on this equation for their solutions. To emphasize its importance, Equation (3.1) is often called the *law of compound interest*. Since this equation is fundamentally important, we devote this entire chapter (Chapter 3) to discussing it. We need to thoroughly understand the relationships among the parameters of this equation.[4]

---

[3]A comparison of the derivation procedure for the basic equation and the use of first principles in Example 1.2 illustrates their equality.

[4]Note that when i is zero, Equation (3.1) reduces to $F = P$, meaning that the investment P remains unchanged. The situation is similar to *burying money in the backyard* or keeping it *under the pillow,* i.e., not investing the principal. In the business world, $i = 0$ makes no sense. But in the personal world, some people offer loans to members of their religious group at no interest. We also hear about some generous, rich people offering interest-free loans to their relatives, close friends, or charity organizations.

3.1.1 Interest Earned

We can use Equation (3.1) to determine the total interest earned from the investment P for n periods. It is evaluated simply by subtracting the P from the final sum F. Thus,

$$I = F - P$$

where F is given by Equation (3.1). One can simply use this subtraction to determine I, with no need for an equation.

Alternatively, an equation can be derived by substituting for F from Equation (3.1) as

$$\begin{aligned} I &= F - P \\ &= P(1 + i)^n - P \\ &= P[(1 + i)^n - 1] \end{aligned} \tag{3.2}$$

Note that Equation (3.2) is a *derived* formula. It is in fact redundant as long as one is aware of the simple fact that total interest I is simply the difference between F and P.

3.1.2 Exponentiality of Compounding

In Equation (3.1) the time period n appears as an exponent. This results in an exponential growth in the function $(1 + i)^n$ as n increases, and therefore in the investment P to which this function is a multiplier. Such a growth is illustrated in Table 3.1, as an example for $P = \$100$ and $i = 10\%$.

As seen in this table, a single one-time investment of merely \$100 at 10% yearly interest rate, compounded annually, becomes over a million dollars in 100 years. If you want your grandchild's grandchild to be a millionaire, then just invest today a meager sum of \$100! (Assuming 25 years as the span of a generation, your grandchild's grand-

Table 3.1 Exponential Growth Due To Compounding ($P = \$100, i = 10\%$)

Year	F (year end)
0 (now)	100(= P)
1	110
2	121
3	133
⋮	⋮
10	259
⋮	⋮
50	11,739
⋮	⋮
100	1,378,061

child will follow you 100 years from now.) Isn't this the easiest way to ensure a millionaire in your future family!

3.1.3 Compounding Benefits the Lender

Compounding enables the interest earned for a given period to become part of the principal at period end and thus to begin earning interest during the subsequent period. In other words, under compounding, *interest earns interest*. This results in exponential growth in the principal. It is this exponentiality that ensures a millionaire great-great-grandchild to a person who is wise enough to invest \$100 today.

For a given problem, the difference in the values of I obtained from Equations (3.2) and (1.1) yields the additional interest earned due to compounding (over simple interest). Since compounding earns additional interest, it is beneficial to the lender or investor, as illustrated in Example 3.1.

EXAMPLE 3.1

After finishing his undergraduate education in construction engineering, John decides to start his own business. His rich, generous uncle has agreed to loan him \$80,000 at an annually compounded interest rate of 4% per year. John will pay back his uncle the accumulated sum F on the fifth loan anniversary. How much will John have to pay? How much more does John's uncle earn due to compounding over simple interest? (Example 1.1 is considered simple interest.)

Solution

The relevant formula is Equation (3.1). The value of n is 5, since over the loan life compounding takes place five times. The variables' values are

$$\begin{aligned} P &= \$80{,}000 \\ i &= 4\% \text{ per year} = 0.04 \\ n &= 5 \text{ years} \end{aligned}$$

Substituting these values in Equation (3.1),

$$\begin{aligned} F &= P(1 + i)^5 \\ &= \$80{,}000\,(1 + 0.04)^5 \\ &= \$97{,}332.23 \\ &\approx \$97{,}332 \end{aligned}$$

Thus, John pays back his uncle on the fifth loan anniversary a sum of \$97,332.

For the second part of the answer, we refer to, and compare with, Example 1.1 where simple interest was considered. In Example 1.1, John paid back \$3,200 four times and \$83,200 once as the last payment. So the total payment under simple interest was (\$3,200 × 4) + \$83,200 = \$96,000. Since his payment under compound interest is \$97,332, compounding earned his uncle \$1,332 more (\$97,332 − \$96,000). This \$1,332 represents the additional interest earned due to compounding. We thus see that compounding benefits the lender (obviously a load to the borrower!).

3.2 EQUATION VERSUS FIRST PRINCIPLES

Refer to Example 1.2 and compare the lengthiness of its solution with the comparable first part of Example 3.1. The former is based on first principles and is for a shorter period ($n = 3$); the latter is based on an equation and is for a longer period ($n = 5$). Even then the solution of Example 1.2 is lengthy. It would have been lengthier had the value of n been 5. We thus see the advantage of solution conciseness in using an equation (over the first principles). The rule is that if an appropriate equation exists, and you feel confident using it, then use it—you will normally save time and effort. Otherwise, follow the first principles.

3.3 ANALYSIS OF THE BASIC EQUATION

As an embodiment of the law of compound interest, Equation (3.1) is a very basic formula in engineering economics. Let us therefore study this equation extensively. Note that it involves four variables, and in its usual form it relates F with P, i, and n.

The interest rate i in Equation (3.1) must be expressed as a decimal fraction, not a percentage. For example, if in a problem i is given to be 7.5%, then use 0.075 in the equation.

The time period n as an exponent of the basic equation is always a positive integer. In most problems of engineering economics, n is in the range 3 to 10. The upper limit arises due to (a) the limited useful lives of products and services, and (b) increasing uncertainty in the data as n gets larger. For high-tech products, n is small. For example, in the case of personal computers, n is usually 2 to 4 years due to obsolescence created by feverish ongoing innovations in this product. In some projects, especially those involving established technologies, n may be longer than 10 years. In public projects n may be so large that it is almost infinite; examples are bridges, hospitals, schools, and churches, which offer service perpetually.

Most numerical problems involving compound interest provide the values of three of the variables and ask for the evaluation of the fourth (unknown) variable. There are two ways to solve such problems; the choice depends on the analyst.

1. *Use the basic equation.* Substitute the values of the known variables, rearrange the terms in such a way that the unknown is on the left side of the equal sign, and then evaluate the unknown.
2. *Use the derived formula.* The derived formula[5] will look like this:

$$\text{Unknown} = f(\text{known variables})$$

where f denotes *"function of."* Substitute the known variables' values in the derived formula to evaluate the unknown.

---

[5]Such formulas are derived from the basic equation. The derivation rearranges the variables in such a way that the unknown is on the left of the equal sign and the knowns are on the right.

Here are the four possible analysis situations arising from the basic equation. These include the derived formulas for those who prefer alternative 2 above.

1. *Finding F, Given P, i, n.* Use Equation 3.1.

2. *Finding P, Given F, i, n.* Equation (3.1) can be rearranged as

$$P = \frac{F}{(1+i)^n} \tag{3.3}$$

3. *Finding i, Given P, F, n.* Equation (3.1) can be rearranged as

$$(1+i)^n = \frac{F}{P}$$

$$1 + i = (F/P)^{1/n}$$

$$i = (F/P)^{1/n} - 1 \tag{3.4}$$

4. *Finding n, Given P, F, i.* Equation (3.1) can be rearranged as

$$(1+i)^n = \frac{F}{P}$$

Taking the ℓog of both sides,

$$\ell\text{og}\,(1+i)^n = \ell\text{og}(F/P)$$

$$n\,\ell\text{og}\,(1+i) = \ell\text{og}(F/P)$$

$$n = \frac{\ell\text{og}(F/P)}{\ell\text{og}(1+i)} \tag{3.5}$$

Note that Equations (3.3) to (3.5) are simply rearrangements of the same basic equation. For solving a problem, one can ignore these three formulas altogether, begin with Equation (3.1), and rearrange the terms as the solution progresses. However, those who prefer to save the time of rearranging the terms begin with the appropriate formula. Note that this involves having to deal with three more equations.

3.4 SOLUTION METHODS

In this section, we discuss the various ways the basic equation, or its rearrangements as embodied in Equations (3.3) to (3.5), can be used to solve problems involving compounding.

3.4.1 Classical Approach

By classical approach we mean the use of "old-fashioned" paper and pencil, and a basic knowledge of algebra; it assumes the absence of even a scientific calculator. Any of the four variables can, as an unknown, be evaluated if the values of the other three are given. Begin the solution with the basic equation, or use one of the derived formulas, Equations (3.3) to (3.5). Be careful about the compatibility of units. Due to the exponential term, evaluation of the unknown may require the use of ℓog tables. However, since n is normally an integer and its value small, the evaluation of $(1 + i)^n$ can be done simply by multiplying the value of $(1 + i)$ by itself n times. This does away with ℓog tables when i and n are the known variables, that is, where P or F is the unknown. If i or n happens to be the unknown, the use of ℓog tables becomes necessary.

Since a scientific calculator is ubiquitous today, engineers almost never follow the classical approach. They use instead one of the following three methods (next three subsections) in solving for P, F, i, or n, or for problems discussed in the other chapters.

3.4.2 Scientific Calculator

A scientific calculator is widely available today; its use requires the analyst to

a. Select the appropriate formula for the unknown, and
b. Key in the values of the known variables in the recommended sequence.

The result is displayed almost instantly. However, be careful in using any calculator, since the keys and sequencing of data entry are not the same in all calculators. Scientific calculators invariably have an exponent key for evaluating the exponential function $(1 + i)^n$, where i and n are known. Follow the instructions carefully on how to use your calculator. It is always a good practice to confirm the calculator results through sample calculations, worked out approximately in your head if possible or on paper following the classical approach (Subsection 3.4.1).

If you decide to begin the solution with the basic equation, then rearrange the terms carefully to get the unknown on the left of the equal sign. The alternative is to use the appropriate derived formula from among Equations (3.3) to (3.5), whose terms are already rearranged.

The four examples[6] that follow illustrate the use of a scientific calculator in evaluating the unknowns—F in Example 3.2, P in Example 3.3, i in Example 3.4, and n in Example 3.5.

---

[6]These examples, and several others in the text, relate to personal financing. They may leave you wondering whether we are learning about engineering economics. Yes indeed, we are. Many personal financing examples/problems are like engineering economics problems. As an illustration, Example 3.2 has been rephrased here to become an engineering economics problem. Note that the rephrasing has made the problem statement unnecessarily lengthy. That is why personal-financing examples have been preferred to illustrate a concept. Moreover, they are easier to comprehend, since they relate to everyday life.

Example 3.2

Jenny, a plant supervisor, has at the end of this fiscal year been left with \$300 in her department's expense account. Her company's policy is that a sum under \$500 does not have to be returned to the business office. Such sums may be used by the department heads to enhance productivity in any way deemed appropriate, either in the fiscal year or later. If the money is to be used later it must be invested to earn reasonable interest. She decides to use the \$300 left this year to replace a coordinate measuring machine 8 years later, which is estimated to cost \$10,000. So she invests this money in a CD at an interest rate of 6% per year, compounded annually. How much additional capital will be required to replace the machine?

EXAMPLE 3.2

Jenny deposits \$300 now in her account and leaves it there for 8 years to earn interest at the rate of 6% per year, compounded annually. How much does she get at the end of the eighth year?

Solution
The given data are $P = \$300, n = 8$ years, $i = 6\%$ (=0.06) per year. We need to determine F, for which the appropriate formula is Equation (3.1), wherefrom

$$\begin{aligned} F &= P(1 + i)^n \\ &= 300(1 + 0.06)^8 \\ &= 300(1.06)^8 \\ &= 300 \times 1.5938 \\ &\approx 478 \end{aligned}$$

In most problems of engineering economics, cents are normally rounded to the nearest dollar, as done here. Thus, Jenny gets \$478 at the end of the eighth year.[7]

EXAMPLE 3.3

John will need \$5,000 in four years as down payment for an apartment he plans to buy after getting a job as an environmental engineer. How much should he save now in a CD that pays 7% yearly interest, compounded annually?

Solution
The given data are $F = \$5{,}000, n = 4$ years, $i = 7\%$ (= 0.07) per year. Sketch a cash-flow diagram like Fig. 3.1, if that is helpful in comprehending the problem. We need to determine P, for which we can adopt one of the following two approaches. Use the one you feel more comfortable with.

1. *Basic Equation.* Substitute the given values in Equation (3.1) and carry out the calculations, rearranging the terms to get the unknown on the left side of the equal sign. This leads to

$$\begin{aligned} F &= P(1 + i)^n \\ 5{,}000 &= P(1 + 0.07)^4 \\ &= P(1.07)^4 \\ &= P \times 1.3108 \end{aligned}$$

---

[7]The calculator makes it easier to evaluate $(1.06)^8$ in this example. The alternative classical approach would have required multiplying 1.06 by itself eight times.

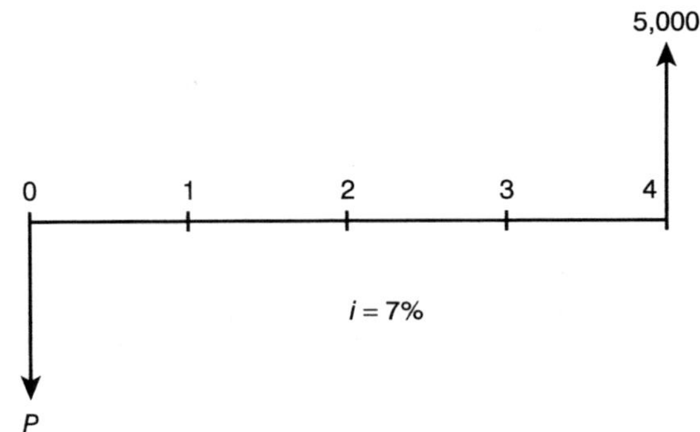

Figure 3.1 Diagram with P as the Unknown (Example 3.3)

Therefore,

$$P = 5{,}000/1.3108$$
$$= 3{,}814$$

Thus, John needs to save $3,814 now.

2. *Derived Formula.* For P the unknown, the appropriate formula is Equation (3.3), wherefrom

$$\begin{aligned} P &= F/(1 + i)^n \\ &= 5{,}000/(1 + 0.07)^4 \\ &= 5{,}000/(1.07)^4 \\ &= 5{,}000/1.3108 \\ &= 3{,}814 \end{aligned}$$

Thus, John needs to save $3,814 now.

EXAMPLE 3.4

Joe gets back $4,000 for a one-time deposit of $3,200 made six years ago in a money market account. What yearly interest rate did he earn if compounding was annual?

Solution

Here the given data[8] are $P = \$3{,}200$, $F = \$4{,}000$, $n = 6$ years. We need to determine i, for which we can adopt one of the following two approaches discussed in Section 3.3. Use the one you feel comfortable with.

---

[8]We may at times be confused by the meanings of *present* denoted by P and *future* denoted by F, as in this example. We are using P to represent a time in the past—six years ago—and F to represent the present. Note that the concept of present and future is *logical* and *relative* rather than *real*, since we are concerned simply with the cashflow timings. Thus, P or F does not have to represent the *real* present or the *real* future. Concentrate instead on the historicity of cashflows. We are primarily concerned with the time gap, in terms of compounding periods, between the investment and its maturity.

1. *Basic Equation.* Substitute the given values in Equation (3.1) and carry out the calculations while rearranging the terms to get i on the left side of the equal sign. This leads to the answer, as below.

$$\begin{aligned} F &= P(1 + i)^n \\ 4{,}000 &= 3{,}200(1 + i)^6 \\ (1 + i)^6 &= 4{,}000/3{,}200 \\ &= 1.25 \\ (1 + i) &= 1.25^{1/6} \\ &= 1.03789 \\ i &= 1.03789 - 1 \\ &= 0.03789 \\ &= 3.789\% \\ &\approx 3.8\% \end{aligned}$$

   Thus, Joe's money market fund earned an annual interest of 3.8%.

2. *Derived Formula.* The appropriate formula for i as the unknown is Equation (3.4), wherefrom

$$\begin{aligned} i &= (F/P)^{1/n} - 1 \\ &= (4{,}000/3{,}200)^{1/6} - 1 \\ &= 1.25^{1/6} - 1 \\ &= 1.03789 - 1 \\ &= 0.03789 \\ &\approx 3.8\% \end{aligned}$$

Thus, Joe earned an yearly interest of 3.8%.

EXAMPLE 3.5

Jaya plans to surprise her son with a cash gift of $1,000 that he can use as down payment to buy his first car. She can afford to save $500 now in a bank account that will earn an annual interest of 6%, compounded each year. How long will Jaya have to wait to surprise her son?

Solution

The given data are $P = \$500, F = \$1{,}000, i = 6\%$ $(= 0.06)$ per year. We need to determine n, for which we can adopt one of the following two approaches.

1. *Basic Equation.* Substitute the given values in Equation (3.1) and carry out the calculations, rearranging the terms concurrently to get the unknown on the left side of the equal sign. This leads to the answer as

$$
\begin{aligned}
F &= P(1+i)^n \\
1{,}000 &= 500(1+0.06)^n \\
(1+0.06)^n &= 1{,}000/500 = 2 \\
\ell og(1+0.06)^n &= \ell og\, 2 \\
n \times \ell og(1.06) &= \ell og\, 2 \\
n &= \ell og\,(2)/\ell og\,(1.06) \\
&= 0.3010/0.0253 \\
&= 11.897 \text{ years}
\end{aligned}
$$

Thus, Jaya will have to wait almost 12 years to be able to surprise her son.

2. *Derived Formula.* For n as the unknown, the appropriate formula is Equation (3.5), wherefrom

$$
\begin{aligned}
n &= \ell og\,(F/P)/\ell og\,(1+i) \\
&= \ell og\,(1{,}000/500)/\ell og\,(1+0.06) \\
&= \ell og\,(2)/\ell og\,(1.06) \\
&= 0.3010/0.0253 \\
&= 11.897 \text{ years}
\end{aligned}
$$

Thus, Jaya will have to wait almost 12 years to surprise her son.

From these four examples it is obvious that the derived formulas yield concise solutions. This is because such formulas are already in the rearranged form, ready for use. The use of the basic equation, on the other hand, requires manipulation of the terms each time it is used, as the solution progresses. However, this approach does away with the need for formulas.

The lengthiness of the basic-equation-based solutions in the preceding four examples can easily be curtailed by those proficient in algebra. The detailed calculation steps, illustrated there, can be collapsed into fewer, yielding brevity.

3.4.3 Software

The software-based method assumes that a computer is available along with a suitable engineering economics application program. The program may be canned or simply be spreadsheet based. Computer systems, especially PCs, are widely available to students and practicing engineers in the industrialized countries. However, the same is not true in the developing countries.

If you have access to a computer system and are skilled in using it, the software-based method is probably the most suitable for solving engineering economics problems. Most software offers graphics capability that can display the results in an extremely presentable form. This capability is even more desirable in complex problems to enhance result visualization.

In the software-based method, the values of the known variables are input into the system using keys, a mouse, or other similar devices. The software evaluates the unknown almost instantly and displays the result on the screen. However, be careful in using the computer for solving problems. Remember GIGO (garbage in, garbage out)! You must have a "feel" for the expected result. It is always a good practice to earn confidence in computer-based results through sample calculations with a scientific calculator.

3.4.4 Functional Notation

The functional-notation method is popular with practicing engineers and technologists, and that is why we discuss it extensively in the next section. In fact, we encourage this method throughout the text.

As an alternative to evaluating the exponential function $(1 + i)^n$ its values are made available in a table for different sets of i and n. Here is a portion of such a table as an example:

i	n	$(1 + i)^n$
5%	3	1.157625
	4	1.215506
	5	1.276282
	⋮	⋮
	8	1.477455
8%	3	1.259712
	4	1.360489
	5	1.469328
	⋮	⋮
	8	1.850930
11%	3	1.367631
	4	1.518070
	5	1.685058
	⋮	⋮
	8	2.304538

If such a table is given, the analyst simply has to note the value of $(1 + i)^n$ from the table corresponding to i and n for a problem. Once this value is known, Equation (3.1) reduces to the simple relationship:

$$F = P \times \text{Exponential function's value}$$

Note that the table has freed the analyst from having to calculate the value of $(1 + i)^n$. Instead, he or she simply reads the value off the table. Since this approach is based on tables, it may be called the *tabular method*. But it is commonly called the *functional-notation* method, for the reason explained in the next section.

3.5 FUNCTIONAL-NOTATION METHOD

The functional-notation approach, commonly used by practicing engineers and technologists, does away with the need to calculate the value of the exponential factor $(1 + i)^n$. Rather than calculate, its value is read off the tables for the specific interest rate i and time period n. The exponential and other associated factors are expressed in a special way, called *functional notations*, to minimize the confusion in using them. Tables containing values of functional notations are known as *compound interest tables*. Such tables are sandwiched in between Chapters 8 and 9.

Equation (3.1), given by $F = P(1 + i)^n$, can be rewritten as

$$F = P(F/P, i, n) \tag{3.6a}$$

where

$$(F/P, i, n) = (1 + i)^n \tag{3.6b}$$

Note that the notation $(F/P, i, n)$ is simply another way to express the exponential term $(1 + i)^n$. It too is a function of i and n. It is used for evaluating F by multiplying its value with P, as is obvious in Equation (3.6a). We call it the *compound amount factor*, since it actualizes the effect of compounding on a given sum P. The factor $(F/P, i, n)$ is read as *Find F given P, i, and n*. Note that the notation has been so composed that the unknown F is on the left side of the slash, while the knowns P, i, and n are on the right. This is similar to the way we write algebraic equations with the unknown on the left and the knowns on the right.

For conciseness, no algebraic sign exists between P and $(F/P, i, n)$ in Equation (3.6a). In reality, however, there is a sign, and it is multiplication—the same as in Equation (3.1) between P and the exponential factor $(1 + i)^n$.

The compound amount factor $(F/P, i, n)$ is used to evaluate F, given P, i, and n. In a similar way, it is possible to evaluate P for given values of F, i, and n. We do this by rewriting Equation (3.3) as

$$P = F(P/F, i, n) \tag{3.7a}$$

where

$$\begin{aligned}(P/F, i, n) &= 1/(1 + i)^n \\ &= (1 + i)^{-n}\end{aligned} \tag{3.7b}$$

The notation $(P/F, i, n)$ is called the *present worth factor,* since it is used to evaluate the present worth P of a future sum F.

To get more conversant with the functional notations, refer to the pages of compound interest tables at the end of Chapter 8. There is a page of data for one interest rate. Each page contains several columns. The first column is always for the time period n. For the time being, we limit ourselves to the second and third columns, under *Single Payment*. From these we can read off the value of $(F/P, i, n)$ or $(P/F, i, n)$ for a given n. As can be seen, the tables contain values for other factors too, which we shall discuss later.

In using the table while solving a problem, simply reach for the page for the particular interest rate, then look for the value of n under the first column and read off the factor's value under the appropriate column. Can you find the value of the F/P factor[9] for $i = 8\%$ and $n = 10$ to be 2.159? You can use your calculator to confirm that $(1 + 0.08)^{10}$ is indeed 2.159. Note that some sacrifice is made in the accuracy of a factor's value by rounding it to three decimal places in the table, and therefore in the final result, in comparison to its exact value.

By the time we have learned about all the functional factors, eight in number (see the columns in any table), confusion may arise as to which column's value to use for a problem. The confusion can be minimized by looking at the notation closely. Let us consider the expression

$$F = P(F/P, i, n)$$

We are interested in checking whether the functional notation $(F/P, i, n)$ used in this expression is the correct one. Should it have been $(P/F, i, n)$? To do this, we compare the two sides of the equal sign. On the left we have variable F. So we should get F on the right side too. On the right, the variable P is logically and visually multiplied with F/P. This multiplication $(P \times F/P)$ does indeed yield F, since the numerator P cancels out the denominator P, leaving behind F. Thus, on the right side of the equal sign we also get F, which is what we have on the left. We have therefore checked[10] that the functional notation $(F/P, i, n)$ used in the expression is the correct one.

In case you are not sure of the functional factor[11] to use in a problem, mentally follow the procedure explained in the preceding paragraph.

---

[9] F/P is a short form of $(F/P, i, n)$; both mean the same.

[10] There is another way to check for this. Mentally consider an equal sign in place of / in the parentheses of the notation. In this case, with = replacing the /, we get $F = P$ within the parentheses, which is the same as outside, so the functional notation is correct. This comes from the way the functional notations have been composed.

[11] This is an attractive feature of the functional-notation approach to solving engineering economics problems. Users do not have to remember any mathematical equation. They only need to be skillful in selecting the correct notation. For example, for determining F for a given P, the user simply needs to know that the appropriate functional factor is $(F/P, i, n)$. With this knowledge he reads off the factor's value in the table and multiplies it with the known P to determine the unknown F.

We shall come across other functional notations later. They are all based on the same logic and are therefore *user friendly* in comparison to methods requiring calculation of the exponential term.

Note that the words *notation* and *factor* are being used interchangeably.

To illustrate the use of functional notations, we redo the four examples (Examples 3.2 to 3.5) presented earlier. For evaluating F we use Equation (3.6a) and for P Equation (3.7a), as illustrated in Examples (3.6) and (3.7). There are no direct formulas for evaluating i and n using functional notations. Instead, we manipulate the data and rearrange the terms, as illustrated in Examples (3.8) and (3.9).

EXAMPLE 3.6

Jenny deposits $300 in an account to earn annually compounded interest of 6% per year. How much does she get at the end of the eighth year?

Solution

We are given $P = \$300$, $i = 6\%$ per year, and $n = 8$ years, and asked to determine F. Select the functional notation formula where the unknown F is on the left side of the equal sign. It is Equation (3.6a), as

$$F = P(F/P, i, n)$$

On substitution of the given values, we get the following: within the parentheses the values of only i and/or n are substituted.

$$F = 300(F/P, 6\%, 8)$$

Note that i has been expressed in percent, since the tables are titled in percent. However, the factors' values in the table have been calculated using fractional values of i, as required in the basic equation.

From the tables, we need to evaluate $(F/P, 6\%, 8)$. Turn the table pages to locate the one for 6%. Run your finger on the first column under n to reach 8. Read off the value of F/P in the second column (Find F Given P) as 1.594. Use this in the preceding expression to determine F as

$$\begin{aligned} F &= 300 \times 1.594 \\ &= \$478.20 \end{aligned}$$

Let us compare the notation-based result with that using a calculator to see how much accuracy has been sacrificed by using the tables. We need to evaluate the term $(1 + i)^n$, which is what $(F/P, i, n)$ represents. Recalling that i must be in a fraction in the exponential form, we find that $(1 + 0.06)^8 = 1.593848$. With this value, $F = \$300 \times 1.593848 = \$478.15$. The difference in the results for F is five cents ($478.20 – $478.15), which for all practical purposes is negligible.

EXAMPLE 3.7

John will need $5,000 in four years as down payment for an apartment he plans to buy after getting a job. How much should he invest now in a CD that pays 7% yearly interest, compounded annually?

Solution

We are given F, i, and n, and asked to determine P. So select the functional notation formula with the unknown P on the left side of the equal sign. It is Equation (3.7a), that is,

$$P = F(P/F, i, n)$$

On proper substitution of the given values, we get

$$P = 5{,}000(P/F, 7\%, 4)$$

Turn to locate the page for 7%, and run your finger on the first column (under n) to reach the entry for 4. Read the value of P/F in the third column (Find P Given F) as 0.7629. Use this in the preceding expression to evaluate P as

$$\begin{aligned} F &= 5{,}000 \times 0.7629 \\ &\approx \$3{,}815 \end{aligned}$$

Thus, John needs to invest $3,815 now.

When the interest rate i or the time period n is the unknown, the functional-notation approach requires interpolation, as illustrated in Examples 3.8 and 3.9.

EXAMPLE 3.8

Joe gets back $4,000 for a one-time deposit of $3,200 made six years ago in a money market account. What yearly interest rate did he earn if compounding was annual?

Solution

The given data are $P = \$3{,}200, F = \$4{,}000, n = 6$ years. We need to determine the value of the interest rate i. In the functional-notation approach there is no formula for evaluating i directly, and therefore we manipulate the basic relationship. We can use either Equation (3.6a) or (3.7a). The former is

$$F = P(F/P, i, n)$$

By substituting the given values and rearranging the terms, we get

$$\begin{aligned} 4{,}000 &= 3{,}200(F/P, i, 6) \\ (F/P, i, 6) &= 4{,}000/3{,}200 \\ &= 1.25 \end{aligned}$$

Our task now is to search the tables to locate the interest rate i, corresponding to $n = 6$, for which the value of F/P is 1.25.

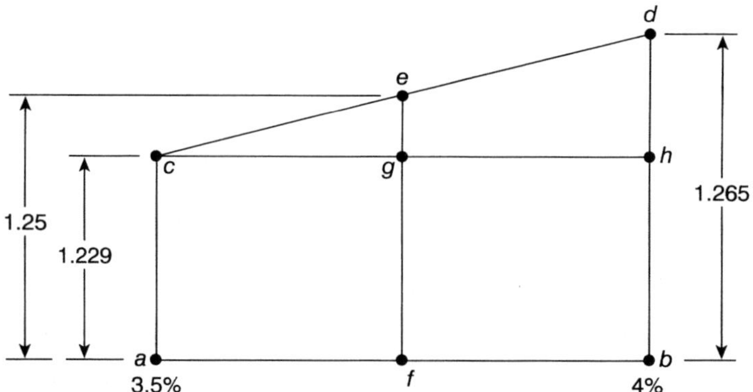

Figure 3.2 Interpolation

To do this, turn to any page of the tables. Let us say this happens to be the page for $i = 6\%$. On that page, for $n = 6$, $(F/P, 6\%, 6)$ is 1.419. This value is greater than 1.25, so our search continues. But shall we search in the pages of higher interest rates or the lower ones? Let us try the adjoining page of 7%, wherefrom $(F/P, 7\%, 6) = 1.501$. Since 1.501 is greater than 1.419—the F/P value corresponding to 6%—we are heading in the wrong direction, since our search is for 1.25. So turn toward lower interest rates. Let us say we check the page for 5%. For this interest rate, the F/P factor[12] for $n = 6$, being 1.34, is still greater than 1.25, but we are getting closer. For $i = 4\%$, the F/P factor corresponding to $n = 6$ is 1.265; and for $i = 3.5\%$ it is 1.229. So the interest rate for which $(F/P, i, 6)$ will be 1.25 is somewhere in between 3.5% and 4%. To find the exact value of that interest rate, an interpolation within this range is essential.

We set out the data to be interpolated in a sketch. Referring to Fig. 3.2, the F/P values are plotted vertically and the interest values horizontally. The unknown i will be found on the horizontal axis. The height ac represents 1.229, the value of F/P corresponding to 3.5%, while bd represents 1.265, corresponding to 4%. We want to know the value of i that will yield F/P as 1.25. Let us say 1.25 is represented[13] by ef. So we need to know how far point f is from point a, that is, the length of line af. We draw a horizontal line ch, parallel to ab, to generate the two triangles cge and chd. Since these triangles[14] are similar,

$$\frac{eg}{dh} = \frac{cg}{ch}$$

$$\frac{ef - gf}{bd - bh} = \frac{af}{ab} \qquad (\text{since } cg = af \quad \text{and} \quad ch = ab)$$

[12] *F/P factor* is another way to express $(F/P, i, n)$. Likewise, we use P/F factor for $(P/F, i, n)$.

[13] Since the value of ef lies *between* the known boundaries of $ac = 1.229$ and $bd = 1.265$, the process of determining the unknown is called *interpolation*.

[14] This interpolation is based on first principles. Some analysts derive an equation for the purpose of interpolation using $x_{\min}$, $y_{\min}$, etc. I prefer the first-principles approach, since it avoids an additional equation. Moreover, this approach requires nothing to remember—just a grasp of the interpolation concept.

$$\frac{1.25 - ac}{1.265 - ac} = \frac{af}{4 - 3.5} \qquad (\text{since } gf = ac \quad \text{and} \quad bh = ac)$$

$$\frac{1.25 - 1.229}{1.265 - 1.229} = \frac{af}{0.5}$$

$$\frac{0.021}{0.036} = \frac{af}{0.5}$$

$$af = \frac{0.021 \times 0.5}{0.036}$$

$$\approx 0.292$$

The interest rate represented by point f will be the value corresponding to point a plus the value represented by length af, that is,

$$\begin{aligned} i &= 3.5\% + 0.292\% \\ &= 3.792\% \end{aligned}$$

Thus, Joe earned a yearly interest of 3.792%. This is slightly higher than the calculator-based value 3.789% in Example (3.4). The minute difference is due to the rounded F/P values in the table, which are limited to three decimal places.

EXAMPLE 3.9

Jaya is thinking of surprising his son with a cash gift of $1,000 that he can use as down payment to buy his first car. She can afford to save $500 now in a bank account that will earn an annual interest rate of 6%, compounded per year. How long will Jaya have to wait to surprise her son?

Solution

Here the given data are $P = \$500, F = \$1{,}000, i = 6\%$. We need to determine the value of n. In the functional-notation approach there is no formula for evaluating n directly. We use Equation (3.6a) or (3.7a) and manipulate the terms. With the former, we have

$$F = P(F/P, i, n)$$

Substituting the given values,

$$\begin{aligned} 1{,}000 &= 500(F/P, 6\%, n) \\ (F/P, 6\%, n) &= 1{,}000/500 \\ &= 2 \end{aligned}$$

Our task is to determine from the tables the value of n for which $(F/P, 6\%, n)$ is 2. Turn to the page for the 6% interest. Search down under the F/P column for a value of 2. Notice that for $n = 11$, F/P is 1.898, and for $n = 12$ it is 2.012. So the value of n for which F/P will be 2 is somewhere in between 11 and 12. To determine that value of n we interpolate in a way similar to that in Example 3.8.

Set out the data to be interpolated in a sketch similar to Fig. 3.2. Refer to that figure for the discussion that follows. Set the F/P values vertically and the n values horizontally. In this case, ac represents 1.898, corresponding to $n = 11$, while bd represents 2.012, corresponding to $n = 12$. We want to know the value of n that will yield F/P as 2. Let us say 2 is represented by ef. We need to know how far point f is from point a, that is, the length of line af. Draw a line ch to generate the two triangles ceg and cdh. Since these triangles are similar,

$$\frac{eg}{dh} = \frac{cg}{ch}$$

$$\frac{ef - gf}{bd - bh} = \frac{af}{ab} \qquad (\text{since } cg = af \quad \text{and} \quad ch = ab)$$

$$\frac{2 - ac}{2.012 - ac} = \frac{af}{12 - 11} \qquad (\text{since } gf = ac \quad \text{and} \quad bh = ac)$$

$$\frac{2 - 1.898}{2.012 - 1.898} = \frac{af}{1}$$

$$\frac{0.102}{0.114} = af$$

$$af = \frac{0.102}{0.114} \approx 0.895$$

The value of n represented by point f corresponds to point a plus the value represented by length af. Thus,

$$n = 11 + 0.895 = 11.895 \text{ years}$$

Jaya will have to wait 11.895 years to surprise her son. (This value of n is slightly lower than 11.897 years in Example 3.5. The minute difference is due to the rounding of F/P values in the table, which are limited to only three places of decimal.)

The evaluation of i or n using the interest table involves search and interpolation,[15] as was seen in Examples 3.8 and 3.9. The interpolation seems lengthy and cumbersome, but with practice it becomes easier.

---

[15]In the interpolations discussed in this section, and also elsewhere in the text, we make an implicit assumption. We assume that the variation between the two variables being interpolated is *linear*. This linearity assumption introduces some error if the variables have nonlinear relationships, as with the exponential function $(1 + i)^n$ or the associated functional notations. As long as the error is not large, it is acceptable for all practical purposes. To minimize the error, keep the boundaries of the variable on the x-axis as close as possible.

A final note on interpolation. In most problems the interpolation may have to be done the way it was explained in Examples 3.8 and 3.9. There are times, however, when interpolation may be done in your head; for example when the value being searched for is just halfway between the two boundaries. Suppose that in Example 3.8 the value of F is \$3,990 instead of \$4,000. Then $(F/P, i, n)$ would be $3{,}990/3{,}200 = 1.247$. This value happens to be exactly in between 1.229 corresponding to 3.5% and 1.265 corresponding to 4%. Therefore, the unknown i would simply be exactly in between 3.5% and 4%, that is, 3.75%. Such mental interpolations are possible also in problems where the point being searched happens to be at the quarters, thirds, or tenths of the range.

3.6 BASIC EQUATION'S DIAGRAM[16]

We have learned about cashflow diagrams in Chapter 2. How does the basic relationship, Equation (3.1), where the initial sum P and the final sum F are the two cashflows, look as a cashflow diagram?

As discussed in Chapter 2, a cashflow diagram looks different depending on whose viewpoint is considered. From the lender's or investor's viewpoint, the cashflow diagram of the basic equation will look as in Fig. 3.3(a). Here, the initial sum P is invested now ($n = 0$). Its vector is located at time marker 0, and the direction is *downward*, since it is a cost[17] to the investor. At the end of the nth period, a final sum F is received back (a benefit) by the investor. Since F will then increase (move *up*) the investor's account balance, the F vector is *upward* at time marker n. The timeline is shown broken, since n is a variable in the basic equation, whose value can be any positive integer.

From the borrower's viewpoint, the cashflow diagram will be as shown in Fig. 3.3(b). This diagram is like Fig. 3.3(a) except that the directions of P and F are reversed. The direction convention, however, remains the same: P is upward since the borrowed sum P is a benefit to the borrower, while F is downward since its payment to the lender at the end of the nth period is a cost.

Since the final sum F is greater than the initial sum P due to compounding, in both the diagrams of Fig. 3.3 F is longer than P. The ratio of their lengths is equal to the exponential function $(1 + i)^n$, according to Equation (3.1). It is this ratio (F/P) that is given in column 2 of the interest tables, while the ratio P/F is in column 3. It may be noted that any value in column 3 is the reciprocal of the corresponding value in column 2 for all sets of i and n. This is also obvious from a comparison of Equations (3.6b) and (3.7b). Thus,

$$(P/F, i, n) \times (F/P, i, n) = 1$$

---

[16]The cashflow table discussed in Chapter 2 is not of much use in single-payment problems, and hence not used here. The usefulness of cashflow tables becomes evident in problems involving multiple cashflows, as in later chapters.

[17]It brings *down* his account balance; refer to Section 2.2 (Chapter 2) for vector convention, if necessary.

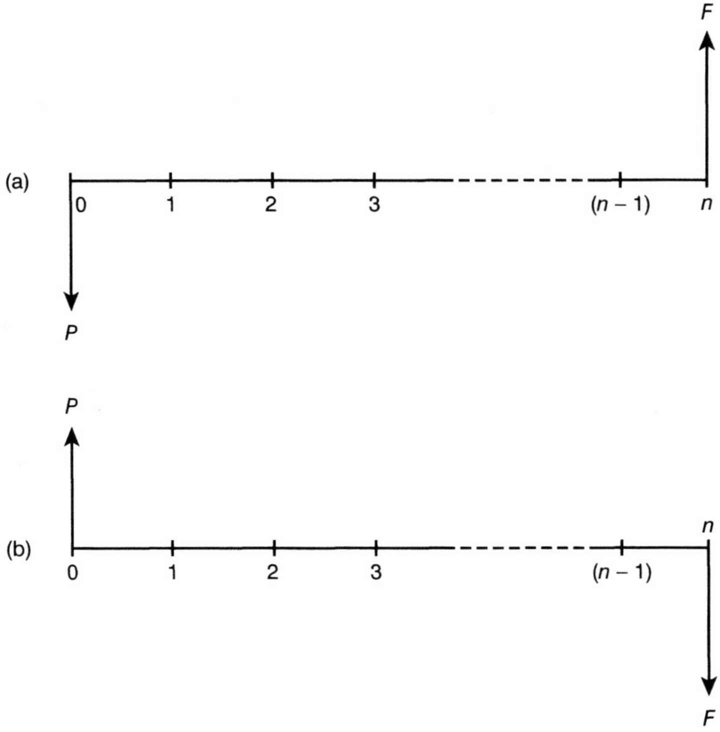

Figure 3.3 Basic Equation's Cashflow Diagram

Problem statements involving P, F, i, and n may at times be difficult to comprehend. In such cases cashflow diagrams should be sketched to facilitate comprehension, as illustrated later through Examples 3.10 and 3.11.

3.7 COMPOUNDING PERIOD

As discussed earlier, compounding is beneficial to the lender. It enables the principal to accumulate faster than under simple interest. Obviously, the more frequent the compounding, the faster is the growth of the principal for a given interest rate. In engineering projects, compounding is usually annual. However, it may take place more frequently, such as quarterly or monthly, especially in cases of personal borrowings and investments.

Suppose Rita deposits \$2,000 in an account for two years at an annually compounded interest rate of 5% per year. At the end of the first year, due to compounding, her savings become \$2,100, comprising the initial deposit of \$2,000 plus the interest of \$100 (= \$2,000 × 0.05) for the year. During the second year the interest earned is \$2,100 × 0.05 = \$105. Thus, at the end of the two years, Rita's account balance is \$2,100 + \$105 = \$2,205. This could have been obtained from functional notation as[18]

[18]The difference of \$1 between \$2,205 and \$2,204 is due to the rounding off of the functional factor's value in the table.

$$\begin{aligned} F &= 2{,}000(F/P, 5\%, 2) \\ &= 2{,}000 \times 1.102 \\ &= 2{,}204 \end{aligned}$$

Now, consider that the compounding is done every six months. Then,

$$\begin{aligned} &\text{Principal at the end of the first six months} \\ &= \$2{,}000 + \text{Interest for the first six months} \\ &= \$2{,}000 + (\$2{,}000 \times 0.05 \times 6/12) \\ &= \$2{,}050 \end{aligned}$$

The factor 6/12 in the calculation accounts for the fact that interest is earned only during 6 of the 12 months of the year.

This $2,050 now earns interest for the second six months. Thus,

$$\begin{aligned} \text{Principal at the end of the second six months} &= \$2{,}050 + (\$2{,}050 \times 0.05 \times 6/12) \\ &= \$2{,}050 + \$51.25 \\ &= \$2{,}101.25 \end{aligned}$$

Similarly, the principal at the end of the third six months, that is, the first six months of the second year, is

$$\begin{aligned} \$2{,}101.25 + (\$2{,}101.25 \times 0.05 \times 6/12) &= \$2{,}101.25 + \$52.53 \\ &= \$2{,}153.78 \end{aligned}$$

And finally, the principal at the end of the last six months, that is, at the end of the second year, is

$$\begin{aligned} \$2{,}153.78 + (2{,}153.78 \times 0.05 \times 6/12) &= \$2{,}153.78 + \$53.84 \\ &= \$2{,}207.62 \end{aligned}$$

Thus, the six-monthly compounding enables Rita's $2,000 savings to become $2,207.62 at the end of two years. Compared to the $2,205 with annual compounding, this is $2.62 more for Rita. While this $2.62 may sound little, for large initial investments or more frequent compoundings the difference can be quite significant.

In order to compete, financial institutions offer savers compoundings more frequently than yearly. Half-yearly, quarterly, or monthly compoundings are prevalent in the financial marketplace. In some cases compounding may even be daily, and in the extreme it may be *continuous*—every moment of the time.

How can we account for more frequent compoundings without having to labor through the first principles? This is done simply by considering *periods*, rather than *years* or *months* or other units of time, as compounding intervals.

The interest rate usually quoted is yearly; we call it *nominal*[19] interest rate and denote by r. The symbol i is used for *interest rate per period*. If compounding is annual, then $i = r$. However, if it is more frequent, then apportion the nominal rate to reflect

---

[19]Financial institutions use APR (annual percentage rate) for nominal (annual) interest rate r. Sometimes they quote a daily rate by dividing the APR by 365, the number of days in a year. The daily rate is quoted to convey to the borrower the feeling of a lower interest rate.

the interest rate for the period. This will yield the i that is used in the formulas. Since for annual compounding $i = r$, we did not bother to make this distinction earlier between i and r.

In Rita's case, r is 5% per year. Since the compounding is done every six months, that is, twice a year, the interest rate i for the period (6 months) is $5\%/2 = 2.5\%$. Also, over the two years of her investment there will be four periods of compounding (four six months in two years). So the value of n is 4. We can use Equation (3.6a) to evaluate the final sum F for Rita provided we use these values of i and n, as follows:

$$\begin{aligned} F &= P(F/P, i, n) \\ &= 2{,}000\,(F/P, 2.5\%, 4) \\ &= 2{,}000 \times 1.104 \\ &= \$2{,}208 \end{aligned}$$

This answer is approximately the same as that obtained earlier ($2,207.62) from first principles. The difference of 38 cents is due to the limited number of decimal places in the F/P value.

EXAMPLE 3.10

Sam owns and operates a small electronics repair business that he set up on graduation from the local engineering college. Five years ago he had a very profitable fiscal year. From the profits that year he invested $5,000 in a 10-year CD at 5%, compounded semiannually. The last few years have been bad, but this year he has made a decent profit and decides to invest $4,000 for five years in another CD at 6%, compounded quarterly. At the end of five years, he intends to buy a small CNC milling machine at an expected cost of $25,000. After using the proceeds of the two CDs, how much more will he have to spend on the machine?

Solution

This example is slightly more complex. Let us use a cashflow diagram to comprehend the problem. Based on what we learned in Chapter 2, the cashflow diagram is drawn in Fig. 3.4. Referring to the diagram, P_1 ($5,000 deposit) accumulates to F_1, while P_2 ($4,000) accumulates to F_2. Thus, the proceeds from the two CDs maturing at the same time will be the total of F_1 and F_2. Sam will have to spend the difference between $25,000 and the total proceeds.

As discussed in Chapter 2, the diagram should show, as in Fig. 3.4, the number of compounding periods, rather than the years. The value of n for the first CD (n_1) is 20, since two compoundings take place each year over ten years. For the second CD too, the value of n (n_2) is 20, since four compoundings take place each year over five years.

Also note that the applicable interest rates will be different from those given, due to compoundings other than annual. Since the first investment of $5,000 is compounded twice a year, i_1 is $5\%/2 = 2.5\%$. For the second investment of $4,000 compounded quarterly, i_2 is $6\%/4 = 1.5\%$.

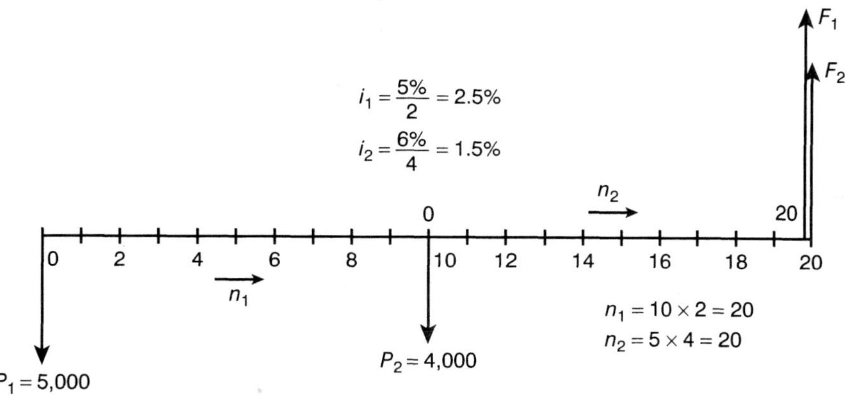

Figure 3.4 Two Pairs of Cashflows

Once the problem has been fully comprehended using the cashflow diagram, the next step is to solve for the unknowns, which in this case are F_1 and F_2. We can use any of the four methods discussed earlier to evaluate these two future sums. Let us use the (preferable) functional-notation method for its simplicity. You should be able to write the following relationship *without referring to any equation whatsoever, if you have mastered the fundamentals presented* in the text so far.

$$\begin{aligned}\text{Total proceeds} &= F_1 + F_2\\ &= 5{,}000\,(F/P, 2.5\%, 20) + 4{,}000\,(F/P, 1.5\%, 20)\\ &= 5{,}000 \times 1.639 + 4{,}000 \times 1.347\\ &= 8{,}195 + 5{,}388\\ &= \$13{,}583\end{aligned}$$

Since the milling machine costs \$25,000, Sam will have to spend \$11,417 more, the difference between \$25,000 and the \$13,583 proceeds from the two CDs.

EXAMPLE 3.11

Javed, a civil engineer, owns and operates a small consulting firm. He invested \$5,000 five years ago in a six-year CD at a 5.5% annual rate, compounded semiannually. This investment will mature next year. He had planned at the time of investment to purchase finite element software for design work when the CD would mature, but the software company is offering an attractive deal this year. Future updates of the software will be free if purchased this year. He is considering closing the CD and using the proceeds to purchase the software this year rather than next year as originally planned.

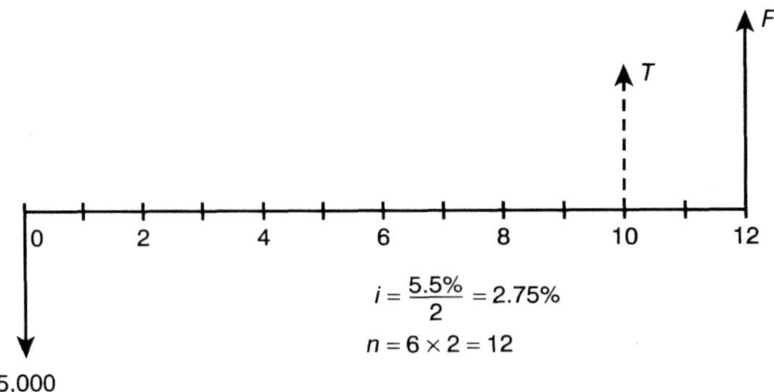

Figure 3.5 Penalty for Premature CD Closing

There is a penalty for premature closing, however. Not only does he lose the interest for the current (sixth) year, he also foregoes 1% of the total interest earned. How much will he get if he closes the CD, and how much is his loss from premature closing?

Solution

The problem statement seems complex and may be difficult to comprehend. So, let us draw a cashflow diagram with the hope that it will facilitate comprehension.

Based on what we learned in Chapter 2, the cashflow diagram is sketched in Fig. 3.5. Due to the semiannual compounding, $n = 6 \times 2 = 12$. For the same reason, $i = 5.5\%/2 = 2.75\%$ per period. As stipulated, the $5,000 investment is due as F at $n = 12$. But, since Javed is considering buying the software now (period 10), a year before the CD is due, he will get[20] T. We need to evaluate T and consider the penalties. The first penalty is that he does not get the interest for the sixth year, that is, for the last two periods. This is taken care of by T, since it corresponds to $n = 10$. In addition, he will lose 1% of the interest earned, which is simply the difference between T and the original $5,000 investment.

Again we use the functional notation method. The appropriate formula is Equation (3.6a), wherefrom we have

$$T = 5{,}000(F/P, 2.75\%, 10)$$

Now, we search for the value of this F/P factor in the tables. Since there is no page for 2.75%, we interpolate the values for 2.5% and 3%. Being exactly between 2.5% and 3%, for which F/P values are 1.280 and 1.344, the F/P value for 2.75% is their mean, as

[20]The use of symbol T for future sum should not be confusing. It is really a future sum with reference to period 0 when the investment was made; but since we are already using F, we had to use another symbol. We chose T; someone could have chosen K or any other symbol. As an alternative to using symbols other than F for future sums, one can use F_1, F_2, etc. This might minimize confusion but introduces subscripts.

$$
\begin{aligned}
(F/P, 2.75\%, 10) &= ((F/P, 2.5\%, 10) + (F/P, 3\%, 10)]/2 \\
&= (1.280 + 1.344)/2 \\
&= 1.312
\end{aligned}
$$

Thus,

$$
\begin{aligned}
T &= 5{,}000 \times 1.312 \\
&= \$6{,}560
\end{aligned}
$$

Therefore, Javed earns total interest of

$$
\begin{aligned}
I &= T - P \\
&= \$6{,}560 - \$5{,}000 \\
&= \$1{,}560
\end{aligned}
$$

He loses 1% of this, that is, \$15.60 as another penalty. Thus, Javed will get back \$6,560 − \$15.60 ≈ \$6,544 on closing the CD. This is the first part of the answer.

Were the CD to run its full term of six years, the proceeds would have been F, where

$$F = 5{,}000(F/P, 2.75\%, 12)$$

Evaluating in the same way, we find

$$
\begin{aligned}
(F/P, 2.75\%, 12) &= [(F/P, 2.5\%, 12) + (F/P, 3\%, 12)]/2 \\
&= (1.345 + 1.426)/2 \\
&= 1.3855
\end{aligned}
$$

Thus,

$$
\begin{aligned}
F &= 5{,}000 \times 1.3855 \\
&\approx \$6{,}928
\end{aligned}
$$

Therefore, the loss from premature closing of the CD is

$$
\begin{aligned}
\text{Proceeds if not closed} - \text{Proceeds if closed} &= \text{F} - \$6{,}544 \\
&= \$6{,}928 - \$6{,}544 \\
&= \$384
\end{aligned}
$$

3.8 EFFECTIVE INTEREST RATE

The interest rate quoted most often is annual, also called the nominal rate. For a given nominal interest rate, compounding more often than yearly yields higher earnings than with annual compounding. This in effect means that the *actual* annual interest rate is higher than the nominal rate. We distinguish the two by labeling the quoted annual rate as *nominal* interest rate, denoted by r, and the actual annual rate as *effective* interest rate, denoted by i_{eff}. The effective interest rate is also termed *annual yield.*

Let us reconsider Rita's case discussed in the previous section to understand the difference between nominal and effective interest rates. Her 5% yearly interest is the *nominal* interest rate r. Rita's final sum F at the end of two years was $2,205 under annual compounding. With semiannual compounding it was $2,208, that is $3 more. We want to know the *effective* annual rate the semiannual compounding yielded. In other words, we want to know what annual interest rate would have allowed Rita's $2,000 investment to become $2,208, had the compounding been yearly? This rate will be i_{eff}. Let us work it out using functional notation. The solution is very similar to that in Example 3.8. Following the procedure explained there, we have

$$\begin{aligned} F &= P(F/P, i, n) \\ 2{,}208 &= 2{,}000\,(F/P, i, 2) \\ (F/P, i, 2) &= 2{,}208/2{,}000 \\ &= 1.104 \end{aligned}$$

From the tables, the adjoining[21] interest rates for this value of F/P for $n = 2$ are 5% with a value of 1.102 and 6% with a value of 1.124. Using the relationship based on similarity of triangles, as explained in Example 3.8 through Fig. 3.2, we get

$$\begin{aligned} \frac{1.104 - 1.102}{1.124 - 1.102} &= \frac{af}{6 - 5} \\ \frac{0.002}{0.022} &= \frac{af}{1} \\ af &= \frac{0.002}{0.022} \\ &= 0.091 \end{aligned}$$

The interest rate corresponding to the F/P value of 1.104 is thus

$$5\% + 0.091 \approx 5.09\%$$

Thus the *effective annual* interest rate i_{eff} is 5.09%. In other words, the increase of $2,000, invested for two years at 5% *nominal* annual rate under semiannual compounding, to $2,208 is equivalent to an *effective* annual rate of 5.09%. Remember that the effective annual interest rate will always be higher than the nominal for compoundings more frequent than yearly.

Instead of the first principles, the following formula can be used to determine the *effective annual* interest rate i_{eff} for a given *nominal* interest rate r and number of compoundings m per year.

---

[21]Interpolation should be done only between the adjoining set of data. Otherwise the nonlinear nature of the basic compounding equation introduces more error.

$$i_{\text{eff}} = \left(1 + \frac{r}{m}\right)^m - 1 \tag{3.8}$$

Recall that when compounding is annual ($m = 1$), $i = r$. We find this to be true in Equation (3.8); with $m = 1$, i_{eff} becomes r which equals the i we have been using. Thus, for annual compounding, $i = r = i_{\text{eff}}$.

Applying Equation (3.8) to Rita's case where $r = 5\%(= 0.05)$ and $m = 2$, we find

$$\begin{aligned} i_{\text{eff}} &= (1 + 0.05/2)^2 - 1 \\ &= 1.025^2 - 1 \\ &= 0.0506 \\ &= 5.06\% \end{aligned}$$

This value of i_{eff} is slightly lower than the 5.09% determined earlier, due to rounding errors.

Why use three kinds of interest rates, which may be confusing? Because each serves a different purpose. Of the three interest rates discussed in this section,

The *nominal interest rate* r is often the quoted one. It is expressed on an annual basis and is used for crude comparison.

The *interest rate i corresponding to the compounding period* is the one that is used in calculations and on which the interest tables are based. It is simply the nominal interest rate r divided by m, the number of compoundings per year.

The *effective annual interest rate* i_{eff} is used for comparison of "*real*" interest rates offered by financial institutions. The nominal interest rate r does not account for the effect of compounding, if other than annual, on the growth of investment; i_{eff} does this.

Lending institutions quote the interest rate they charge their customers. They usually like to quote the nominal interest rate r. To make the interest rate seem low, they sometimes quote r on a monthly or daily basis. For $r = 18\%$ per year, the equivalent monthly rate of 1.5% (18/12) seems so attractive, the daily rate at 0.05% being even more so.

It should be clear by now that what is really important in deciding from where to borrow is the effective annual interest rate i_{eff}, the lowest being the best. It is also called *annual percentage yield* (APY). Lenders are required by law to quote the APY so that borrowers can compare the cost of borrowing. The up-front finance charges in securing a loan and other costs of financing, such as insurance, should all be accounted for in such a comparison.

EXAMPLE 3.12

According to its brochures the local credit union offers to its savers an annual interest rate of 8%, compounded quarterly. What are the values of r, i, and i_{eff}?

Solution

a. The given annual interest rate is the nominal interest. Hence $r = 8\%$.

b. Since there are four compoundings per year, $m = 4$; the interest rate per compounding period

$$\begin{aligned} i &= r/m \\ &= 8\%/4 \\ &= 2\% \end{aligned}$$

c. We determine i_{eff} from Equation (3.8) for $r = 8\%$ ($= 0.08$), and $m = 4$, as

$$\begin{aligned} i_{\text{eff}} &= (1 + 0.08/4)^4 - 1 \\ &= 1.02^4 - 1 \\ &= 1.08243 - 1 \\ &= 0.08243 \\ &= 8.24\% \end{aligned}$$

3.9 DOUBLING TIME

We have learned by now that an investment increases exponentially under compound interest. We are sometimes interested in knowing how long it takes an investment to double. This doubling time is of interest in other areas too, for example in predicting how long it will take for the population of a country, or the wealth of an individual, to double.

The doubling time of an investment P can be determined from the basic equation:

$$F = P(1 + i)^n$$

For the future sum F to be double of P, we have

$$(1 + i)^n = F/P = 2$$

To know the value of n that will satisfy this relationship, we can take the natural logarithm (ℓn) of both sides, which yields

$$\begin{aligned} \ell\text{n}(1 + i)^n &= \ell\text{n}\, 2 \\ n\, \ell\text{n}(1 + i) &= 0.693 \\ n &= 0.693/\ell\text{n}(1 + i) \end{aligned} \tag{3.9}$$

Equation (3.9) is used to determine the doubling time for a given interest rate i, which should be expressed as a fraction. Note that it involves evaluation of the natural logarithm of $(1 + i)$. For example, for an interest rate of 5% per year, $\ell\text{n}(1 + i) = \ell\text{n}(1 + 0.05) = 0.04879$, and therefore the doubling time will be $0.693/0.04879 = 14.2$ years.

There is a simpler way to determine the doubling time, which does away with the need to use a logarithm. The term $\ell n(1 + i)$ can be expressed in its series as

$$\ell n(1 + i) = i - i^2/2 + i^3/2 - i^4/2 + \cdots$$

The value of i is usually small. Even for an interest rate as high as 20%, the fractional value of i is 0.2 only. For values of i much less than one, the terms $i^2/2,\ i^3/2,\ i^4/2,\ \ldots$ on the right side of the above relationship will be much smaller in comparison to the first term i; they can therefore be ignored. This reduces $\ell n(1 + i)$ to i. Thus, from Equation (3.9)

$$n = 0.693/i$$

where i is in fraction.

If 0.693 is approximated by 0.7, then the relationship becomes

$$n = 70/i \tag{3.10}$$

where i is in percent.

Equation (3.10) is handy in predicting the doubling time of an investment. For the previous illustration, where i was 5%, the doubling time from this equation = 70/5 = 14 years, which is very close to what we got (14.2 years) using Equation (3.9). Remember, however, that the smaller the value of i, the closer the result from Equation (3.10) is to that from Equation (3.9). This is because we made such an assumption earlier by ignoring the other higher terms of the logarithmic series of $\ell n(1 + i)$.

Equation (3.10) can be used to determine the doubling time in other areas too. For example, the world human population of 6 billion, under the current growth rate of 1.4%, will take 70/1.4 = 50 years to double. Consider another case. If you are investing in the stock market and want your investment to double in five years, then from Equation (3.10) your investment must appreciate annually by 14% ($i = 70/n = 70/5 = 14\%$).

SUMMARY

Compounding of interest is the basis of all investments, lendings, and borrowings in the business world. It is embodied in the basic equation $F = P(1 + i)^n$, sometimes called the *law of compounding*. Interest earned under compounding is more than that earned under simple interest. The exponential nature of the basic equation yields faster growth in investment nearer the end of the investment duration. For example, a one-time \$100 investment for 100 years at a 10% interest rate, compounded annually, will bring in over half a million dollars as interest during the last five years of the investment. Compounding benefits the lender at the cost of the borrower.

In solving single-payment problems, the focus of Chapter 3, one can use either the basic equation or its derived formulas. The discussions emphasized the fundamentals of interpolation, as well as of solving problems. The functional notation method has been discussed, and the compound interest tables have been introduced. The concept of effective annual interest rate has been presented. To determine the doubling time of an investment, use $70/i$, where i is the interest rate in percent.

EXERCISES

Discussion Questions

3.1 Why has Equation (3.1) been given so much prominence in this chapter?

3.2 Explain the law of compound interest to a layperson, using an example from everyday life.

3.3 Of the four methods that can be used to solve problems involving compound interest, which one do you prefer and why?

3.4 Why might practicing engineers prefer the functional notation method?

3.5 Explain the difference between nominal and effective annual interest rates.

3.6 Give an example, not the one in the text, where the doubling time relationship, Equation (3.10), can be applied.

Multiple-Choice Questions (Circle the *best* answer.)

3.7 The equation representing the law of compound interest is
a. $I = Pin$.
b. $F = P(1 + in)$.
c. $F = P(1 + i)^n$.
d. none of the above

3.8 Compounding is beneficial to
a. the lender.
b. the borrower.
c. both, depending on the interest rate.
d. either, depending on the frequency of compounding.

3.9 The present worth factor is given by
a. $(F/P, i, n)$.
b. $(P/F, i, n)$.
c. $(F/P, i, n)/(P/F, i, n)$.
d. $(P/F, i, n)/(F/P, i, n)$.

3.10 The use of interest tables is ______ for solving problems involving compound interest.
a. absolutely essential
b. essential
c. somewhat essential
d. not essential

3.11 You borrow $200 at 6% annual interest rate, compounded quarterly, and plan to pay back the loan and the interest together at the end of fifth year. The sum to be paid can be determined from
a. $F = 200(F/P, 6\%, 5)$
b. $F = 200(F/P, 1.5\%, 5)$
c. $F = 200(F/P, 6\%, 20)$
d. none of the above

3.12 For compoundings more frequent than annual, the effective interest rate
a. is higher than the nominal rate.
b. is lower than the nominal rate.
c. depends on the amount borrowed.
d. equals the nominal rate.

3.13 If a bank charges 1.5% monthly interest for credit card loans, the nominal interest rate is
a. 1.5%.
b. 15%.
c. 18%.
d. 180%.

Numerical Problems[22]

3.14 Better Mousetrap Incorporated had an excellent fiscal year, with profit far more than expected. It decides to invest $12,000 of its profit as a 2-year CD. How much will this investment mature to if the yearly interest rate is 6%, compounded annually?

3.15 If compounding is quarterly in Problem 3.14, what is the answer?

3.16 Determine through interpolation the value of $(P/F,\ 8.4\%, 10)$.

3.17 How much does one have to invest today for a lump-sum payback of $10,000 in five years if the annual interest rate is 4%, compounded quarterly?

3.18 If John is paid back $14,500 for an investment of $12,750 made three years ago under compound interest, what was the interest rate, assuming annual compounding?

3.19 Visionary Robotics invests $10,000 in the local credit union, which pays back $11,500 at the end of three years. What is the nominal interest rate if compounding is quarterly?

3.20 How long will it take a $4,500 investment to mature to $6,525 at a yearly interest rate of 5%, compounded annually?

3.21 Monica receives a cash gift of $1,000 on her tenth birthday. She decides to invest it in a five-year CD whose proceeds she plans to use when she is 15 as a down payment to buy her first car. She has heard that credit unions usually pay higher interest than banks. She can invest with one of the two local credit unions, one affiliated to her father's company and the other to her mother's. The former offers 12.5% interest per year compounded annually, the latter 12% per year compounded monthly. Which credit union should she invest with?

3.22 Myra received her federal education loan of $3,000 for the current semester on August 31. She deposited it the same day with the local bank at 3% annual interest, compounded monthly. For her expenses, she withdrew $800, $500, and $450 on the last days of September, October, and November respectively. How much will her proceeds be if she closes the account on December 31?

---

[22]Use the methods learned in this chapter to solve these problems.

3.23 Savita owns and operates a small construction business she inherited from her parents. Her biggest surprise on the first day at work is a capital gift from her parents' past savings. Ten years ago her dad invested $8,000 in a CD that compounded interest semiannually. Also, six years ago her mom saved $2,000 for her in an account at 6% annual interest compounded monthly. The two savings matured on her first day of work with a total proceeds of $15,540. What interest rate did her dad's CD earn?

3.24 How many years will it take a sum to double if it earns annually compounded interest of 9% per year. How much sooner will the doubling occur under monthly compounding? Assume the 9% interest rate to be nominal.

3.25 Engineering Unlimited takes a loan at a 16% yearly interest rate, compounded quarterly, for purchasing a robotic system. The company pays back the loan (capital and interest) through two payments of $50,000 at the end of six months and $65,000 on the first loan anniversary. How much was the loan?

CHAPTER FOUR

Multiple Payments

IN THIS CHAPTER YOU WILL LEARN TO

- Comprehend problems involving multiple cashflows
- Further appreciate the importance of the basic equation
- Solve both regular and irregular cashflow problems
- Derive equations for uniform, arithmetic, and geometric series cashflows
- Solve uniform, arithmetic, and geometric series cashflow problems
- "Tailor" the cashflows into the three known patterns

In the previous chapter we discussed the effect of compounding on a single payment. We saw how an investment P increases under compounding to become future[1] sum F. We learned that both the interest rate i and the time period n affect the relationship between F and P. We analyzed problems of single payments, which in fact involve a pair of cashflows—P by one party and F by the other. In this chapter, we extend our discussions to multiple payments involving more than one pair of cashflows.

The pair of cashflows in single-payment problems are two one-time transactions. They relate mostly to personal financing. Engineering economics problems usually involve multiple payments. The investment in a resource (machine, equipment, or similar item) comprises not only the initial (purchase) cost but also the recurring costs of maintenance. Besides, there are costs of utilities and insurance that must be paid regularly, such as monthly. In other words, industrial projects encompass several costs and benefits, giving rise to multiple payment problems.

---

[1]The concept of *present* and *future* in engineering economics is relative, not absolute. Consider that you deposited \$500 in a bank account five years ago that paid you \$575 last year as principal plus the interest. As learned in the previous chapter, in this case the present sum P is \$500 and the future sum F is \$575, though both payments have been made in the past. The present was logically the time four years ago, when \$500 was deposited, with reference to the future which occurred last year with the \$575 payment. Thus, the term *present* does not necessarily mean the present time in the physical sense. It is rather a *relative* time that can occur anywhere on the timeline. The same is true for the term *future*. The *future*, however, always comes after the present. Note that such connotations of *present* and *future* render the word *past* almost meaningless.

Even in personal financing there are situations that involve multiple payments, for example regular payments for loans on homes, cars, and so on. Leasing of products and services also requires multiple periodic payments.

Multiple payment problems arise because time has been "sliced" into convenient durations of weeks, months, and years. Wages and salaries are paid weekly or monthly, investment incomes are reported quarterly, and the budgets are drawn annually. These result in engineering projects with cashflows that occur several times.

Multipayment problems fall into one of the three categories:

1. One Cost,[2] Several Benefits
2. Several Costs, One Benefit
3. Several Costs, Several Benefits

The first two categories are in fact special cases of the third. One-time investment on a machine requiring no maintenance is a problem of the first category with recurring benefits, while regular savings to accumulate a capital is of the second. Problems involving several costs and several benefits are numerous and encountered most frequently.

Multipayment cashflows may be *irregular* or *regular*. Irregular cashflows are erratic; they occur inconsistently. Examples are occasional deposits in and withdrawals from a savings account, or repair costs of a machine when it breaks down. We discuss problems of this type in Section 4.1. Cashflows are *regular* if they are repetitive, occurring at each time period. Examples are monthly home mortgage payments or annual payments for the leased machine. We discuss such cases in Section 4.2.

Regular cashflows may have some definite pattern. In one common pattern they are constant, such as with a monthly mortgage payment. Constant cashflows form a *uniform series*; we discuss them in Section 4.3. In another common pattern, the cashflows increase with the time period by a fixed amount, forming an *arithmetic series*. A machine whose maintenance cost increases each year by the same amount is a good example. Such problems are analyzed in Section 4.4. In some cases the cashflows may form a *geometric series*, increasing each period by a certain percentage. Such cases are covered in Section 4.5. Finally, in Section 4.6 we discuss cashflows that are *approximately* uniform or series-type, and learn how to "tailor" them to simplify the solution.

As we shall see in the following sections, the law of compounding embodied in Equation (3.1) continues to be the basis of all the formulas in this chapter, as well as elsewhere.

4.1 IRREGULAR CASHFLOWS

In problems involving irregular payments the cashflows can occur at any time period. A bank savings account in which money is deposited when possible, and withdrawn when needed, leads to irregular cashflows. The cashflow diagram in such a case displays vectors here and there—not at each time period.

---

[2]As discussed in Chapter 2,the term *cost* is synonymous with *disbursement, expenditure,* or *cash outflow*. By the same token, *benefit* is synonymous with *receipt, revenue,* or *cash inflow*.

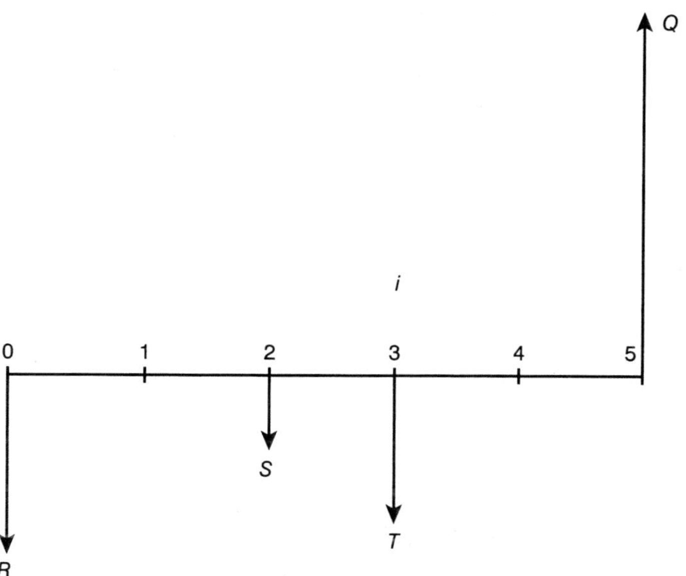

Figure 4.1 Irregular Cashflows

Consider a problem whose cashflow diagram is Fig. 4.1. A cost R is incurred now, none at period 1, S at period 2, T at period 3, and none at period 4. Since the cost does not occur at each period, its cashflows[3] are *irregular*. The payback of these *irregular* costs occurs together at period 5 as Q. Such several-costs, one-benefit problems are similar to several-deposits, one-withdrawal problems.

Problems like the one in Fig. 4.1 are easily solved by determining the F value for each cost. One can use the basic equation or its functional notation. Using Equation (3.6a), and noting that Q must equal the total of the three cost F values, we get

$$Q = R(F/P, i, 5) + S(F/P, i, 3) + T(F/P, i, 2)$$

Again, since we followed the functional notation approach there was no need to remember any formula. This made it easier to set up the expression. The basic skill lies in being able to use the appropriate notation and the correct value of n, for example 5 for R. The other skill in evaluating Q is to be able to search for the F/P values in the tables and carry out the interpolation if any.

Irregular cashflow problems of the one-cost, several-benefits or several-costs, several-benefits type are also solved the same way, by accounting for each cashflow separately. The following example illustrates the procedure.

---

[3]With reference to the timing of Q, symbols R, S, and T depict "present" sums at their time periods, and are so treated later. Instead of these symbols, we could have used P for present sum and labeled them as P_1, P_2, P_3. Since they are present sums, the appropriate functional notation to determine their future values remains $(F/P, i, n)$. Within the parentheses P is left unchanged to signify "present," what R, S, and T logically are.

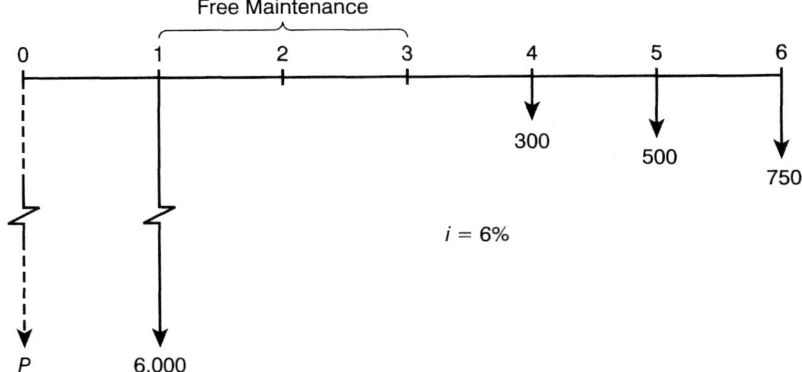

Figure 4.2 Diagram for Example 4.1

EXAMPLE 4.1

A new machine with expected useful life of five years is being planned for acquisition next year, when it will cost $6,000. This cost includes maintenance for the first two years of the purchase. The maintenance for the subsequent three years of the useful life is projected to cost $300, $500, and $750 respectively. How much should be budgeted now for the purchase and maintenance of this machine if the yearly interest rate is 6%, compounded annually?

Solution

Let us sketch a cashflow diagram to help comprehend the problem. Such a diagram is shown in Fig. 4.2. If you have difficulty understanding this diagram, then reread Chapter 2.

The diagram shows that the machine will be purchased next year for $6,000 (vector down at period 1, since it represents a cost). There are no cashflows at periods 2 and 3, since the purchase cost includes maintenance for the first two years. The maintenance costs of the subsequent three years are displayed as cashflows at periods 4, 5, and 6. Note that the $300 cost during the third year, following the period-end convention, is posted at period 4, which marks the end of period 3 as well as the beginning of period 4. So are the other two maintenance costs.

Altogether there are four costs, which will be paid for from the budget this year, that is, from P. Thus, P must be *equivalent* to these costs. Note that the vector P is shown[4] dotted to highlight that in itself it is *not a cost*, but a budgeted sum that should be enough to pay for the four future costs.

---

[4]Why isn't vector P up? This is because we are setting up the problem in a way where P is the equivalent cost now ($t = 0$) for the four future costs. Some analysts draw P vertically up, in which case it represents the one-time cash inflow into the machine account, to be used for its purchase and maintenance. Thus, P can be vertically up or down depending on the sense of its representation. In either case, the equivalency prevails, yielding the same result.

The equivalency between P and the four costs suggests that P can be determined simply by adding the P value of each of these costs. Which equation is appropriate for finding P for a given F? It is Equation (3.7a). Thus,

$$P = 6000(P/F, 6\%, 1) + 300(P/F, 6\%, 4) + 500(P/F, 6\%, 5) + 750(P/F, 6\%, 6)$$

Substituting the values of the functional factors from the interest table for 6%, we have

$$\begin{aligned} P &= 6{,}000 \times 0.9434 + 300 \times 0.7921 + 500 \times 0.7473 + 750 \times 0.7050 \\ &= 5{,}660.40 + 237.63 + 373.65 + 528.75 \\ &\approx \$6{,}800 \end{aligned}$$

Thus, the machine should be budgeted this year for $6,800. At the end of the first year, it will be $\$6{,}800(1 + 0.06) = \$7{,}208$, of which $6,000 will be used for the purchase. The remaining balance of $1,208 earning interest during the project life will fully pay for the three maintenance costs, as and when required.

4.2 REGULAR CASHFLOWS

Several situations arise, both in personal financing and engineering economics, where costs and/or benefits occur regularly at each period. This makes the cashflows[5] *regular*.[6] The cashflow diagram in such a case displays vectors at each time period. A savings account in which money is deposited regularly, say on the first day of each month, is a good example of regular cashflows. Another example is the dividend received each quarter from investment in stocks.

Consider a savings account illustrated by its cashflow diagram in Fig. 4.3. Savings are made regularly: A at period 1, B at period 2, . . . , and G at period 7. The balance in the account is withdrawn at period 7 as Q. This is similar to a several-costs, one-benefit problem.

Problems like the one in Fig. 4.3 are solved in the way illustrated earlier in Example 4.1. For determining Q, we have in this case

$$Q = A(F/P, i, 6) + B(F/P, i, 5) + \cdots + F(F/P, i, 1) + G$$

The solution requires evaluating each term on the right of this expression. Again, in this solution approach—based on functional notation—there is no need to remember any formula.

Regular, multipayment problems may not display any pattern in the cashflow magnitudes, as in Fig. 4.3. However, several such problems possess a certain pattern in their cashflow magnitudes. For example, the payments may be the same at each period, or

[5]Throughout the text, the words *cashflow* and *payment* are used interchangeably.

[6]The regularity is *either* of the costs or benefits, *not of the two together*.

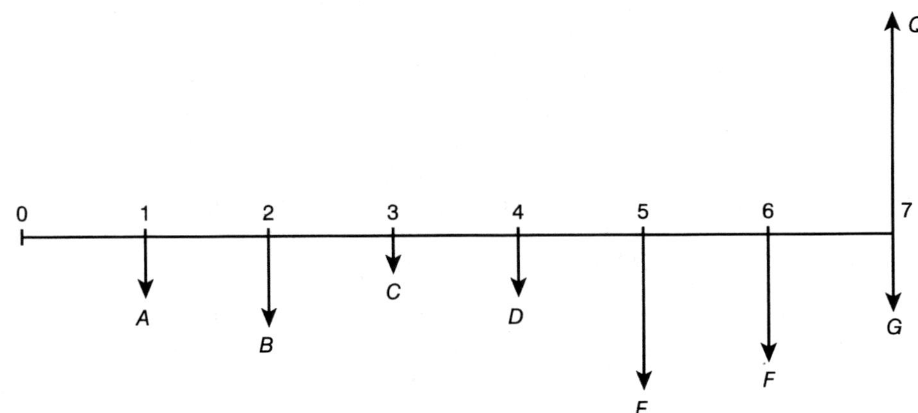

Figure 4.3 Regular Cashflows

they may increase or decrease in a certain way. The existence of a pattern may be exploited to avoid evaluating each cashflow separately, thus achieving a more efficient solution. We discuss three cashflow patterns, commonly encountered in engineering economics, in the next three sections (4.3, 4.4, and 4.5).

4.3 UNIFORM SERIES

Engineering projects often entail uniform series cashflows. The term *uniform series* means that the cashflow exists at each period and its magnitude is the same. Referring to Fig. 4.3, it means that $A = B = \cdots = G$. Home-mortgage-loan and car-lease payments are good examples of uniform series. In these cases, the benefits of owning a home or car are realized by taking a loan now and regularly paying a fixed sum over a certain length of time. Such payments generate a *series* of cashflows of constant or *uniform* magnitude.[7]

Uniform series problems may involve loan P now and equal payments A over a predetermined number of periods n. They may also be of the opposite type, that is, regular deposits of sum A to accumulate capital F at the end of n periods. In a third type of problem, a sum P is invested now to earn fixed regular income of A. All these problem types are *conceptually* the same (uniform series), and therefore analyzed the same way.

---

[7]Here are two other examples:

1. Most of us find it difficult to save for retirement once we receive our paychecks. Employers help in circumventing this difficulty by offering pension or retirement plans with deduction at the source. Self-employed people can set up their own retirement plans, usually encouraged by government through tax exemptions. In such situations, a certain sum A is saved regularly over n periods. The savings are invested for the "long haul," earning better than short-term interest. The balance in the retirement account is realized in several ways: In one the retiree receives a lump sum, and in another a regular sum throughout his life.

2. Companies sometimes raise future capital from within by saving each year out of the profit.

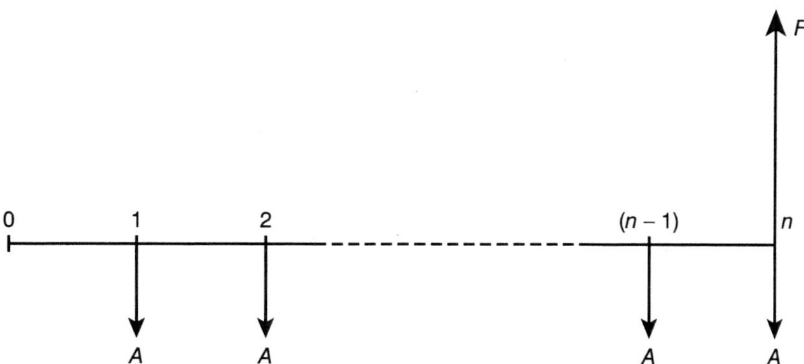

Figure 4.4 Regular Saving A Accumulates to F

We consider below two types of uniform series problems. In one, uniform savings A accumulate to a future sum F at the end of n periods (Fig. 4.4); our aim is to know how A and F are related. In the other, a present loan P is returned through payments of A over n periods (Fig. 4.5a). Here we seek a relationship between P and A. Juxtaposed to Fig. 4.5(a) is Fig. 4.5(b) where a sum P invested now (cost) brings uniform income (benefit) of A over n periods.

4.3.1 Relationship between A and F

Consider the cashflow diagram in Fig. 4.4 to represent, for example, a retirement plan with A as the regular contribution. Note that there is no contribution at period 0 (now), which may be considered the beginning of employment when the employee joins the retirement plan by signing in. The first contribution is made at period 1 (deducted from the first paycheck). The same amount A is deducted each period, which may be week, month, or year, and invested to earn compound interest. As discussed in Chapter 2, the use of the term *period* avoids any confusion about the unit of time duration, as long as i and n are in compatible units. Also note in this diagram that the last deduction is made, from the last paycheck, on the day (period n) the lump sum F is collected.

4.3.1.1 Finding F for Given A, i, n

Increasing under compounding, each regular saving A contributes toward F. Thus, F is the total of these contributions along with their interests. Let us assume that the first contribution A at period 1 accumulates to become F_1 at period n. In the same way the second A accumulates to F_2, and so on. Thus, F must equal the total of $F_1, F_2, \ldots, F_n$. In other words,

$$F = F_1 + F_2 + \cdots + F_{n-1} + F_n$$

Thus, in Fig. 4.4, the sum F is basically the collection of its components $F_1, F_2, \ldots,$ F_{n-1}, F_n—all bundled together at period n.

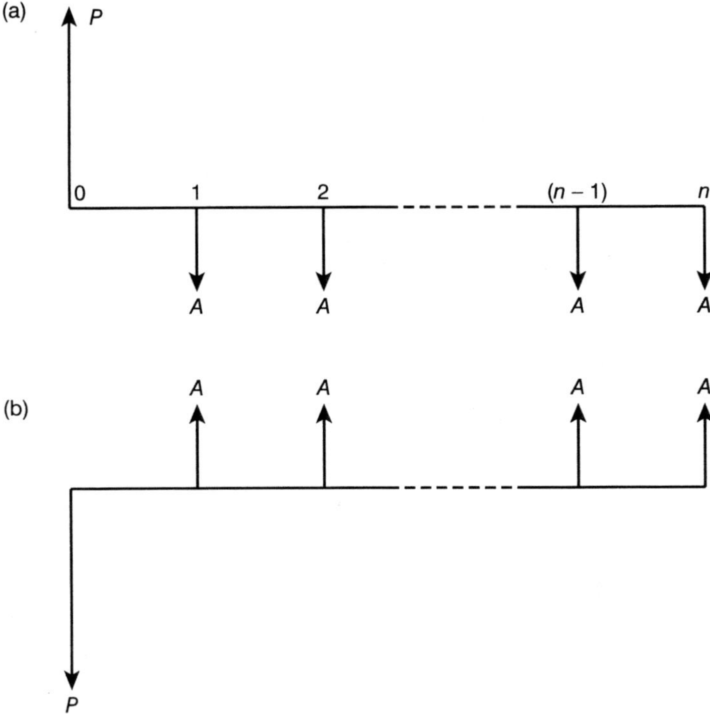

Figure 4.5 (a) Home Mortgage Loan
(b) Lump-Sum Investment as Source of Regular Income

We can consider the first contribution A and F_1 as a pair of single-payment cash-flows, in which A is like the present sum (P) and F_1 is the final sum. We can therefore apply the basic equation, Equation (3.1), to this pair, noting that the value of the time duration for this pair is $(n - 1)$. The relationship between the first A and F_1 is thus

$$F_1 = A(1 + i)^{n-1}$$

We can treat the other contributions in the same way and write for each a relationship similar to the above, taking care of the value for the exponent. Thus, $F = F_1 + F_2 + \cdots + F_{n-1} + F_n$ becomes

$$F = A(1 + i)^{n-1} + A(1 + i)^{n-2} + \cdots + A(1 + i)^2 + A(1 + i) + A$$

Multiplying both sides of this relationship by $(1 + i)$, we get

$$F(1 + i) = A(1 + i)^n + A(1 + i)^{n-1} + A(1 + i)^{n-2} + \cdots + A(1 + i)^2 + A(1 + i)$$

Let us now subtract the relationship for F from that for $F(1 + i)$. This results in

$$\begin{aligned} F(1 + i) - F &= A(1 + i)^n + A(1 + i)^{n-1} + A(1 + i)^{n-2} + \cdots + A(1 + i)^2 + A(1 + i) \\ &\quad - A(1 + i)^{n-1} - A(1 + i)^{n-2} - \cdots - A(1 + i)^2 - A(1 + i) - A \\ &= A(1 + i)^n - A \end{aligned}$$

Note that the subtraction canceled all but two of the terms on the right side. The expression can be rearranged as

$$F(1 + i - 1) = A[(1 + i)^n - 1]$$

$$F = A\frac{(1 + i)^n - 1}{i} \tag{4.1a}$$

Equation (4.1a) is used to analyze uniform series cashflows, specifically to determine F for given values of A, i, and n. In fact, any of the four variables can be evaluated for given values of the other three. Though any of the four methods discussed in Chapter 3 can be used for this, let us continue with the functional notation method, for which Equation (4.1a) can be expressed[8] as

$$F = A(F/A, i, n) \tag{4.1b}$$

where $$(F/A, i, n) = \frac{(1 + i)^n - 1}{i}$$

The notation $(F/A, i, n)$ is called the *compound amount factor,* since it is used as a multiplier to find the effect of compounding on uniform series payments A. Note that we used a term by this name earlier too, in Chapter 3 under single-payment analysis. One should therefore be careful with this term, since depending on the cashflow type, *compound amount factor* can mean either $(F/A, i, n)$ of Equation (4.1b) or $(F/P, i, n)$ of Equation (3.6b). We prefer in this text to avoid the use of functional notations' names, since they are long and difficult to remember; although each name has a distinct meaning based on what the notation really does. Instead of using the names, we use terms like *F/A factor*, meaning the factor used to find F for a given A. Sometimes, we also express it as *F factor for a given A*.

One can read the value of the *F/A* factor off the tables under *Uniform Payment Series*. Note that there are four columns under Uniform Payment Series; we have been introduced to one just now. The other three are discussed later in this chapter.

4.3.1.2 Finding A for Given F, i, n

To facilitate the evaluation of A for given values of F, i, and n, Equation (4.1a) can be rearranged as

$$A = F\frac{i}{(1 + i)^n - 1} \tag{4.2a}$$

Though any of the four methods discussed in Chapter 3 can be used to determine A using this equation, as a preference for the functional notation method we rewrite Equation (4.2a) as

[8]If you experience difficulty following functional notation, reread Chapter 3 where it has been discussed.

$$A = F(A/F, i, n) \tag{4.2b}$$

where $$(A/F, i, n) = \frac{i}{(1+i)^n - 1}$$

The notation[9] $(A/F, i, n)$ is called the *sinking fund factor,* since it eventually *sinks* F a uniform series of regular[10] payments A through F. Again, rather than use the long name for the functional factor, we use the terminology *A/F factor,* or *A factor for a given F.*

Looking closely at what they represent, the two factors $(A/F, i, n)$ and $(F/A, i, n)$ are reciprocals[11], that is, $(A/F, i, n)(F/A, i, n) = 1$. This can also be confirmed from their values for a specific n on any page of the interest tables.

EXAMPLE 4.2

Rani owns and operates a small electronic repair shop. A new instrument, called Rapid Diagnosist, that facilitates faster diagnosis of the items to be repaired has been marketed recently. Being technologically new, it is expensive. Based on the past, costs of such instruments decline fast. Rani predicts that Rapid Diagnosist would sell for $5,000 in the next five years. She decides to retain a fixed sum each year from annual profit to raise this capital five years hence. The retained sum will be invested regularly in the local bank at a 6.5% annually compounded interest rate. How much should be retained each year?

Solution

First, you need to decide whether to draw a cashflow diagram for comprehending the problem statement. This depends on how well you understand the problem. Do you know what is being asked for, what the unknown is, and what variables are known? If you can comprehend all these without the cashflow diagram, you can avoid sketching it. Let us assume that to be the case. We still need to have a mental picture of the problem.

Note that it is a multiple payment problem with regular deposits. It involves A and F, with A being the unknown. The relevant formula is Equation (4.2a) or (4.2b). Use the latter if you favor the functional notation method.

Next, we gather the values of the variables and proceed with the solution. The given data are $F = \$5{,}000$, $i = 6.5\%$ (per year), and $n = 5$ (years). Let us decide to use the functional notation method. From Equation (4.2b) we have

$$A = 5{,}000(A/F, 6.5\%, 5)$$

---

[9] If you find yourself confused about whether to use A/F or F/A in Equation (4.2b), reread Chapter 3, where this point has been explained. Basically, the factor is correct if A and F follow the same sequence within the parentheses as outside. Here, A/F has the same sequence as $A = F$, that is, A precedes F; hence the notation used is correct.

[10] There may be cashflows that are regular but not uniform.

[11] The reciprocity is also true between single-payment factors $(F/P, i, n)$ and $(P/F, i, n)$, as pointed out in Chapter 3, by what they represent, and also by their tabular values for a given set of i and n.

Look for the value of the A/F factor in the tables. Since there is no table for 6.5%, look for the values corresponding to 6% and 7% and determine their average (since 6.5% is the average of 6% and 7%). This works out to be $(0.1774 + 0.1739)/2 = 0.17565$, yielding

$$A = 5{,}000 \times 0.17565$$
$$\approx \$878$$

Thus, Rani must retain $878 each year from the annual profit.

4.3.1.3 Finding i for Given F, A, n Finding n for Given F, A, i

Any of the four, namely Equation (4.1a), (4.1b), (4.2a), or (4.2b), can be used to evaluate unknown i for given values of F, A, and n, or unknown n for given values of F, A, and i. Equations (4.1a) and (4.2a) are formatted for the use of calculators, Equations (4.1b) and (4.2b) for the use of compound interest tables. As illustrated in Chapter 3, the use of tables to evaluate i or n involves interpolation whose basic procedure has been explained in Examples 3.8 and 3.9.

4.3.2 Relationship between A and P

Several engineering economics problems involve a present loan P that is paid back through uniform series disbursements A. Though home-mortgage or car-lease loans quickly come to mind as examples, this scenario is common in industry too, where capital is borrowed to purchase equipment. As the equipment is used and profits (benefits) realized, a certain fixed sum A is paid back to the lender regularly.

The home-mortgage-loan problem is represented by the cashflow diagram in Fig. 4.5(a), where P is the loan (received now, vector up) and A is the regular payment each period during the loan duration. An obvious question is, what value of A will fully pay back the loan P and the associated interests? The lending manager usually determines A by referring to a booklet whose entries are based on the following formula.

4.3.2.1 Finding A for Given P, i, n

There are two ways to derive the relationship between A and P. One is to follow the first principles, as was done earlier in Section 4.3.1.1 for the case of A and F. This derivation is given[12] in Box 4.1 for interested readers.

A simpler approach is to combine the expression in Equation (4.2a) with that in Equation (3.1). This saves time and effort and is preferred by those less interested in theoretical aspects of engineering economics.

On substitution in Equation (4.2a) for F from Equation (3.1), we get

---

[12]Some derivations have been enclosed in boxes to avoid interruption to the discussions. They may be skipped by practicing engineers interested primarily in the application aspects of equations. Students are, however, encouraged to go through the first-principles-based derivations to hone their analytical skills.

Box 4.1

Two cashflow diagrams are shown in Fig. 4.5. In (a), a benefit P is received now and returned over n periods as uniform payments A, while in (b) an investment P is made now and the benefits realized over n periods as uniform A. From the analysis viewpoint the two diagrams represent the same problem.

Let us consider the case corresponding to Fig. 4.5(a). Here, sum P is borrowed now at interest rate i. This loan along with the interest earned under compounding is paid back over n periods through regular payments A. We are interested in deriving a relationship between P and A for given values of i and n.

Each A pays back a portion of P and the accrued interest. We can therefore consider P logically divided in n portions $P_1, P_2, \ldots, P_{n-1}, P_n$. The payment A at period 1 pays for P_1 and its accrued interest, that at period 2 pays for P_2 and its accrued interest, and so on. Thus, in Fig. 4.5(a), instead of sum P, we can conceptually consider at period 0 small vectors $P_1, P_2, \ldots, P_{n-1}, P_n$—all bundled together adding to P. Thus, mathematically

$$P = P_1 + P_2 + \cdots + P_{n-1} + P_n$$

We can consider P_1 and the first payment A at period 1 together as a pair, in which A is like the future sum F for P_1. Using the basic formula, Equation (3.1), where $F = A$, $P = P_1$, and $n = 1$, we can write a relationship between P_1 and the first A as

$$A = P_1(1 + i)^1$$

Thus, $P_1 = A(1 + i)^{-1}$

We can look at the other portions of P the same way, as a pair with the corresponding A, and write relationships similar to the above. By adding the resulting relationships, we get

$$\begin{aligned} P &= P_1 + P_2 + \cdots + P_{n-1} + P_n \\ &= A(1 + i)^{-1} + A(1 + i)^{-2} + \cdots + A(1 + i)^{-(n-1)} + A(1 + i)^{-n} \end{aligned}$$

Multiplying both sides of this relationship by $(1 + i)^n$, we get

$$P(1 + i)^n = A(1 + i)^{n-1} + A(1 + i)^{n-2} + \cdots + A(1 + i) + A$$

Next, multiply both sides of this relationship by $(1 + i)$ to get

$$P(1 + i)^{n+1} = A(1 + i)^n + A(1 + i)^{n-1} + A(1 + i)^{n-2} + \cdots + A(1 + i)$$

Now, subtract $P(1 + i)^n$ from $P(1 + i)^{n+1}$. This subtraction cancels all the terms on the right-hand side of the above two expressions, except the two terms $A(1 + i)^n$ and A. The net result is

$$P(1 + i)^{n+1} - P(1 + i)^n = A(1 + i)^n - A$$

This can be rearranged as

$$P(1 + i)^n [(1 + i) - 1] = A[(1 + i)^n - 1]$$

$$Pi(1 + i)^n = A[(1 + i)^n - 1]$$

$$P = A\frac{(1 + i)^n - 1}{i(1 + i)^n}$$

This is Equation (4.4a), or as rearranged Equation (4.3a).

$$A = P(1 + i)^n \frac{i}{(1 + i)^n - 1}$$

Thus,

$$A = P\frac{i(1 + i)^n}{(1 + i)^n - 1} \qquad (4.3a)$$

Again, any of the four methods discussed in Chapter 3 can determine A using Equation (4.3a). In the functional-notation method, we express this equation as

$$A = P(A/P, i, n) \qquad (4.3b)$$

where $(A/P, i, n) = \dfrac{i(1 + i)^n}{(1 + i)^n - 1}$

The notation $(A/P, i, n)$ is called the *capital recovery factor,* since it is used as a multiplier to recover (get back) the invested[13] capital P through n number of uniform payments A. Rather than use this long name for the factor, we often call it the *A/P factor,* or the *A factor for a given P.* The values of the *A/P* factor are given in the interest tables in one of the columns under Uniform Payment Series.

4.3.2.2 Finding P for Given A, i, n

Equation (4.3a) can be rearranged to determine P for given values of A, i, and n, as

$$P = A\frac{(1 + i)^n - 1}{i(1 + i)^n} \qquad (4.4a)$$

Any of the four methods discussed in Chapter 3 can be used to evaluate P using this equation. Since we prefer the functional notation method, we rewrite Equation (4.4a) as

$$P = A(P/A, i, n) \qquad (4.4b)$$

[13] A home loan represents investment capital from the lending institution's viewpoint.

$$\text{where} \quad (P/A, i, n) = \frac{(1+i)^n - 1}{i(1+i)^n}$$

The notation $(P/A, i, n)$ is called the *present worth factor*[14], since it determines the present value of a series of future cashflows A. Rather than use its long name, we use the terminology *P/A factor* or *P factor for a given A*.

Again, as is obvious from what they represent in terms of i and n, the two factors $(P/A, i, n)$ and $(A/P, i, n)$ are reciprocal of each other. This can also be confirmed from their values for a specific n on any page of the interest tables. For example, $(P/A, 10\%, 8)$ is 5.335, which is the reciprocal of 0.1874—the value of $(A/P, 10\%, 8)$.

4.3.2.3 Finding i for Given P, A, n / Finding n for Given P, A, i

Any of the four equations, namely (4.3a), (4.3b), (4.4a), or (4.4b), can be used to determine i for given values of P, A, and n, or n for given values of P, A, and i. Equations (4.3a) and (4.4a) are in forms suitable for the use of scientific calculators; Equations (4.3b) and (4.4b) are for the use of tabular functional factors. As discussed in Chapter 3, the use of functional factors to evaluate i or n involves interpolation, whose basic procedure has been explained in Examples 3.8 and 3.9. Example 4.3 illustrates the evaluation of interest rate i for known P, A, and n. The procedure for evaluating n is very similar.

EXAMPLE 4.3

Ray's company borrows $5,000 from the local bank to invest in a personal computer for the quality control department on the term that the company will pay back $1,200 each year for the next five years. The first payment is due on the first loan anniversary. What interest rate is the bank charging, if compounding is annual?

Solution

First, we need to decide whether to sketch a cashflow diagram to facilitate problem comprehension. If you can comprehend without the diagram, there is no need to sketch it, since it is not being specifically asked for. However, if the diagram is deemed to be helpful, then sketch it. Let us say that we can do without the diagram.[15]

Note that the problem involves six cashflows; in other words, the multipayment analysis of this chapter is applicable. Next, check whether these cashflows are irregular or regular. The five $1,200 payments are regular, each being annual. Next, do they follow a pattern? The answer is yes, and the pattern is a uniform series, since the payments are equal. Thus, the discussions of Section 4.3.2 are applicable, with Equation (4.3b) or (4.4b) being the relevant formula for the functional notation method.

---

[14]Note the same name under single payment; so be careful in differentiating the two.

[15]The diagram may be physically absent but conceptually present—instead of being on paper, it may be in your head! Get into the habit of developing a mental picture of the diagram before you begin the solution.

We next gather the given values of the parameters and proceed. These are $P = \$5{,}000, A = \$1{,}200, n = 5$. With n in years, the evaluated unknown i will be the annual interest rate. Thus, from Equation (4.3b),

$$\begin{aligned} A &= P(A/P, i, 5) \\ 1{,}200 &= 5{,}000(A/P, i, 5) \\ (A/P, i, 5) &= 1{,}200/5{,}000 \\ &= 0.24 \end{aligned}$$

Searching in the tables under the proper column following the procedure explained in Example 3.8, we find that

$$\begin{aligned} (A/P, i, 5) &= 0.2374 \quad \text{for} \quad i = 6\%, \quad \text{and} \\ &= 0.2439 \quad \text{for} \quad i = 7\% \end{aligned}$$

Conducting the interpolation as explained in Example 3.8, we get

$$\begin{aligned} \frac{0.24 - 0.2374}{0.2439 - 0.2374} &= \frac{af}{7 - 6} \\ af &= \frac{0.0026}{0.0065} \\ &= 0.4 \end{aligned}$$

Therefore,

$$\begin{aligned} i &= 6\% + 0.4 \\ &= 6.4\% \end{aligned}$$

Thus, the bank is charging Ray's company an annual interest rate of 6.4%.

4.4 ARITHMETIC SERIES

In the previous section we treated regular cashflows that are constant, forming uniform series. In this and the next section, we consider cases where the regular cashflows increase with the time period. When the increase is constant, such cashflows generate an *arithmetic series*. Consider payments of \$100 at period 1, \$120 at period 2, \$140 at period 3, and so on. In here, the increase in successive payments is constant, being \$20. Thus, the cashflows \$100, \$120, \$140, \$160, . . . form an arithmetic series. You must have come across such a series[16] in algebra.

[16]The cashflows \$800, \$700, \$600, . . . , \$100 also form an arithmetic series with a decrease of \$100 (or an increase of −\$100) in successive cashflows.

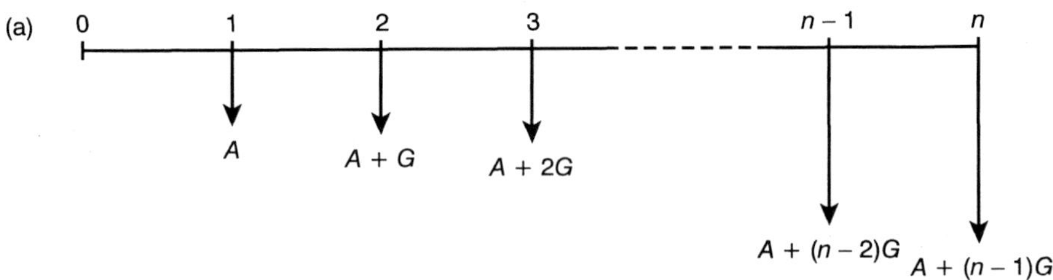

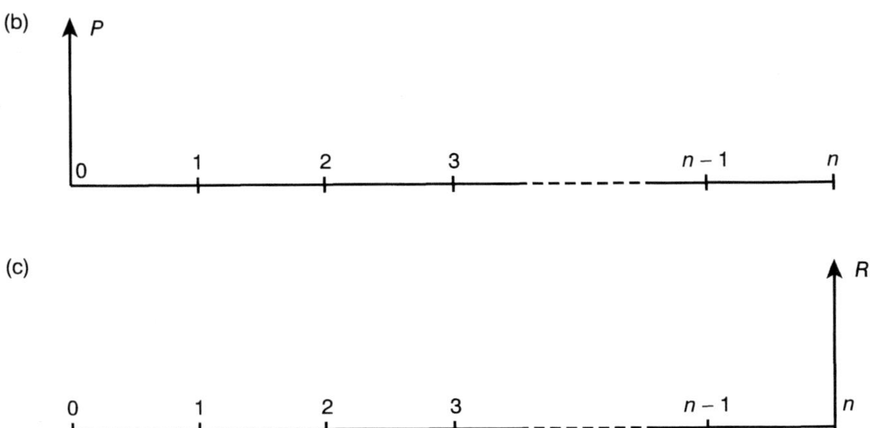

Figure 4.6 Arithmetic Series

An appropriate example of an arithmetic series is maintenance cost that increases each year by a constant amount. Increasing utility costs, insurance costs, and other similar costs of doing business may also form arithmetic series.

In most problems involving an arithmetic series, we are interested in knowing the equivalent present cost of the future costs. This helps the engineer in budgeting and making decisions based on the minimum present cost of the alternatives.

Let us consider the cashflow diagram of Fig. 4.6. In (a), the first cashflow A is at period 1, which may represent the first-year maintenance cost of the machine. The cost increases each year by G, forming the series of Fig. 4.6(a). At period 2 the cost is $A + G$, at period 3 it is $A + 2G$, and so on; finally at the nth period it is $A + (n - 1)G$. Note that the uniform series discussed in Section 4.3 can be considered a special case of arithmetic series where $G = 0$.

What we are interested in is to know the present cost implication of the increasing future costs. In other words, how much do we budget now in a maintenance account, that is, the value of P in Fig. 4.6(b), to pay for the increasing future costs of machine maintenance (Fig. 4.6a). We need to derive an expression that relates P with the arithmetic series payments A, $A + G$, $A + 2G$, . . . , $A + (n - 1)G$. We may also be interested in knowing the equivalent future value R of such payments (costs), as shown in Fig. 4.6(c).

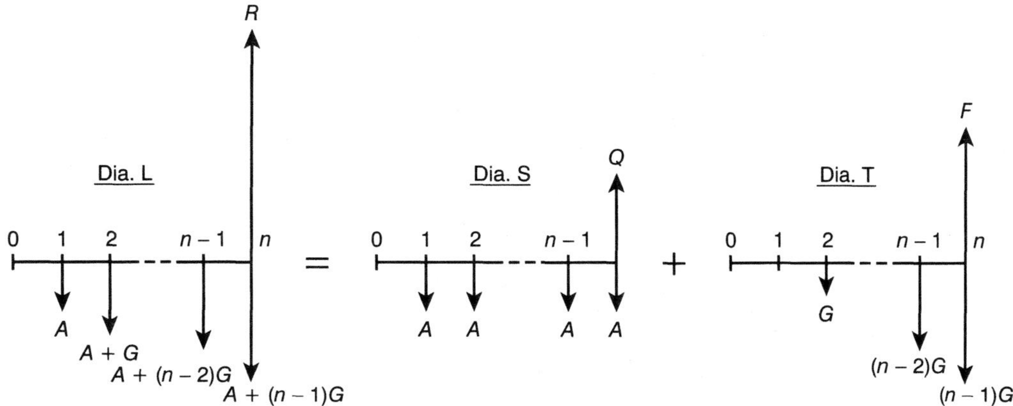

Figure 4.7 Two Components of an Arithmetic Series

4.4.1 Relationship between P and G

Consider the cashflow diagram combining Fig. 4.6(a) and Fig. 4.6(c). This is shown in Fig. 4.7 as cashflow diagram L, where R is the future balancing equivalent of the arithmetic series payments. We are interested to know the value of R at period n for the payments A, $A + G$, $A + 2G$, $\ldots$, $A + (n - 2)G$, and $A + (n - 1)G$. Since A is common in all the cashflows of this series, we can consider diagram L as composed of two diagrams S and T. Diagram S is based on uniform payments A, which are equivalent to Q at period n. Diagram T is based on increasing G, with the series being equivalent to F. Note that there is no cashflow in diagram T at period 1, since the entire cashflow A at this period has been accounted for in diagram S. Since, at each period, the total of the cashflows in diagrams S and T equals those in L—the original cashflow diagram—we can say that $Q + F$ must equal R.

4.4.1.1 Finding P for Given G, i, n

Diagram S, for which we have already derived equations, has been discussed in Section 4.3.1. If we can derive an expression for diagram T, we should have an equation for diagram L, that is, for the arithmetic series.

Diagram T requiring derivation is shown on its own in Fig. 4.8. The cashflow begins at period 2 with G, increasing by G and ending at n as $(n - 1)G$. The derivation linking F with G, i, and n is given in Box 4.2, wherefrom

$$F = \frac{G}{i}\left[\frac{(1 + i)^n - 1}{i} - n\right]$$

Substituting for F from Equation (3.1), the expression reduces to

$$P(1 + i)^n = G\frac{(1 + i)^n - ni - 1}{i^2}$$

$$P = G\frac{(1 + i)^n - ni - 1}{i^2(1 + i)^n} \tag{4.5a}$$

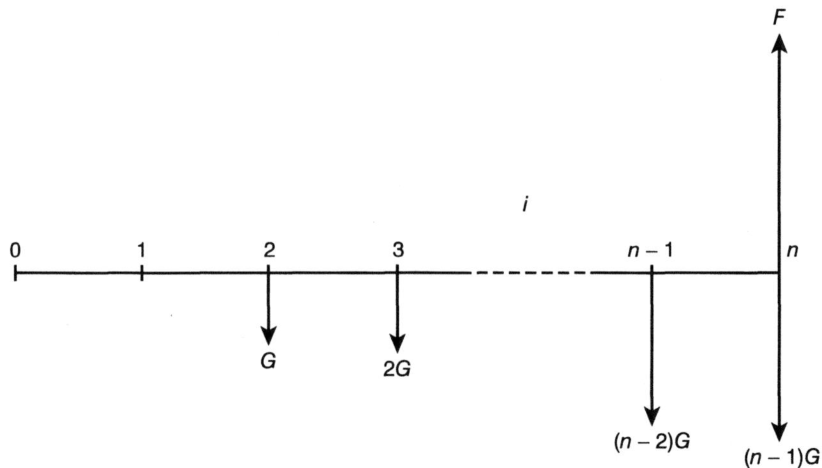

Figure 4.8 Analysis of Arithmetic Series

Any of the four methods discussed in Chapter 3 can be used to find P using this equation. However, since we prefer the functional notation method, Equation (4.5a) is expressed as

$$P = G(P/G,\ i,\ n) \tag{4.5b}$$

where $$(P/G,\ i,\ n) = \frac{(1 + i)^n - ni - 1}{i^2(1 + i)^n}$$

The notation $(P/G, i, n)$ is called the *gradient present worth factor,* since it is used to determine the present value of a series of future cashflows that increase arithmetically by G. Rather than use its long name, we can use the terminology *P/G factor*, or *P factor for a given G*. The values of this factor are given in the interest tables under *Arithmetic Gradient*. The following example illustrates the solution procedure involving arithmetic series cashflows.

EXAMPLE 4.4

Under a government environment incentive plan, companies installing air-pollution control equipment are entitled to a 50% grant to offset the purchase and maintenance costs. XYZ decides to avail itself of this opportunity by acquiring equipment costing $50,000. The maintenance cost is estimated to be $4,000 in the first year, increasing thereafter annually by $300. If the equipment's useful life is 10 years and the yearly interest rate is 6% compounded annually, how much will the grant be?

Solution

If you can comprehend the problem, there is no need to sketch the cashflow diagram. Note that the maintenance cost increases each year by a fixed amount, so the costs form an arithmetic series, for which a formula is available. We need to determine the equivalent present cost, 50% of which will be the government grant.

BOX 4.2

Refer to Fig. 4.8 and note that there is no cashflow at period 1. The arithmetic series cashflows increase by a constant amount G. We are interested in determining the future sum F that is equivalent to the cashflows in this series. Note that a payment $(n - 1)G$ is made on the day F is due. The derivation should aim at relating F with the series payments $G, 2G, \ldots, (n - 1)G$.

Each cashflow of the series contributes toward the final sum F. For example, G at period 2 increases under compounding to become, let us say, F' at period n. Similarly, $2G$ becomes F'', and so on. Thus, F will be equal to $F' + F'' + \cdots$. We can pair each series cashflow with its contributed portion and apply the basic expression, Equation (3.1), to each pair. Thus,

$$F' = G(1 + i)^{n-2}$$
$$F'' = 2G(1 + i)^{n-3}$$
$$\vdots$$

Once each payment has been paired we can write

$$F = F' + F'' + \cdots$$

or

$$F = G(1 + i)^{n-2} + 2G(1 + i)^{n-3} + \cdots + (n - 2)G(1 + i) + (n - 1)G$$

Multiplying both sides by $(1 + i)$, we get

$$(1 + i)F = G(1 + i)^{n-1} + 2G(1 + i)^{n-2} + \cdots + (n - 2)G(1 + i)^2 + (n - 1)G(1 + i)$$

Subtracting the expression for F from that for $(1 + i)F$, we get

$$\begin{aligned}(1 + i)F - F = {} & G(1 + i)^{n-1} + 2G(1 + i)^{n-2} + \cdots + (n - 2)G(1 + i)^2 \\ & + (n - 1)G(1 + i) - G(1 + i)^{n-2} - \cdots \\ & - (n - 3)G(1 + i)^2 - (n - 2)G(1 + i) - (n - 1)G \\ iF = {} & G(1 + i)^{n-1} + G(1 + i)^{n-2} + \cdots + G(1 + i)^2 + G(1 + i) - (n - 1)G \\ = {} & G(1 + i)^{n-1} + G(1 + i)^{n-2} + \cdots + G(1 + i)^2 + G(1 + i) + G - nG \\ = {} & G[(1 + i)^{n-1} + (1 + i)^{n-2} + \cdots + (1 + i)^2 + (1 + i) + 1] - nG\end{aligned}$$

The terms within add[17] to $\dfrac{(1 + i)^n - 1}{i}$, hence

$$iF = G\frac{(1 + i)^n - 1}{i} - nG$$

[17] If interested, see any book on college algebra.

Dividing both sides by i,

$$F = \frac{G}{i}\left[\frac{(1+i)^n - 1}{i} - n\right]$$

Replacing F as per Equation (3.1), we get

$$P(1+i)^n = G\frac{(1+i)^n - ni - 1}{i^2}$$

$$P = G\frac{(1+i)^n - ni - 1}{i^2(1+i)^n}$$

This relationship is Equation (4.5a)

The equivalent present cost comprises the purchase price and the present value of all the maintenance costs over 10 years. The latter is the equivalent P value of arithmetic-series cashflows with $A = \$4{,}000$ and $G = \$300$. For determining it, we can use Equation (4.5a) or (4.5b). Let us opt for the functional-notation-based Equation (4.5b). Compounding is annual with $i = 6\%$, and $n = 10$. For these data, the equivalent present cost is

Purchase cost + Present value of maintenance costs[18]
$= 50{,}000 + [A(P/A, i, n) + G(P/G, i, n)]$
$= 50{,}000 + 4{,}000(P/A, 6\%, 10) + 300(P/G, 6\%, 10)$
$= 50{,}000 + 4{,}000 \times 7.360 + 300 \times 29.602$
$= 50{,}000 + 29{,}440 + 8{,}881$
$= \$88{,}321$

Being 50% of the above cost, the grant will be \$44,160.50.

4.4.1.2 Finding *G* for Given *P*, *i*, *n*

The determination of G rarely arises in practice. But, if it does use Equation (4.5a) or (4.5b) to evaluate G, by substituting the known values of P, i, and n, and manipulating the terms.

4.4.1.3 Finding *i* for Given *P*, *G*, *n*
Finding *n* for Given *P*, *G*, *i*

Equation (4.5a) or (4.5b) is used to evaluate unknown i for given values of P, G, and n, or unknown n for given values of P, G, and i. Equation (4.5a) is formatted for the use of scientific calculators, Equation (4.5b) for interest tables. As explained in Chapter 3, use of functional factors to determine i or n involves interpolation, which has been illustrated in Examples 3.8 and 3.9.

[18]See Fig. 4.7.

4.4.2 Relationship between A and G

Substitution of P in terms of A, i, and n from Equation (4.4a) modifies Equation (4.5a) to

$$A\frac{(1+i)^n-1}{i(1+i)^n}=G\frac{(1+i)^n-ni-1}{i^2(1+i)^n}$$

$$A=G\frac{(1+i)^n-ni-1}{i[(1+i)^n-1]} \tag{4.6a}$$

4.4.2.1 Finding A for Given G, i, n

Again, any of the four methods discussed in Chapter 3 can be used to determine A using this equation. Since we prefer the functional notation method, we rewrite Equation (4.6a) as

$$A=G(A/G,i,n) \tag{4.6b}$$

where $$(A/G,i,n)=\frac{(1+i)^n-ni-1}{i[(1+i)^n-1]}$$

The notation $(A/G, i, n)$ is called the *gradient uniform series factor,* since it is used to determine the equivalent uniform cashflow A for a series of cashflows increasing arithmetically by G. We use the terminology *A/G factor*, or *A factor for a given G*. Values of this factor too are given in the interest tables under *Arithmetic Gradient.*

4.4.2.2 Finding G for Given A, i, n

The determination of G for given values of A, i, and n rarely arises in practice. If it does, use Equation (4.6a) or (4.6b), substitute the known values, and manipulate the terms.

4.4.2.3 Finding i for Given A, G, n
Finding n for Given A, G, i

Equation (4.6a) or (4.6b) can be used to evaluate the unknown i for given values of A, G, and n, or n for given values of A, G, and i. Equation (4.6a) is suitable for the use of scientific calculators, Equation (4.6b) for tabular functional factors. As explained in Chapter 3, the use of functional factors to evaluate i or n does involve some interpolation, as illustrated in Examples 3.8 and 3.9.

4.5 GEOMETRIC SERIES

In the previous section we discussed cashflows that form arithmetic series. In this section, we consider situations where they form *geometric series*. You must have studied geometric series in college algebra. In geometric series, the cashflows vary in such a way that the magnitude at any period is equal to that at the previous period multiplied by certain factor, which is the same for all cashflows of the series. Consider cashflows of \$100 at period 1, \$120 at period 2, \$144 at period 3, and so on. These cashflows increase by a factor of 1.2 (\$120 is 1.2 times the previous cashflow of \$100; similarly \$144 is 1.2 times of \$120). The factor by which cashflows change is called the *common ratio*, or *rate*, and is denoted by g. In the cashflows \$40, \$60, \$90, \$135, . . . forming a geometric series, g is 1.5. Costs on maintenance, utilities, and insurance may increase geometrically. If g is less than one, as can happen in a deflationary economy, the cashflows decrease. Note that the uniform series discussed in Section 4.3 can be considered a special case of geometric series where $g = 1$.

For geometric series cashflows too, appropriate equations can be derived, as in Box 4.3. Summarized below are the formulas applicable to problems involving geometric series. Note that A_1 is the first cashflow of the series, as shown in Fig. 4.9.

$$P = A_1 \frac{1 - (1 + g)^n (1 + i)^{-n}}{i - g} \qquad \text{when} \qquad i \neq g \tag{4.7}$$

When the interest rate i and the common ratio g are equal, the appropriate relationship is

$$P = A_1 \frac{n}{1 + i} \qquad \text{when} \qquad i = g \tag{4.8}$$

Representation of Equations (4.7) and (4.8) into functional factors so that tables can be used, as done earlier for other equations, becomes clumsy due to the additional variable g. Therefore, we use these two equations as they are to solve geometric series problems. Depending on whether $i = g$ or not, one of the two is used to evaluate any of the variables P, A_1, i, g, or n provided the values of the others are given. Also note that once A_1 has been evaluated (or is known as given), any of the cashflows in the geometric series can be determined using $A_n = A_1(1 + g)^{n-1}$, by substituting the particular value of n.

EXAMPLE 4.5

An international company appoints a 30-year Harvard MBA as its new president on a five-year contract at a salary of \$450,000. The salary is to increase each year by 9%. On the first day of the job, a serious past misconduct is discovered about the incumbent. The board of directors therefore decides to dismiss him the same day. However, as stipulated in the employment contract, he must be paid the present value of all the future salaries along with a golden handshake in the sum of \$500,000. If $i = 10\%$, how much is he paid?

Solution

The analysis period is five years. We can draw a cashflow diagram from either the president's viewpoint or the company's. Fig. 4.10 shows the company's viewpoint with cashflows as costs. The president's initial salary of \$450,000 increases at 9% per year. So the value of A_1 is \$450,000, and $g = 9\% = 0.09$. The P value of the salaries plus the \$500,000 golden handshake is the total payment to the dismissed president.

There are two ways to determine the P value of the salaries. In one, we determine the present value of A_1 through A_5 individually, using the basic relationship $F = P(1 + i)^n$, or functional factor, and add them. In the second approach, we exploit the geometric series the cashflows form, using the equation derived in this section. Let us opt for the latter.

Box 4.3

Refer to Fig. 4.9, where cash outflows $A_1, A_2, \ldots, A_{n-1}, A_n$ form a geometric series. They are related to the first cashflow A_1 through g and n, as

$$A_2 = A_1 + A_1 g = A_1(1 + g)$$
$$A_3 = A_2 + A_2 g = A_2(1 + g) = A_1(1 + g)(1 + g) = A_1(1 + g)^2$$
$$\vdots$$
$$A_n = A_1(1 + g)^{n-1}$$

With a proper value of n, any intermediate cashflow can be related to A_1, the first cashflow. For example, $A_4 = A_1(1 + g)^3$ and $A_9 = A_1(1 + g)^8$.

Our task is to determine P that is equivalent to $A_1, A_2, \ldots, A_{n-1}, A_n$. In other words, how much do we budget now to pay for the n future costs increasing geometrically?

Conceptually, each payment of the series is made out of P. The sum P can thus be conceived to comprise portions $P', P'', \ldots$ such that P equals the total of these portions. In other words,

$$P = P' + P'' + \cdots$$

We can pair each cashflow with its corresponding portion of P and apply the basic expression, Equation (3.1), to each pair. For A_1 and P', we have

$$A_1 = P'(1 + i)$$

or

$$P' = A_1(1 + i)^{-1}$$
$$= a$$

where $a = A_1(1 + i)^{-1}$

Similarly,

$$A_2 = P''(1 + i)^2$$

or

$$P'' = A_2(1 + i)^{-2}$$

Since $A_2 = A_1(1 + g)$, due to the geometric relationship illustrated earlier, its substitution gives

$$P'' = A_1(1 + g)(1 + i)^{-2}$$
$$= A_1(1 + i)^{-1}(1 + g)/(1 + i)$$
$$= ab$$

where $b = (1 + g)/(1 + i)$

Similarly,

$$\begin{aligned} P''' &= A_3(1 + i)^{-3} \\ &= A_1(1 + g)^2(1 + i)^{-3} \\ &= A_1(1 + i)^{-1}\left(\frac{1 + g}{1 + i}\right)^2 \\ &= ab^2 \\ &\vdots \end{aligned}$$

Once each portion of P has been paired with the corresponding cashflow of the series, $P = P' + P'' + \cdots$ gives

$$P = a + ab + ab^2 + \cdots + ab^{n-2} + ab^{n-1}$$

Multiplying both sides by b, we get

$$bP = ab + ab^2 + ab^3 + \cdots + ab^{n-1} + ab^n$$

Subtracting the expression for bP from that for P, we get

$$\begin{aligned} P - bP &= a + ab + ab^2 + \cdots + ab^{n-2} + ab^{n-1} \\ &\quad - ab - ab^2 - \cdots - ab^{n-2} - ab^{n-1} - ab^n \\ (1 - b)P &= a - ab^n \end{aligned}$$

Rearranging,

$$P = a\frac{1 - b^n}{1 - b}$$

On substituting the values of a and b, this reduces to

$$P = \frac{A_1}{1 + i} \times \frac{1 - \left(\frac{1 + g}{1 + i}\right)^n}{1 - \frac{1 + g}{1 + i}}$$

$$P = A_1\frac{1 - (1 + g)^n(1 + i)^{-n}}{i - g}$$

This is Equation (4.7).

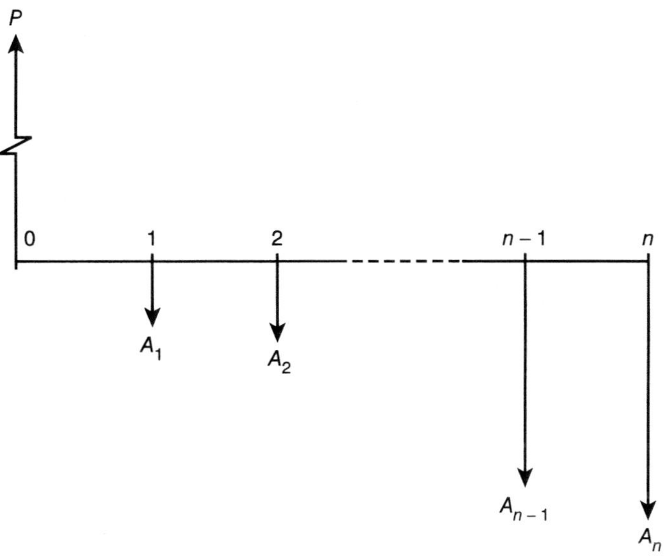

Figure 4.9 Geometric Series

With $i = 0.10$ and $g = 0.09, i \neq g$, and hence Equation (4.7) is applicable. This equation will yield the P value of $A_1, A_2 \ldots, A_5$ at period 0. With $n = 5$, the P value of the future salaries is

$$
\begin{aligned}
&A_1 \frac{1 - (1+g)^5(1+i)^{-5}}{i-g} \\
&= 450{,}000 \frac{1 - [(1+0.09)^5(1+0.10)^{-5}]}{0.10 - 0.09} \\
&= 450{,}000 \frac{1 - (1.538623955 \times 0.620921323)}{0.01} \\
&= 450{,}000 \times 4.4635578 \\
&= \$2{,}008{,}601
\end{aligned}
$$

$$
\begin{aligned}
\text{Total payment} &= P \text{ value of salaries} + \text{Golden handshake} \\
&= \$2,\ 008{,}601 + \$500{,}000 \\
&= \$2{,}508{,}601
\end{aligned}
$$

The discharged president is paid over \$2.5 million.

4.6 TAILORING

In the preceding sections we developed equations for three types of cashflow series: uniform, arithmetic, and geometric. Whenever cashflows follow these patterns, equations are usually efficient.

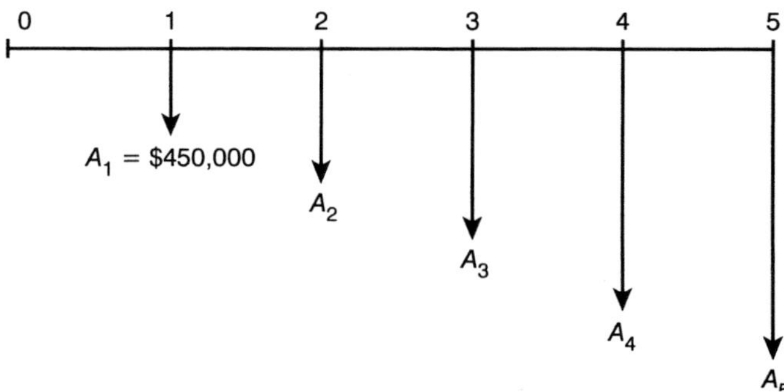

Figure 4.10 Diagram for Example 4.5

However, we often come across cashflows that do not follow the pattern exactly. Such *approximate patterns* can be "tailored" to become one of the three. Tailoring is done to achieve the efficiency of solution offered by the series equations, as illustrated through Examples 4.6 and 4.7.

EXAMPLE 4.6

Evaluate P for the cashflows diagramed in Fig. 4.11(a), if $i = 8\%$.

Solution

In the given cashflow diagram, P is the present sum while the other cashflows are future sums. The evaluation of P involves converting each future sum in today's value and adding them vectorially. Vectorial addition means taking care of the cashflow direction, that is, whether the cashflows are above or below the timeline. In this case, the present value of 40 (at period 5) must be added to P, with the resulting total becoming the present worth of costs. This total is equated with the present worth of benefits by adding the P values of all the cash inflows.

The concept of cost–benefit equivalency can be simplified by considering a cashflow diagram as a classical weighing balance. The cashflows may be considered as weights, provided the time value of money is taken into account. The upward vectors, all assembled at a period, may be thought of as being on one pan of the balance, while the downward vectors assembled at the same period are on the other pan. The cashflows' equivalency is analogous to the balancing of the two pans. This weighing-balance analogy makes it easier to set up the relationship among the cashflows. In the present illustration, the balancing is done at period 0; as weights, the cashflow P and P value of 40 are on one pan while the P values of the other cashflows are on the other pan. Thus,

$$P + P\text{ value of } 40 = \text{Total of the } P \text{ values of other cashflows}$$

Evaluating the P values using Equation (3.7a),

(a)

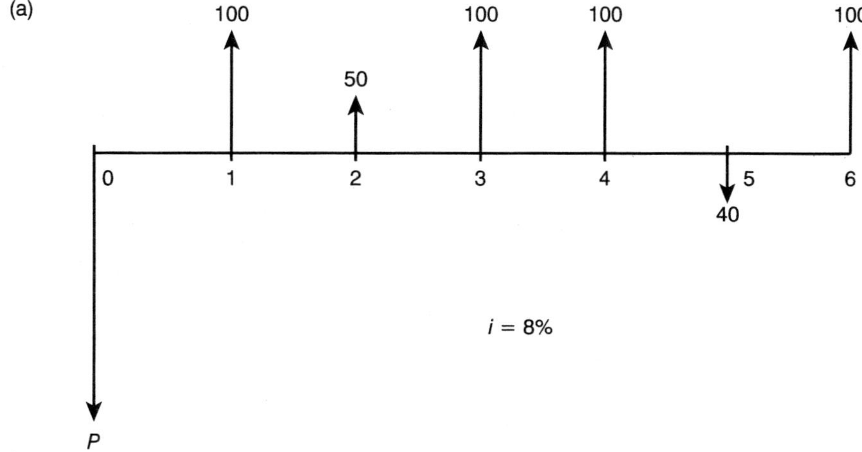

(b)

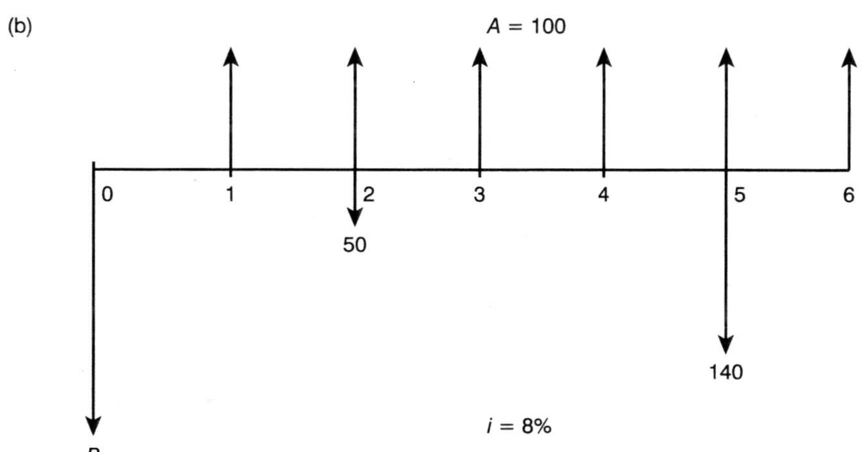

Figure 4.11 Tailoring

$$P + 40(P/F, 8\%, 5) = 100(P/F, 8\%, 1) + 50(P/F, 8\%, 2) + 100(P/F, 8\%, 3) + 100(P/F, 8\%, 4) + 100(P/F, 8\%, 6)$$

Such a solution approach involves evaluation of six functional factors in this example. While perfectly alright, this *straightforward approach* makes no attempt to achieve the solution conciseness offered by the given cashflow pattern. Note that the pattern displays a "touch" of uniformity—it does look like a uniform series. It seems that this approximate series can be tailored.

Tailoring the Diagram

A closer look at the cashflow pattern in Fig. 4.11(a) suggests that it is approximately of the uniform type discussed in Section 4.3. Of the six future cashflows, four are of the same magnitude, 100 each. If we modify the other two, those, at periods 2 and 5, to be 100 each and vertically up, then we can use the equation for uniform series.

Let us apply the weighing-balance concept mentioned earlier. At period 2 the vector magnitude is 50 up. Let us increase it to 100. Since we are increasing it up by fifty, that is, placing a weight of 50 on one pan of the balance, we need to put the same weight on the other pan, that is, add a downward vector of 50 at period 2. In other words, the given upward vector 50 at period 2 can be treated as two vectors, one of 100 upward and the other of 50 downward. The modified vectors, 100 up and 50 down, are equivalent to the original 50 up, since $100\uparrow - 50\downarrow = 50\uparrow$.

Applying the same concept at period 5, the given downward vector 40 is equivalent to two vectors: 100 up and 140 down. With these modifications at periods 2 and 5, the *tailored* cashflow diagram looks like Fig. 4.11(b). The two cashflow diagrams in Fig. 4.11 are thus equivalent; any solution based on Fig. 4.11(b) must yield the same result as that based on Fig. 4.11(a).

To appreciate the advantage of tailoring, let us now evaluate P using the modified cashflow diagram (Fig. 4.11b). The uniform series cashflows streamline the conversion of the upward vectors using just one P/A factor. We still have to convert the two downward cashflows (50 and 140) using P/F factors. Taking care of the vector directions, from the weighing-balance analogy we get

$$P + 50(P/F, 8\%, 2) + 140(P/F, 8\%, 5) = 100\,(P/A, 8\%, 6)$$

Compare this expression with the previous one, based on the straightforward approach. Due to tailoring we now have to evaluate only three functional factors, not six as earlier. Thus, tailoring has enhanced solution efficiency.

Substituting the values of the factors from the interest tables in the preceding expression, we get

$$P + 50 \times 0.8573 + 140 \times 0.6806 = 100 \times 4.623$$

Thus,

$$\begin{aligned} P &= 100 \times 4.623 - 50 \times 0.8573 - 140 \times 0.6806 \\ &= 462.30 - 42.87 - 95.28 \\ &= 324.15 \end{aligned}$$

In this example, tailoring proved advantageous, since it saved time and effort. Had the number of future cashflows been fewer (three or less), tailoring would not have been better—the straightforward approach would have been equally efficient. In general, the larger the number of cashflows in the approximate series and the fewer the required modifications, the more advantageous is the tailoring.

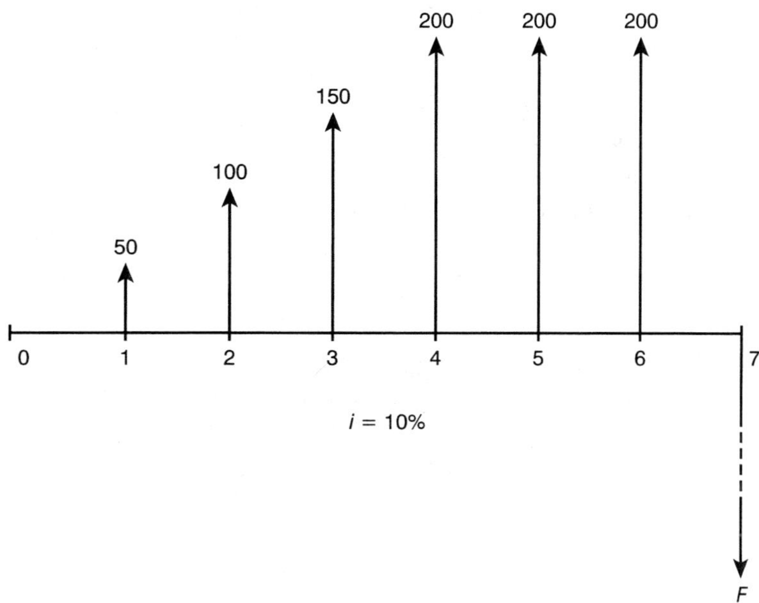

Figure 4.12 Diagram for Example 4.7

EXAMPLE 4.7

Evaluate F in Fig. 4.12 for $i = 10\%$.

Solution

Again, there are two approaches to evaluating F in Fig. 4.12. With the straightforward approach each cashflow can be converted to its F value, using Equation (3.6a). The total of these F values will yield the answer. Since there are six cashflows, this approach will involve six functional factors.

The other approach, based on tailoring to exploit any pattern in the cashflows, begins by looking closely at the diagram to find what patterns, if any, exist. In this case, the cashflows at periods 1, 2, and 3 form an arithmetic series, while those at periods 4, 5, and 6 form a uniform series[19]. It seems that tailoring of the diagram is advantageous.

Tailoring the Diagram

Referring to the diagrams pertaining to uniform and arithmetic series (Figs. 4.4 and 4.6a), we notice in Fig. 4.12 that

1. An equivalent cashflow located one period left of period 0 can easily be found for the arithmetic-series cashflows by using the P/G factor. Let us call this equivalent cashflow P_1.

---

[19]One can consider the cashflow at period 4 with the arithmetic series.

2. Also, an equivalent cashflow located[20] at period 3 can easily be determined for the uniform-series cashflows at periods 4, 5, and 6 by using the P/A factor. Let us call this equivalent cashflow P_2.

Once P_1 and P_2 have been determined, they can then be transferred to period 7 as F values using F/P factors. The total of these F values will equal the unknown F at period 7.

Following this logic, for $G = 50$ and $n = 4$, we get

$$\begin{aligned} P_1 &= 50(P/G, 10\%, 4) \\ &= 50 \times 4.378 \\ &= 218.9 \end{aligned}$$

Note that this P_1 is located one period left of period 0.

Next, for the cashflows forming a uniform series, $A = 200$. Noting that n is 3 for the series, we get

$$\begin{aligned} P_2 &= 200(P/A, 10\%, 3) \\ &= 200 \times 2.487 \\ &= 497.4 \end{aligned}$$

Now we transfer P_1 and P_2 to their equivalent values at period 7 and add the transferred values to determine F. For the former $n = 8$ (since P_1 is a period left of 0), and for the latter $n = 4$. Thus,

$$\begin{aligned} F &= 218.9(F/P, 10\%, 8) + 497.4\,(F/P, 10\%, 4) \\ &= 218.9 \times 2.144 + 497.4 \times 1.464 \\ &= 469.32 + 728.19 \\ &\approx 1{,}198 \end{aligned}$$

Note that the tailoring involved evaluation of four functional factors altogether; the straightforward approach would have involved six. Thus, tailoring has been beneficial.

SUMMARY

Most engineering economics problems involve multiple cashflows. They may be one of the three basic types: one cost, several benefits; several costs, one benefit; or several costs, several benefits. Basic considerations in analyzing such problems have been discussed in Chapter 4. Multiple cashflows may be regular or irregular. They are regular if they occur at each period. Regular cashflows may or may not display patterns, as evidenced in their diagrams or tables.

Problems with no pattern in their cashflows are analyzed by treating each cashflow separately, using the basic formula, Equation (3.1), or functional notations. Those with patterns are analyzed using the relevant equations.

Most common patterns in cashflows form one of the three series: uniform, arithmetic, or geometric. These three series and their equations have been discussed in

[20]The formula we have derived earlier for the P/A factor yields P just left of the first A, in this case at period 3.

Chapter 4. For the uniform and arithmetic series cashflows, the functional-notation method is available to the analyst. For the geometric series cashflows the functional-notation method is cumbersome, and hence only formulas are used.

When the cashflows deviate slightly from a pattern, there are two choices. One is to ignore their proximity to the pattern and use the basic equation or its variations. This straightforward approach is usually lengthy. The other is to *tailor* the given cashflows to modify them to fit in one of the patterns, and then use the equation for that pattern. The analyst has to judge which of the two choices makes sense from the solution efficiency viewpoint. Section 4.6 presented discussions and examples on *tailoring*, which is one of the challenging and creative tasks in engineering economics.

It may be noted that all the derivations in this chapter were based on Equation (3.1), emphasizing once again that this equation is really the most *basic* formula in engineering economics.

EXERCISES

Discussion Questions

4.1 Give an example from recent personal financing where multipayments were involved.

4.2 In regular cashflow problems, what is *regular*: cashflow, the time period, or both? Explain your answer with an example.

4.3 Which pattern—arithmetic or geometric series—is more likely in industrial problems, and why? Consider maintenance and utility costs as examples.

4.4 In evaluating i or n in a problem, which one—functional notation or the calculator method—do you prefer, and why?

4.5 Why is the functional-notation approach not pursued for geometric-series cashflows?

Multiple-Choice Questions (Circle the *best* answer.)

4.6 In regular cashflow problems,
- a. cashflows must be the same at each time period.
- b. a cashflow must occur at each time period.
- c. cashflows cannot be unequal.
- d. none of the above

4.7 Multipayment problems are always
- a. regular.
- b. irregular.
- c. in an arithmetic series.
- d. none of the above

4.8 Referring to Fig. 4.1, the present value of cashflow T is
- a. $T(P/T, i, 3)$.
- b. $T(F/P, i, 3)$.
- c. $T(P/F, i, 3)$.
- d. $T(T/P, i, 3)$.

4.9 If you set aside 5% of your fixed earnings into a retirement plan, the cashflows will form
a. a uniform series.
b. an arithmetic series.
c. a geometric series.
d. a logarithmic series.

4.10 Which of the following equations is correct?
a. $F = P(P/F, i, n)$
b. $P = A(A/P, i, n)$
c. $F = P(F/A, i, n)$
d. $P = A(P/A, i, n)$

4.11 For a geometric series with $A_1 = \$100$ and $g = 10\%$, the value of A_6 in dollars will approximately be
a. 60.
b. 110.
c. 160.
d. 210.

4.12 For the following cashflows, the correct relationship is

Year	Cashflow
0	− $300
1	80
2	120
3	160
4	200
5	25

a. $300 = 80(P/A, i, 4) + 25(P/F, i, 5)$.
b. $300 = 80(P/A, i, 5) - 215(P/F, i, 5)$.
c. $300 = 80(P/A, i, 5) + 105(P/F, i, 5)$.
d. none of the above

4.13 The uniform series is a special case of geometric series with
a. $g = 0$.
b. $g = 1$.
c. $g = -1$.
d. none of the above

Numerical Problems

4.14 Evaluate F in Fig. 4.13 by the straightforward approach (no tailoring) assuming $i = 10\%$.

4.15 Monica wins one million dollars in a state lottery that cost her a dollar for the ticket. She has two options: to receive the million dollars now, or $80,000 per year for the rest of her life. If she expects to live for another 25 years, which option is better for Monica? Assume that the applicable annually compounded interest rate is 10% per year.

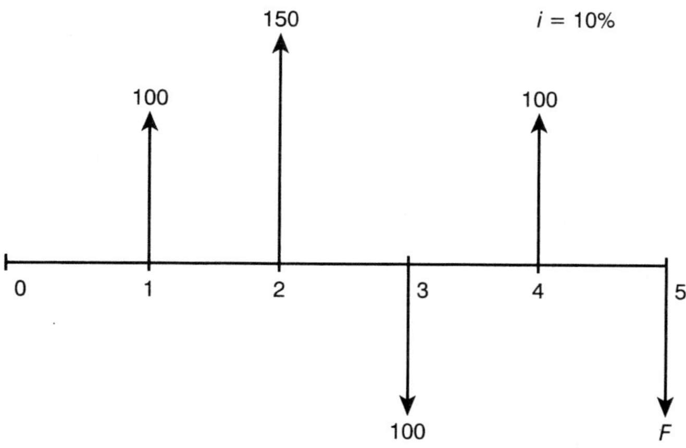

Figure 4.13 Diagram for Problems 4.14 and 4.21

4.16 For a mortgage loan of $100,000 Ismayel's quarterly payment for the last 20 years has been $2,000. If he has to continue to pay this amount for another 10 years to fully pay back the loan, what is the nominal interest rate?

4.17 The purchasing department is planning to buy a PC system for $10,000. The system is expected to have a useful life of 5 years during which its updating and maintenance costs will be

Year	Cost
1	$400
2	500
3	600
4	700
5	800

How much should be budgeted for both the purchase and upkeep of the system if yearly interest rate is 12%, compounded annually?

4.18 John began a retirement plan 15 years ago by investing $2,000 at year 1. Each subsequent year's investment was $200 more than the previous year's. Due to excessive health care expenditure John could not invest in the fifth and seventh year. The investments for years 6 and 8 were $3,000 and $3,400 respectively. What is the account balance after this year's investment? Assume an annually compounded interest of 8% per year.

4.19 Twenty-First Century Manufacturers are planning to acquire a new machine that will cost $5,000. The machine's estimated life is 10 years. It is expected to be trouble-free during the first year; the maintenance costs for the subsequent four years are likely to be $100, $200, $300, and $400 respectively. It will require an overhaul halfway through its life at a cost of $2,000. The maintenance costs following the overhaul are expected to be $200, $220, $242, $266.20, and $292.82.

How much should be budgeted to pay for the purchase and future overhaul and maintenance of the machine? Assume an interest rate of 10% per year, compounded annually.

4.20 If in Fig. 4.1, $T = 2S$, $R = 2.5S$, and $Q = 8S$, what is the interest rate?

4.21 Evaluate F in Fig. 4.13 by tailoring the diagram to exploit the most appropriate cashflow pattern. Compared to the straightforward approach of Problem 4.14, has tailoring been beneficial? By how much?

4.22 Evaluate D in Fig. 4.14 by tailoring the diagram.

4.23 Evaluate P in Fig. 4.15 by tailoring the diagram.

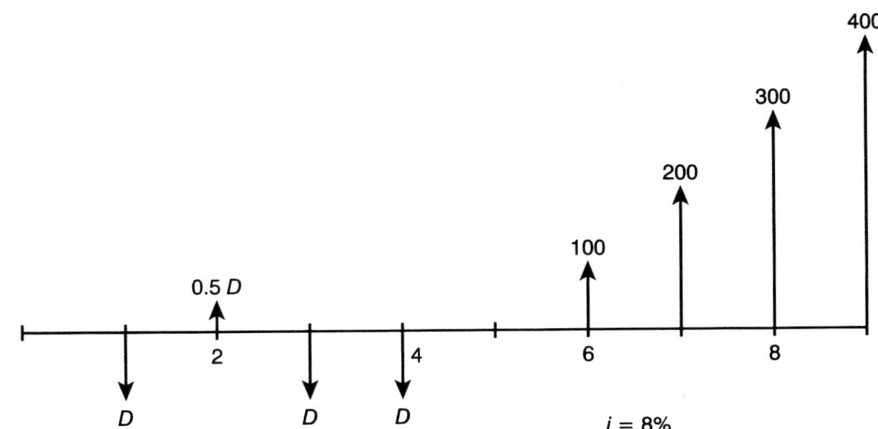

Figure 4.14 Diagram for Problem 4.22

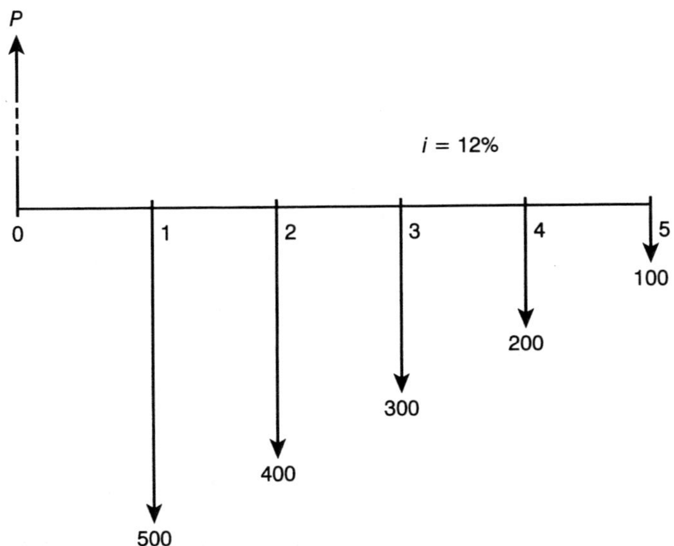

Figure 4.15 Diagram for Problem 4.23

PART II

Criteria

After having laid in Part I the *foundation* of fundamental principles relating to cashflows, we are ready to *build the structure* in Part II. As the structure depends on its foundation, so do the discussions of this part on those of Part I. In the next seven chapters, we discuss the criteria by which the best alternative is selected from among the many feasible ones. Six different criteria are practiced in industry; we discuss each of them in separate chapters: payback period in Chapter 5, present worth in Chapter 6, future worth in Chapter 7, annual worth in Chapter 8, rate of return in Chapter 9, and benefit–cost ratio in Chapter 10. Finally, we compare the six criteria in Chapter 11.

CHAPTER FIVE

Payback Period

IN THIS CHAPTER YOU WILL LEARN ABOUT

- The meaning of *payback period*
- Determination of the payback period
- When to apply the payback criterion
- Application of the payback criterion
- Prevalency in industry
- Limitations of the payback-period method
- Time-valued payback-period analysis

In the four chapters of Part I we discussed the basic principles underlying cashflows and their manipulations. We are now ready to solve engineering economics problems. To do that we need a criterion by which the best alternative will be selected from among the feasible ones. There are six different criteria commonly practiced by industry. In this chapter we learn about one of them, called the *payback period.* The other decision criteria are discussed in the next five chapters.

The payback period method of solving engineering economics problems is unique. This is the only one of the six methods that does not depend on what we have learned in Part I. It differs from the other five, presented in Chapters 6 through 10, in that it ignores the *time value of money* whereas the others don't.

5.1 MEANING

Payback period is the duration in which an investment pays for itself. Investments on equipment and other resources are made with the expectation of profit or net income. The time it takes the generated profit to pay back the capital invested is the payback period. Consider a company that invests in an automatic tool changer at a cost of $5,000. The tool changer is expected to increase production, generating $2,000 of additional profit per year. Thus, the $5,000 investment will be paid back in 2.5 years. This time duration of 2.5 years is the payback period of the investment in the tool changer.

Payback period can thus be expressed as

$$\text{Payback period} = \frac{\text{Investment cost}}{\text{Net income per unit time}} \tag{5.1}$$

In the denominator we have used the term *net income.* This is due to the fact that the invested equipment may involve operating costs, for example for maintenance, insurance, and utilities.[1] The net income is the difference between gross income from the invested equipment and its total operating cost. Note also that it is expressed as *per unit time,* usually per year. Being in the denominator, this unit of time becomes the unit of the payback period. If the net income is expressed as per month, the payback period will be in months. Instead of the term *investment cost* in the numerator of Equation (5.1), other synonymous terms may also be used. For example, for a new machine the term *first cost* is equally appropriate.

EXAMPLE 5.1

A hydraulic machine is being planned for acquisition at a cost of \$5,000. During its useful life of seven years, it is expected to generate an annual gross income of \$3,400. The cost of space to house the machine and of the utilities is \$900 per year. Its maintenance cost is projected to be \$300 per year. Determine the payback period for the machine.

Solution
Here, the first cost (or investment) is \$5,000. The recurring expenditure comprises the space and utility cost of \$900 and the maintenance cost of \$300, both per year. Thus, the total cost in utilizing the machine is \$900 + \$300 = \$1,200 per year. The net income from investment in the machine is obtained by subtracting this total cost from the gross income. This yields a net income of \$3,400 − \$1,200 = \$2,200 per year. Therefore,

$$\text{Payback period} = \frac{\$5{,}000}{\$2{,}200 \text{ per year}}$$
$$= 2.27 \text{ years}$$

Note that there was redundant data in Example 5.1, namely the machine's useful life of seven years. This data was not needed for the solution. Engineering economics problems may sometimes contain redundant data. In illustrative or exercise problems, redundant data may intentionally be provided to judge the depth of your understanding of the subject matter. In industrial problems, redundant data may creep in during data collection for the project. You must be able to identify any redundant data in a problem and boldly ignore them, rather than ponder on their relevance.

[1]This term is more prevalent in the United States. It means services such as electricity, gas, telephone, fax, Internet, and janitors, that are essential to the operation of the business/industry/plant.

Note that in Example 5.1 *we did not account for the time value of money.* We treated the second year's net revenue at par with that of the first year. This is the major weakness of the payback period analysis, since, as was learned in Part I, money does have a time value.

The machine in Example 5.1 has a useful life of 7 years. With a payback period of 2.27 years, it returns the $5,000 investment well before the end of its life. It may continue to be used for another 4.73 years beyond the payback period. The payback period method ignores the benefits and costs beyond the payback period, which is another drawback.

Could the payback period be longer than the useful life of the invested resource? Such a case does not arise normally, since such projects are usually not funded. But if it did, the term *payback period* would lose its meaning, since only a portion of the investment would be recovered. The result will be a *partial payback.*

In Example 5.1, the recurring cost of using the machine and the gross income were given to be constant. In several problems, these vary from year to year, yielding a variable net income. In that case we track the *cumulative net income,* and the payback period is that cut-off duration at which cumulative net income equals the investment.

5.2 VARIABLE CASHFLOWS

When gross income and operating cost data for the investment are variable, a tabular or graphical approach is more suitable. The operating costs vary since costs on maintenance, utilities, insurance, and so on, usually increase with time. In such cases the data are summarized, preferably in a table similar to the cashflow table discussed in Chapter 2. Example 5.2 illustrates how problems with variable costs and benefits are solved using the *tabular approach,* while Example 5.3 illustrates the *graphical approach.*

EXAMPLE 5.2

If for the machine in Example 5.1, expected gross incomes and operating costs are as follows, what is the payback period?

Year	Gross Income	Total Operating Cost
1	$3,500	$1,100
2	3,350	1,200
3	3,600	1,250
4	3,200	1,300
5	3,450	1,400
6	3,400	1,500
7	3,350	1,600

Solution
The investment remains the same, that is, $5,000. Both the gross income and the annual operating cost vary[2]. For the net income, subtract the *total operating cost* from the *gross income.* The results are given in the last column, by extending the data table.

Year	Gross Income	Total Operating Cost	Net Income
1	$3,500	$1,100	$2,400
2	3,350	1,200	2,150
3	3,600	1,250	2,350
4	3,200	1,300	1,900
5	3,450	1,400	2,050
6	3,400	1,500	1,900
7	3,350	1,600	1,750

As evident in this table, net income from the machine is variable. Therefore, Equation (5.1) is not applicable. The solution approach in such a case is to track the *cumulative* value of the net income. The payback period corresponds to the time duration when cumulative net income equals the investment. The calculations can be streamlined by creating another column for the cumulative income[3]. The cumulative income at any point in time is the total of all the previous incomes. As it increases, keep watching its value to check whether it has equaled or exceeded the investment.

Year	Gross Income	Total Cost	Net Income	Cumulative Income
1	$3,500	$1,100	$2,400	$2,400
2	3,350	1,200	2,150	4,550
3	3,600	1,250	2,350	6,900
4	3,200	1,300	1,900	
5	3,450	1,400	2,050	
6	3,400	1,500	1,900	
7	3,350	1,600	1,750	

In the last column, second row, the cumulative income of $4,550 has been obtained by adding the first year's net income of $2,400 to the second year's $2,150. The third year's cumulative income of $6,900 results from adding the third-year income of $2,350 to the cumulative income at the end of second year ($4,550). Note that we stopped computation as soon as cumulative income exceeded the investment.

[2]In rare cases if they vary by the same amount, their difference, i.e., (the net income) will be constant. If so, use Equation (5.1).

[3]Cumulative income means "cumulative net income."

The table shows that by the end of the second year the machine recovers $4,550, which is less than the $5,000 investment. But, by the end of the third year, the machine recovers more than the investment.

So what is the payback period? It definitely is greater than 2 years. Is it 3 years or somewhere in between 2 and 3 years? That depends on how the income is realized. If the profits and costs are accounted for at the end of the year, then the payback period is 3 years. However, if they are realized "as you go," that is, on a continuous basis, then the payback period is under 3 years. In such a case, it can be determined by interpolating between years 2 and 3:

$$\begin{aligned}\text{Payback period} &= 2 + \frac{5{,}000 - 4{,}550}{2{,}350} \\ &= 2 + 0.19 \\ &= 2.19 \text{ years}\end{aligned}$$

Thus, the payback period is 2.19 years.

5.3 USE OF A TABLE AND DIAGRAM

Are cashflow tables and diagrams of much use in payback-period analysis? With some insight we can conceive that evaluation of the payback period, especially where net income from the investment is variable as in Example 5.2, is based on manipulation of the cashflow table. Consider Example 5.1 once again. The problem can be restated through its cashflow table, where cashflows are the net incomes.

Year	Cashflow
0	−$5,000
1	2,200
2	2,200
3	2,200
4	2,200
5	2,200
6	2,200
7	2,200

Here, $2,200 is the yearly net income ($3,400 − $1,200) from the investment. One can use this table and follow the cumulative approach outlined in Example 5.2 to determine the payback period. This first-principles approach does away with the need of any formula, namely Equation (5.1).

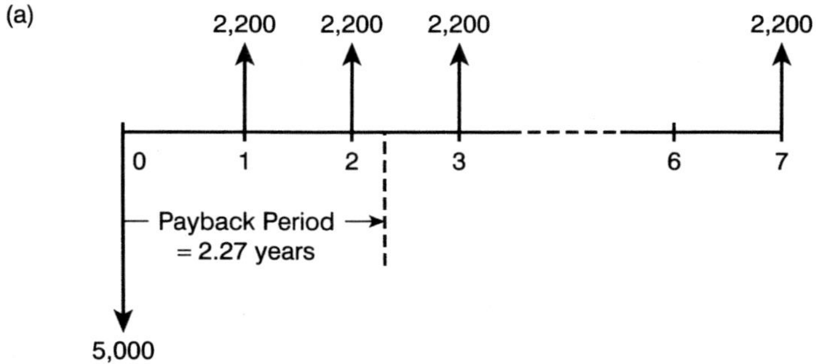

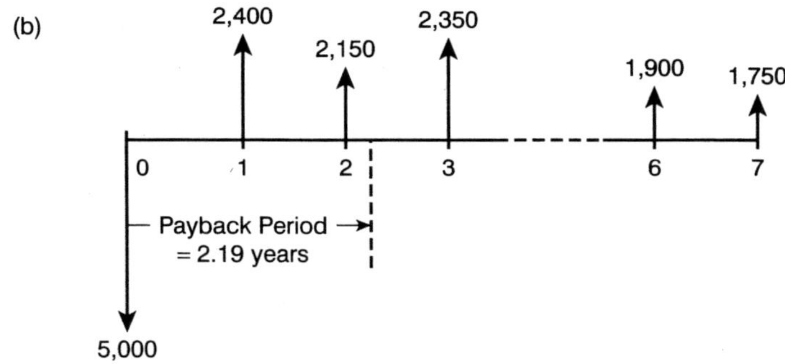

Figure 5.1 Payback Period Concept

The cashflow-table-based approach yields the same result for the payback period, as expected. The choice between this approach and the use of an equation is a matter of preference. For investments yielding constant net income, Equation (5.1) is convenient and therefore preferred. Where incomes from the investment are variable, Equation (5.1) is inapplicable. The tabular approach illustrated in Example 5.2 is the only choice.

Cashflow diagrams too have limited use in solving problems under the payback criterion. However, they help us in understanding the underlying concept of this criterion and its limitations. The cashflow diagrams for Examples 5.1 and 5.2, discussed so far in this chapter, are given in Fig. 5.1. Example 5.1 corresponds to the diagram in Fig. 5.1(a), while Example 5.2 corresponds to Fig. 5.1(b). Note the income vectors' constancy in Fig. 5.1(a), but their variability in Fig. 5.1(b), as dictated by the problem statements. The payback period can be displayed in such diagrams by marking it off the timeline, such as 2.27 years in Fig. 5.1(a) and 2.19 years in Fig. 5.1(b). Note in this figure that the cashflow next to the right of the payback period mark also contributes toward the payback. This is because the cashflows normally occur during the full time period, although we post them together as one lump sum at period ends. For example in Fig. 5.1(a), the cashflow of $2,200 at period 3 contributes during the initial 0.27 year following period 2. It has been sketched at year 3 only because of the period-end-posting-of-cashflows convention.

EXAMPLE 5.3

Do Example 5.2 using the graphical approach.

Solution

The solution for the payback period can also be based on a graphical approach. In that case a graph is drawn between the cumulative income on the y-axis versus time on the x-axis. A horizontal line is drawn to represent the investment. The payback period, corresponding to the intersection of the cumulative and investment lines, is read off the x-axis.

The graphical approach comprises five steps:

1. Set up the cumulative income on the y-axis and the time periods on the x-axis, and plot the data.
2. Join the data points by straight lines to draw the cumulative income line.
3. Draw the horizontal investment line.
4. Project the intersection of the investment and cumulative income lines on the x-axis.
5. Read the payback period off the x-axis.

The graphical approach to solving Example 5.2 has been illustrated in Fig. 5.2, where the cumulative income data have been plotted. The data points have been joined by straight lines, generating the *cumulative income line.* The intersection of this line and the horizontal investment line yields the payback period on the x-axis as 2.19 years.

Take a moment to ponder that the cumulative line for constant net income as in Example 5.1 will be a single straight line with one slope. The slope is a measure of the rate at which investment is recovered.

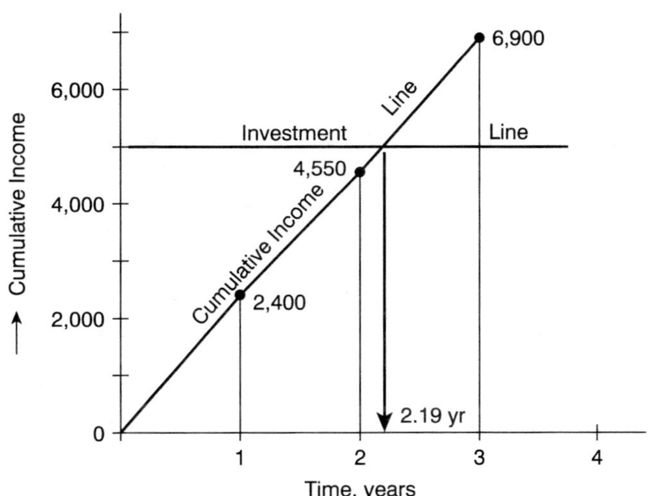

Figure 5.2 Graphical Analysis of Payback Period

The graphical approach to evaluating the payback period, as in Example 5.3, is simply an extension of the tabular approach. In both cases, cumulative income data must be computed. The difference lies in whether we algebraically interpolate the cumulative income data or graph them. The graphical approach offers a superior visualization and is preferred by engineers and technologists, especially if the results are to be communicated to others. Remember, however, that the accuracy of the graphical method's results is influenced by the accuracy of plotting the data and reading the payback period off the x-axis. This is not a major limitation when graphic calculators and CAD systems—quite a common tool of engineers and technologists today—are used.

5.4 SINGLE PROJECTS

In Sections 5.1 through 5.3, we learned the procedure of evaluating payback period. We can now discuss decision making under the payback-period criterion. It is obvious from the preceding discussions that this criterion favors projects with shorter payback periods, since the investment is recovered sooner.

The cut-off payback period for funding a project is usually set by the company management after considering various financial and marketing factors. Based on the number of alternatives, projects may be of three types: single, two-alternative, and multi alternative. Single projects with no alternative are the simplest to analyze. Here the engineering economist's task is to determine the payback period of the project. If the payback period is shorter than that set up by the management, the project is funded. In Example 5.1, investment in the machine with a 2.27-year payback would be approved if the cut-off period is 3 years. Companies normally finance projects with payback periods under 3 years.

5.5 TWO-ALTERNATIVE PROJECTS

A payback-period-based decision on whether to invest in a resource with no alternative is straightforward, as was discussed in Section 5.4. Two- and multialternative problems are comparatively more difficult. We discuss the two-alternative case in this section and the multialterative case in the next section.

When engineers are faced with the task of selecting one of the two alternatives based on payback-period criterion, the procedure involves evaluating the payback periods of the alternatives and choosing the one whose payback period is shorter. The attraction of a shorter-payback-period project is obvious, since it will return the investment sooner. Example 5.4 explains the procedure.

EXAMPLE 5.4

For automating the painting of car bodies, two models of a robotic system—basic and deluxe—are available in the market. For the following cashflows, which one should be selected on the basis of payback period?

Year	Basic	Deluxe
0	−$50,000	−$70,000
1	20,000	30,000
2	20,000	28,500
3	20,000	27,000
4	20,000	25,500
5	20,000	24,000

Solution
Since the cash inflow data are not specific about what they represent, we can assume them to be net incomes. The basic model yields a constant annual income of $20,000. We can therefore find its payback period using the standard formula, Equation (5.1). With an investment of $50,000 and annual income of $20,000, the payback period for the basic model is

$$\$50{,}000/(\$20{,}000 \text{ per year}) = 2.5 \text{ years}$$

For the deluxe model, the incomes are variable. So we follow the tabular or graphical approach, both involving evaluation of cumulative incomes. Following the procedure outlined in Example 5.2, the payback period for the deluxe model is determined to be

$$2 + \frac{70{,}000 - 58{,}500}{27{,}000} = 2.43 \text{ years}$$

Based on the desirability of shorter payback period, the deluxe model should be selected.

Two-alternative problems fall into one of the two categories. In one, a comparison of the alternatives' payback periods is made and the alternative with the shorter period is selected, as in Example 5.4. In this case, it is assumed that the management has not set any cut-off payback period. The alternative with the shorter payback period is selected irrespective of its value.

In the second category of problems, a cut-off payback period has been fixed by the management. The engineer's task is to evaluate the payback periods of both the alternatives and compare each one with the set one. Three situations can arise:

1. Both payback periods are longer than the cut-off payback period. Therefore, neither alternative is selected. However, if the resource is crucial, then the compliance to the cut-off payback period may be waived and the alternative with the shorter payback period is selected.
2. One alternative's payback period is longer than the cut-off period, while the other's is shorter. Obviously, the alternative whose payback period is shorter than the cut-off value is selected.

3. Both alternatives' payback periods are shorter than the cut-off payback period. Obviously, the alternative with the shorter payback period is selected.

In Example 5.4, there was no cut-off payback period. The selection was therefore based on a comparison between the payback periods of the two models. Had the management set the payback period in that example to, say, 2 years, neither model would have been selected.

5.6 MULTIALTERNATIVE PROJECTS

Engineering projects involving more than two alternatives are solved in ways similar to those for two alternatives discussed in Section 5.5. To select the best alternative out of the several feasible ones, the solution procedure comprises: (a) evaluating the payback period for each alternative, and (b) selecting the best alternative on the basis of shortest payback period, considering any constraint imposed by the cut-off payback period set by the management. Example 5.5 illustrates the procedure.

EXAMPLE 5.5

The small drilling machine in the job shop is worn beyond repair. The supervisor has done some research and found that any of the three makes—A, B, or C—will suffice as a replacement. The cashflows for these three candidate machines are given below. If the company funds projects of payback periods only under 3 years, which one should be selected?

Year	Make A	Make B	Make C
0	−$5,000	−$7,000	−$9,000
1	2,000	3,000	4,000
2	2,000	2,500	3,750
3	2,000	2,000	3,500
4	2,000	1,500	3,250
5	2,000	1,000	3,000
6	2,000	500	2,750

Solution

The cash inflows are assumed to be net incomes. We need to follow the two-step procedure of evaluating the payback period of each make and selecting the one with the shortest payback period provided it is less than 3 years.

Make A

Using the standard equation, since incomes are constant, the payback period for make A is

$$\$5{,}000/(\$2{,}000 \text{ per year}) = 2.5 \text{ years}$$

Make B

For make B, the incomes are variable. So follow the tabular or graphical approach; both involve evaluation of the cumulative income. Following the procedure outlined in Example 5.2, the payback period for this make is

$$2 + \frac{7{,}000 - 5{,}500}{2{,}000} = 2.75 \text{ years}$$

Make C

For this make too, the incomes are variable. Following the approach of Example 5.2, the payback period for make C is

$$2 + \frac{9{,}000 - 7{,}750}{3{,}500} = 2.36 \text{ years}$$

Since all the three makes satisfy the set cut-off payback period of 3 years, each one is worthy of selection. However, on the basis of shortest payback period, make C is selected.

Multialternative problems can also be categorized into two types. One involves the evaluation and comparison of alternatives' payback periods, and selection of the alternative with the shortest period. In such problems, no cut-off payback period has been set.

In the second type of problem, a cut-off payback period is set by management, as in Example 5.5. The engineer's task is to evaluate the payback periods of all the alternatives and compare each one with the cut-off payback period. Projects with payback periods above the cut-off are discarded. Of the remaining, the one with the shortest payback period is selected. In Example 5.5, the three alternatives (makes A, B, C) had payback periods lower than the set 3-year period. So the decision was made in favor of the shortest of the three payback periods, resulting in the selection of make C. In case the payback periods of all the alternatives are above the cut-off value, but it is crucial to acquire the resource, the alternative with the shortest payback period is funded.

5.7 MERITS AND LIMITATIONS

As a criterion for economic decision making, the payback period has its supporters and denouncers. The supporters boast about its simplicity, while the denouncers point out its nonconsideration of the time value of money. In response to this criticism, the conventional method is sometimes modified to account for the time value of money, as illustrated in the next section.

The payback period method is suited for small investments by, and for, the operating departments. The final decision to invest rests usually with the department manager, with no need of approval from upper management. Normally, the capital involved is small and within the limit set by the management.

Under what circumstances is the payback-period criterion more appropriate? This is an important question. The usual circumstances are

1. The working capital is low. The company is operating with a short-term planning horizon with tight cashflows. It needs to recover the invested capital as soon as possible for use elsewhere.
2. The investment is associated with a technology that is changing rapidly[4]. Short payback periods ensure that the investment will be recovered well in time—before the invested resource becomes obsolete. The recovered capital may be "recycled" for upgrading.
3. Investment in innovative technologies should also be governed by shorter payback periods because of the higher risk.

Till recently the payback period has been an extremely popular method in industry for both engineers and nonengineers. It required little knowledge of engineering economics. If the economic climate is stable with low interest rates, ignoring the time value of money may not affect the decision too much, especially for short-term projects. However, the other methods discussed in Chapters 6 through 10 have now become popular, primarily because they account for the time value of money and their analyses are facilitated by computer technology.

Payback period method can lead to decisions that contradict those by other methods, such as rate of return method (Chapter 9). Engineering economists must be mindful of this limitation of the payback period. They may use another method to confirm the decision by the payback period criterion. In fact, the payback-period criterion is often used in industry for initial screening of the projects. Promising projects are then subjected to more rigorous analysis by one of the methods discussed in Chapters 6 through 10.

The merits and limitations of the payback period method are summarized here.

Merits

1. Simple to use
2. Suitable for small investments
3. Quicker investment decision, since upper management usually not involved
4. Suitable for projects involving rapidly changing technologies
5. Easily understood, even by noneconomists

Limitations

1. Time value of money ignored
2. An approximate, less exact method

[4]The result may be obsolescence of the invested resource. How much we repent when the PC bought just last year gets superseded by a newer, more powerful one, and that at a lower cost! Short payback periods minimize the negative impact, through obsolescence, of rapidly changing technologies on investment.

3. Can lead to erroneous results
4. Benefits beyond the payback period not considered
5. May need to confirm the results by other method(s)
6. Less attractive to companies with "healthy" working capital and/or operating with long-range investment plans

5.8 TIME-VALUED PAYBACK

So far in this chapter, we have limited our discussions to the usual procedure of payback period analysis in which the time value of money is ignored. This is done for the sake of simplicity of analysis—the hallmark of the payback period method. Some economists, however, prefer to add realism to this method by considering the time value of money. Such realistic approaches render the method more complex, however. Examples 5.6 and 5.7 illustrate time-valued payback-period analysis.

EXAMPLE 5.6

A hydraulic machine is being planned for acquisition at a cost of \$5,000. During its useful life of seven years, it is expected to generate a gross income of \$3,400 per year. Its annual maintenance cost is projected to be \$300. The cost of space to house the machine and of the utilities is \$900 per year. Determine the payback period for the machine *by considering the time value of money.* Assume annual compounding with $i = 10\%$ per year.

Solution

This is Example 5.1, where we consider the time value of money. By subtracting the annual expenditure of \$1,200 (= \$900 + \$300) from the gross income, net income is \$3,400 − \$1,200 = \$2,200 per year. So the cashflow table is

Year	Cashflow
0	−\$5,000
1	2,200
⋮	⋮
7	2,200

Since the \$5,000 cost is incurred now, it has no time value with reference to the present (period 0). However, we need to account for the time values of the future cashflows. The first-year income of \$2,200 gets modified in terms of today's dollars as

$$\begin{aligned} P &= 2{,}200\,(P/F, 10\%, 1) \\ &= 2{,}200 \times 0.9091 \\ &= 2{,}000 \end{aligned}$$

Thus, by the end of the first year, \$2,000 of the investment is recovered. The investment yet to be recovered = \$5,000 − \$2,000 = \$3,000.

The second-year net income of \$2,200 gets modified as

$$\begin{aligned} P &= 2{,}200\,(P/F, 10\%, 2) \\ &= 2{,}200 \times 0.8264 \\ &= 1{,}818 \end{aligned}$$

Thus, by the end of the second year, the investment yet to be recovered = \$3,000 − \$1,818 = \$1,182.

The third-year net income of \$2,200 gets modified as

$$\begin{aligned} P &= 2{,}200\,(P/F, 10\%, 3) \\ &= 2{,}200 \times 0.7513 \\ &= 1{,}653 \end{aligned}$$

Since the third-year time-valued income of \$1,653 is more than the remaining investment to be recovered (\$1,182), the payback period lies in between years 2 and 3. By interpolating within these two limits we can determine the exact payback period. Following the interpolation procedure explained in Chapter 3,

$$\begin{aligned} \text{Payback period} &= 2 + 1182/1653 \\ &= 2 + 0.71 \\ &= 2.71 \text{ years} \end{aligned}$$

The time-valued payback period of 2.71 years is longer than the conventional payback period of 2.27 years determined in Example 5.1. This is expected, since consideration of the time value undermined the future benefits, resulting in longer recovery of the investment.

As can be noted in the preceding example, consideration of the time value changed Example 5.1 from being uniform-income type to variable-income type. That is why we could not use the standard formula, Equation (5.1), in Example 5.6. In rare cases, the consideration of time value may do the opposite, that is, change a variable-income problem to a uniform-income type. This occurs when the rate of increase in the future incomes is the same as the interest rate, as illustrated in Example 5.7.

EXAMPLE 5.7

Considering the *time value of money* and assuming annual compounding with $i = 10\%$ per year, what is the payback period for the following project?

Year	Cashflow
0	−$5,000
1	2,200
2	2,420
3	2,662
4	2,928

Solution

Since the $5,000 investment is made now, its time value with reference to year 0 remains unaffected. As required, the cash inflows should account for the time value of money. The first-year income of $2,200 gets modified as

$$\begin{aligned} P &= 2{,}200\,(P/F,\,10\%,\,1) \\ &= 2{,}200 \times 0.9091 \\ &= 2{,}000 \end{aligned}$$

Thus, by the end of the first year, $2,000 of the investment is recovered. The investment yet to be recovered = $5,000 − $2,000 = $3,000.

For the second year, the $2,420 income gets modified as

$$\begin{aligned} P &= 2{,}420\,(P/F,\,10\%,\,2) \\ &= 2{,}420 \times 0.8264 \\ &= 2{,}000 \end{aligned}$$

Thus, by the end of the second year, the remaining investment yet to be recovered = $3,000 − $2,000 = $1,000.

For the third year, the $2,662 income gets modified as

$$\begin{aligned} P &= 2{,}662\,(P/F,\,10\%,\,3) \\ &= 2{,}662 \times 0.7513 \\ &= 2{,}000 \end{aligned}$$

Since the third-year time-valued income of $2,000 is more than the amount to be recovered ($1,000), the payback period is under 3 years. The interpolation between years 2 and 3 is simple, since the amount to be recovered at the end of second year ($1,000) is exactly half of the recovery during the third year ($2,000). Thus, the payback period is 2.5 years. Alternatively,

$$\begin{aligned} \text{Payback period} &= 2 + 1000/2000 \\ &= 2 + 0.5 \\ &= 2.5 \text{ years} \end{aligned}$$

Note that in this example the given incomes were variable (increasing at 10% each year—the same as the interest rate i). The consideration of time value changed these incomes, rendering them uniform at $2,000 each year. We could therefore have used the formula, Equation (5.1), to evaluate the payback period as

$$\begin{aligned} \text{Payback period} &= \text{Investment/Income per year} \\ &= \$5{,}000/(\$2{,}000 \text{ per year}) \\ &= 2.5 \text{ years} \end{aligned}$$

SUMMARY

The payback period is the duration of time in which an invested capital is fully recovered through net incomes from the investment. The payback-period method usually ignores the time value of money. To be selected under the payback criterion, the project's payback period must be shorter than the cut-off value set by the management. The alternative with the shortest payback period is obviously the best, and hence is selected. The payback-period method is often used in industry, especially for analyzing projects requiring low capital. Companies operating under a short planning horizon, due to tight cashflows, find this method attractive. This method is also useful in projects involving rapidly changing technology. It is also used for initial screening of projects competing for funds that are limited. Its simplicity is its hallmark. That the time value of money is not considered is its greatest weakness. The payback criterion should be used carefully, since decision results may at times contradict those by the other methods. Analysts should in that case use another method, such as rate of return, to confirm the results of payback period analysis. The conventional analysis can be enhanced by considering the time value of money. However, such an enhancement is achieved at the cost of the payback method's simplicity.

EXERCISES

Discussion Questions

5.1 Why does the payback-period method continue to be used in industry in spite of the fact that it ignores the time value of money?

5.2 How would you explain the meaning of payback period to a lay person?

5.3 Now that you know how to make economic decisions based on the payback-period criterion, describe any forthcoming personal project for which you might use it. How will you gather the cashflow data for the project?

5.4 Is payback period dimensionless? If yes, why? If no, what is its unit?

5.5 Can one apply the payback period criterion to purchasing a car? If yes, how? If no, why not?

Multiple-Choice Questions (Circle the *best* answer.)

5.6 The payback period is expressed usually in

a. years.

b. quarters.

c. months.
d. days.

5.7 In the graphical method for evaluating the payback period, the investment line is parallel to the
a. x-axis.
b. y-axis.
c. z-axis.
d. none of the above

5.8 The payback period method
a. is exact.
b. is simple.
c. accounts for the time value of money.
d. considers the salvage value in the analysis.

5.9 The payback-period method is most appropriate for______ projects.
a. short-term
b. long-term
c. complex
d. government

5.10 The payback-period method is most appropriate for projects involving
a. established technologies unlikely to change.
b. technology that is changing rapidly.
c. imported technology.
d. appropriate technology.

5.11 The payback-period method's reliability can be enhanced by
a. evaluating it in months rather than in years.
b. considering the time value of money.
c. both a and b
d. neither a nor b

Numerical Problems (Do Problems 5.12 through 5.17 the usual way, that is, ignoring the time value of money.)

5.12 Determine the payback period of a project whose cashflows are

Year	Cashflow
0	−$3,500
1	1,750
2	1,200
3	1,050
4	950

5.13 Assume that your four-year education toward the BS degree costs you $40,000 over and above what it would have cost you to live if you did not attend college. You also lost annual earnings of $15,000 during these four years. Assuming that the degree will enhance your future earnings potential by $10,000 per year, what is the payback period for this higher education?

5.14 A new machine for the manufacture of plastic cutlery costs $158,000. The machine will increase the monthly production by 20,000 packs. Each pack costs 50 cents to manufacture and is sold for 75 cents. What is the payback period for the machine?

5.15 Determine the payback period of a PC that costs $9,500 to install and $150 per month to operate. The system will free 50% of a secretary's time whose salary is $15,000 per year, and the additional fringe benefits are 60% of the salary.

5.16 Two models of a forklift truck—basic and deluxe—are being planned for acquisition. For the following data, which one should be selected on the basis of payback period? The $3,000 cost for the basic model at year 3 is its required overhauling.

Year	Basic	Deluxe
0	−$50,000	−$60,000
1	15,000	17,500
2	15,000	17,000
3	15,000	16,500
	−3,000	
4	15,000	16,000
5	15,000	15,500

5.17 A toolroom lathe needs to be replaced, for which there are three choices—A, B, and C—in the market. On the basis of payback period, which lathe should be procured if the company approves projects of payback periods only under 2.5 years. The cashflow data are

Year	A	B	C
0	−$35,000	−$40,000	−48,000
1	10,750	17,500	17,000
2	10,750	17,500	16,000
3	10,750	16,500	15,000
4	10,750	14,500	15,000
5	10,750	13,500	15,000

5.18 Do Problem 5.12 by considering the time value of money. Assume annual compounding with $i = 12\%$ per year.

5.19 Do Problem 5.13 by considering the time value of money. Assume annual compounding with $i = 8\%$ per year.

5.20 Do Problem 5.14 by considering the time value of money. Assume annual compounding with $i = 10\%$ per year.

5.21 Do Problem 5.15 by considering the time value of money. Assume annual compounding with $i = 15\%$ per year.

CHAPTER SIX

Present Worth

IN THIS CHAPTER YOU WILL LEARN ABOUT

- The meaning of *present worth* (PW)
- The present worth of costs
- The present worth of benefits
- How to determine the PW of a project
- Application of the PW criterion
- The input–output concept
- Analysis period
- Perpetual-life projects
- Capitalized cost

As discussed in Chapter 5, the greatest weakness of the payback-period method is its nonconsideration of the time value of money. In contrast, the other five analysis methods, to be covered next, account for the time value of money. They are therefore more realistic. In this chapter, we discuss one of the five methods, called present worth (PW), and learn to apply it to engineering projects.

The present worth criterion relies on the sound logic that economic decisions—for that matter all decisions—should be made on the basis of costs and benefits pertaining to the time of decision making. The *past* is over and the *future* is unknown, so relying on the *present* makes sense. It is the present we live in. Most decisions of life are made in the present—looking at the future as far as we can see. Economic decisions should be no exception. In the PW method, the economic data are analyzed through the "window" of the present.

6.1 MEANING

Present worth is the value of something at the current time. Time enters into its meaning because the value of an object is not the same at all times. Consider the worth of an apple now. If you are not hungry or don't feel like eating a fruit, the apple has no worth to you at the present time. In three or four hours, when you will feel like eating something and may not have access to what you would like to eat, the same apple will have worth for you. This of course is a rather simplistic way to explain the worth of something[1] as a function of time.

Decision making based on the PW criterion involves evaluation of the present worth. By *present worth* we mean the present value of all the cashflows pertaining to the project. As discussed in Chapter 2, the cashflows may be costs or benefits. The present worth therefore may be of costs or of benefits. Consider the diagram in Fig. 6.1 where the cashflows are in dollars, and assume i to be the interest rate per period. Following the convention discussed in Chapter 2, a cost of \$500 is incurred now. There are two more costs: \$250 at period 3, and \$300 at period 5. We also see in the diagram three benefits, one each at time periods 1, 2, and 4. For this diagram, the present worth of costs means the *equivalent cost now* of all the costs, which includes the present \$500 cost at period 0. Its determination involves evaluating the present value (P value) of the two future costs and adding to the present \$500 cost. Thus, we need to convert the \$250 cost at period 3 and the \$300 cost at period 5 to their equivalents at period 0. These are

$$P \text{ value of the \$250 cost} = 250(P/F, i, 3)$$
$$P \text{ value of the \$300 cost} = 300(P/F, i, 5)$$

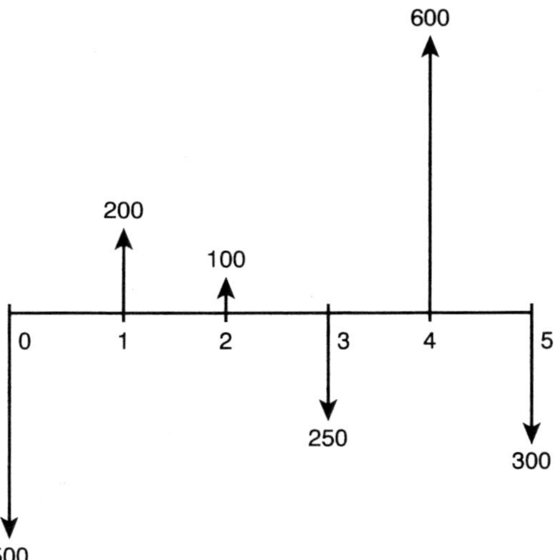

Figure 6.1 Present Worth

[1]In engineering economics, we consider only those items of a project whose value can be expressed in monetary terms. The items are usually cashflows of first cost, benefits, maintenance costs, salvage value, etc. However, sometimes we do consider intangible items of benefit or cost that are difficult to express in monetary terms. Usually we ignore the intangibles during the analysis and consider them subjectively at the final stage of the decision making. If we decide to consider them in the analysis, we guess-estimate their dollar values. Goodwill, for example, whose value may be significant and is usually accounted for while buying or selling an established business, is an intangible item.

Adding these P values to the \$500 cost, the present worth of the costs is

$$PW_{costs} = 500 + 250\,(P/F, i, 3) + 300\,(P/F, i, 5)$$

In the same way we can determine the present worth of the benefits as

$$PW_{benefits} = 200(P/F, i, 1) + 100(P/F, i, 2) + 600(P/F, i, 4)$$

As seen here, in determining the present worth of costs or benefits, all future costs or benefits are transferred to their equivalent values at the present time (period 0). This is how the time value of money is taken into account in the PW method. Consider the following example.

EXAMPLE 6.1

India England is to begin her four years of college to earn a BS degree in environmental engineering. The cost of education during the first year will be \$16,000. It is expected to increase annually by 10%. If the money to be spent on education can earn annually compounded 5% interest per year, what is the present worth of the costs of her education?

Solution

The first-year cost is \$16,000. Since the cost increases each year by 10%, any year's cost is 10% more than the previous year's. Therefore, the second-year cost will be \$17,600 (\$16,000 + 10% of \$16,000). Similarly, the third- and fourth-year costs will be \$19,360 and \$21,296.

We can assume the *present* to be the beginning of the college education. Following the convention that a cashflow is posted at the end of the period, the first-year cost of \$16,000 is incurred at the end of period 1, and so on. Sketch a cashflow diagram if it is helpful.

By transferring the future costs to the present and adding them together, we get the present worth of the costs of India's education as

$$\begin{aligned} PW_{costs} &= 16{,}000\,(P/F, 5\%, 1) + 17{,}600\,(P/F, 5\%, 2) \\ &\quad + 19{,}360\,(P/F, 5\%, 3) + 21{,}296\,(P/F, 5\%, 4) \\ &= 16{,}000 \times 0.9524 + 17{,}600 \times 0.9070 + 19{,}360 \times 0.8638 \\ &\quad + 21{,}296 \times 0.8227 \\ &= 15{,}238 + 15{,}963 + 16{,}723 + 17{,}520 \\ &= \$65{,}444 \end{aligned}$$

To determine the net effect of the costs and benefits on a project, we determine the project's present worth[2] (PW). The present worth of a *project* is the difference between the

[2]In some literature, it is called *net present worth* (NPW) or *net present value* (NPV).

present worth of all the benefits from the project and the present worth of all the costs on the project. Thus,

$$PW = PW_{\text{benefits}} - PW_{\text{costs}} \tag{6.1}$$

Note that we subtract the present worth of costs from the present worth of benefits, not the other way round. This ensures that PW is positive for projects whose present worth of benefits exceeds that of the costs—an obviously desirable aim. Thus, a project is usually funded only if its PW is positive, ensuring a return on investment.

EXAMPLE 6.2

For $i = 6\%$, determine the PW of the project represented by Fig. 6.1. If the PW is negative, explain its significance.

Solution

To evaluate the project PW, we first determine the PW of its benefits and costs. Following the earlier discussion on Fig. 6.1, for $i = 6\%$ we have,

$$\begin{aligned} PW_{\text{benefits}} &= 200\,(P/F, i, 1) + 100\,(P/F, i, 2) + 600\,(P/F, i, 4) \\ &= 200\,(P/F, 6\%, 1) + 100\,(P/F, 6\%, 2) + 600\,(P/F, 6\%, 4) \\ &= 200 \times 0.9434 + 100 \times 0.8900 + 600 \times 0.7921 \\ &= 189 + 89 + 475 \\ &= \$753 \end{aligned}$$

$$\begin{aligned} PW_{\text{costs}} &= 500 + 250\,(P/F, i, 3) + 300\,(P/F, i, 5) \\ &= 500 + 250\,(P/F, 6\%, 3) + 300\,(P/F, 6\%, 5) \\ &= 500 + 200 \times 0.8396 + 300 \times 0.7473 \\ &= 500 + 210 + 224 \\ &= \$934 \end{aligned}$$

The present worth of the project, therefore, is

$$\begin{aligned} PW &= PW_{\text{benefits}} - PW_{\text{costs}} \\ &= \$753 - \$934 \\ &= -\$181 \end{aligned}$$

The present worth is negative, which means that on the basis of the PW criterion the project incurs a loss. It should therefore not be funded, unless there are noneconomic reasons.

6.2 METHOD

The PW method of solving engineering economics problems involves determination of the present worth of the project or its alternatives. Using the criterion of present worth, the better of the two projects, or the best of the three or more projects, is selected. Thus, the application of this method involves two tasks:

1. Evaluation of the present worth(s)
2. Decision on the basis of the PW criterion, according to which project with the highest positive PW is selected.

We discussed the first task in Section 6.1. In this section, we discuss decision making based on the PW criterion.

Engineering economics projects can be categorized into one of the two groups: no-alternative project, or project with alternatives. With a no-alternative project, there is only one candidate—there is no choice. The data for the project are gathered and the PW is determined as discussed in Section 6.1. If the PW is positive and capital is available, the project is funded. No-alternative projects therefore do not involve much decision making beyond the evaluation of the PW.

EXAMPLE 6.3

Determine the present worth of a 10-year project whose first cost is $12,000, net annual benefit is $2,500, and salvage value is $2,000. Assume an annually compounded interest rate of 8% a year.

Solution

If you can visualize the problem without sketching its cashflow diagram, then proceed directly to the calculations. Otherwise, sketch a diagram, as in Fig. 6.2, to comprehend the problem. Note that the salvage value, recovered at the end of useful life, is a cash inflow to the project. It can therefore be considered a benefit.

Based on Equation (6.1) or the definition of a project's present worth and the associated functional notations, PW is given[3] by

$$\begin{aligned} PW &= PW_{\text{benefits}} - PW_{\text{costs}} \\ &= 2{,}500\,(P/A, 8\%, 10) + 2{,}000\,(P/F, 8\%, 10) - 12{,}000 \\ &= 2{,}500 \times 6.710 + 2{,}000 \times 0.4632 - 12{,}000 \\ &= 16{,}775 + 926 - 12{,}000 \\ &= \$5{,}701 \end{aligned}$$

In multialternative projects, there are several feasible alternatives or candidates. The decision making involves selecting the alternative whose PW is the most favorable, that is, the highest of all the positive PWs. For example, if the alternatives' PWs are $300, $650, $268, and $987, then select the one whose PW is $987. If their PWs are all negative, then none should be selected in a normal situation, since negative PW means that the present worth of costs exceeds that of the benefits. If there is an operational necessity so that one of the alternatives must be selected, then select the one whose

[3] If you have difficulty understanding the functional notations used here, reread Chapter 3.

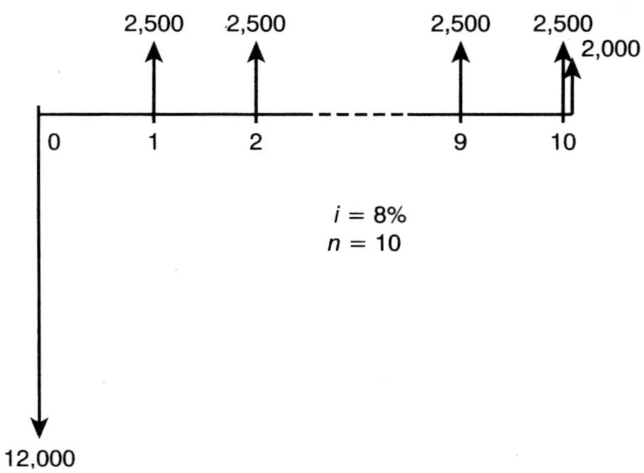

Figure 6.2 Diagram for Example 6.3

negative PW magnitude is the smallest. For example, if the PWs of four alternatives are −$450, −$350, −$896, and −$936, then select the one whose PW is −$350. Being the smallest of the negative PWs, this will minimize the loss and therefore is the most favorable of the four. In case some of the alternatives yield positive PWs and some negative PWs, then ignore those with negative PWs and select, from among those with positive PWs, the one with the highest positive PW. The obvious rule is: Decide in favor of financially the most attractive alternative.

EXAMPLE 6.4

Modern Manufacturers is considering the following five alternative projects as the next year's R&D effort. On the basis of present worth which one should be selected for funding, assuming that the investment must earn an annual return of 12%? All the five projects will take two years to develop and earn income during the five years beyond development.

Project	Annual Cost (during development)	Annual Income (during profitability)
A	$2,000	$3,500
B	1,800	3,200
C	2,200	2,300
D	1,200	2,000
E	3,500	4,800

Solution

The R&D costs are incurred during the two years of development. The benefits are derived during the five years following the development. The decision making involves two steps: determination of the PW for each project and a decision based on the PW criterion.

First, we evaluate the PW of each alternative project. Based on the data similarity, the cashflow diagrams of the projects will be alike. Let us analyze project A, whose cashflow diagram is given in Fig. 6.3, in detail. Its present worth—present being at period 0—can be expressed as

$$PW_A = PW_{benefits} - PW_{costs}$$

To evaluate PW_A, we need to determine project A's $PW_{benefits}$ and PW_{costs}. To do that we need to look at the diagram more closely. One can determine the present worth of the five benefits easily at period 2 as an upward vector, by treating them as a uniform series. This vector at period 2 must be transferred next to period 0 by treating it as a future sum. These operations on the given cash inflows yield

$$\begin{aligned} PW_{benefits} &= 3{,}500(P/A, 12\%, 5)(P/F, 12\%, 2) \\ &= 3{,}500 \times 3.605 \times 0.7972 \\ &= 10{,}059 \end{aligned}$$

The present worth of the costs is easily determined by considering the two cost vectors as a uniform series, yielding

$$\begin{aligned} PW_{costs} &= 2{,}000(P/A, 12\%, 2) \\ &= 2{,}000 \times 1.690 \\ &= 3{,}380 \end{aligned}$$

Thus,

$$\begin{aligned} PW_A &= PW_{benefits} - PW_{costs} \\ &= 10{,}059 - 3{,}380 \\ &= 6{,}679 \end{aligned}$$

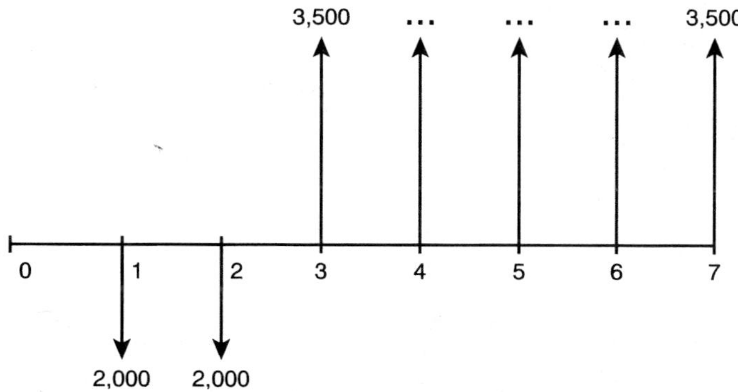

Figure 6.3 Diagram for Example 6.4

There is another way to determine the present worth of project *A*. This is based on the concept of tailoring, learned earlier in Chapter 4. The benefits seem to be an approximate uniform series. We can achieve complete uniformity by adding a $3,500 upward vector at periods 1 and 2. This will necessitate adding similar vectors downward at these two periods. With these changes in the diagram, the modified benefits form a uniform series over the entire analysis period of 7 years. But the modified costs at periods 2 and 3 are now $2,000 + $3,500 = $5,500 (instead of the given $2,000). Thus, from the tailored cashflow diagram, we get

$$\begin{aligned} PW_A &= 3{,}500(P/A, 12\%, 7) - 5{,}500(P/A, 12\%, 2) \\ &= 3{,}500 \times 4.564 - 5{,}500 \times 1.690 \\ &= 15{,}974 - 9{,}295 \\ &= \$6{,}679 \end{aligned}$$

Note that the tailoring required evaluation of only two functional notations, in comparison to the three in the straightforward (nontailored) approach.

Following tailoring, we determine below the present worth of the other four alternative projects in a similar way.

$$\begin{aligned} PW_B &= 3{,}200 \times 4.564 - 5{,}000 \times 1.690 \\ &= 14{,}605 - 8{,}450 \\ &= \$6{,}155 \end{aligned}$$

$$\begin{aligned} PW_C &= 2{,}300 \times 4.564 - 4{,}500 \times 1.690 \\ &= 10{,}497 - 7{,}605 \\ &= \$2{,}892 \end{aligned}$$

$$\begin{aligned} PW_D &= 2{,}000 \times 4.564 - 3{,}200 \times 1.690 \\ &= 9{,}128 - 5{,}408 \\ &= \$3{,}720 \end{aligned}$$

$$\begin{aligned} PW_E &= 4{,}800 \times 4.564 - 8{,}300 \times 1.690 \\ &= 21{,}907 - 14{,}027 \\ &= \$7{,}880 \end{aligned}$$

The second step of the analysis involves deciding which project to select. The decision rule is: Select the one that yields the largest positive PW. In this case, with a present worth of $7,880 it is project *E*, which is therefore selected.

6.3 INPUT–OUTPUT CONCEPT

Engineering economics analysis pertains to a wide variety of industrial projects. Before attempting the analysis, the problem must be fully comprehended. The *concept of input–output* facilitates problem comprehension. According to this concept, a project can be thought of as an input–output model as in Fig. 6.4. The decision box represents the problem and the solution method, including the decision criterion. The inputs to

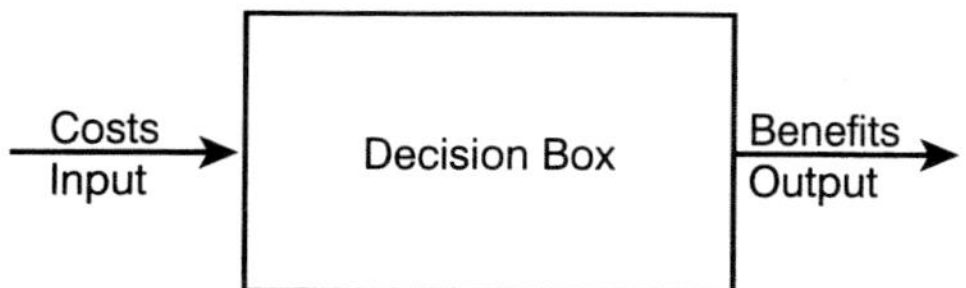

Figure 6.4 Input–Output Model

the box represent all the costs, including project-associated efforts that can be quantified in monetary terms. There may exist other efforts, essential to project completion, that are difficult to quantify in dollars. The outputs of the decision box are all the benefits from the project.

Consider for example research and development (R&D) projects, in which companies invest significant sums to develop new products and services, with the expectation of profits. The expenses for the project are the inputs and the profits through sales are the outputs. The basis of decision making in selecting the best alternative of a project may be one of the following:

1. The output for a given input is the largest.
2. The input for a given output is the smallest.
3. The ratio of the output to input is the largest where both input and output vary from one alternative to another. Such a ratio represents the project's financial or economic efficiency. We discuss this further as *benefit–cost ratio* later in Chapter 10.

On the basis of the input–output concept, engineering economics problems can be grouped into three categories, as discussed in the following subsections. Remember that *input means costs and output means benefits.*

6.3.1 Fixed Output

In many projects the output is fixed, that is, the benefit from each alternative of the project is the same. Consider for example the need to replace the elevator in a building. There may be several candidate elevators from different suppliers as a possible choice. What is the output in this project? The company does not receive any direct monetary benefit by investing in the elevator. The benefits (output) in this case are indirect, namely the convenience to employees who use the elevator and the ease with which goods can be moved in and out of the building. In such a project the output is the same irrespective of which elevator is selected to replace the existing one. In other words, the output of this project is *fixed.* Such projects represent *fixed-output, variable-input* or simply fixed-output, economics problems. The term *variable-input* signifies the fact that the cost of the candidate elevators may be different for each. The decision making involved in buying a car is another example of a fixed-output problem.

EXAMPLE 6.5

Siddharth is a plant engineer. One of the machines in his plant is old and needs replacing. Analyze this situation in light of the input–output concept.

Solution
In applying the input–output concept to Siddharth's decision-making problem of machine replacement, let us first consider the input. The cost of the replacement machine will most likely be budgeted. The purchase cost should include installation and technical training, if any. After receiving quotations from various suppliers, Siddharth will prepare a proposal, for approval by the upper management, that will include justification for replacing the existing machine.

It is not obvious from the problem statement whether the machine is general-purpose or specialized. If the machine is specialized, there may not be much choice of suppliers, sometimes none at all. If that is the case, Siddharth's proposal and budgeting will be straightforward. He will enclose with the proposal a copy of the quotation, asking for approval. On the other hand, if the machine is general-purpose, there may be several choices, yielding what we have called a multialternative project. Siddharth will apply the principles of engineering economics to the various alternatives and select the best based on the company's criterion of investment.

Next, the output. Siddharth may have to consider whether the replacement should just suffice to do what the existing machine does or whether he should seize the opportunity to incorporate current technologies by procuring a more modern machine. Such a machine may yield higher productivity or better product quality, or both, using advanced features such as on-line monitoring, predictive maintenance, ports for interfacing with computers, and so on.

If the machine is of the general-purpose type and several alternatives are available at different costs, then the project's input is variable. All the machines will do the same—replace the current machine. So the alternatives' output is fixed. Thus, the project is represented by a variable-input, fixed-output problem.

The extent of analysis, depicted by the decision box in Fig. 6.4, will depend on the issues relating to the input and output. The PW criterion discussed in this chapter, or any of the other five criteria presented in Part II, can be used to make the final decision.

Fixed-output problems involving multiple alternatives can be solved by comparing only the present worth of the costs, since the benefit from each alternative is the same (fixed). There is no need to analyze the benefits, which will keep the solution shorter and simpler. Obviously, the criterion of decision making in such problems is to *minimize the present worth of costs,* as illustrated in Example 6.6.

EXAMPLE 6.6

Hari has been using an old car he purchased as a freshman (first year of college). During the three years he has worked as an industrial technologist he has saved enough money to buy a new car. He is considering the Honda Accord, Chevy Prism, and Plymouth Acclaim, each with a seven-year useful life. Given the following data which one should he buy if the decision criterion is present worth? Assume an annually compounded interest rate of 9% per year.

Car	First Cost	Annual Cost	Salvage Value
Honda Accord	$20,000	$300	$3,000
Chevy Prism	18,000	500	2,000
Plymouth Acclaim	17,500	750	1,500

Solution

In this problem the benefit—the convenience of owning a car—is the same. The convenience translates into indirect benefits, such as time saved commuting to work, which are intangible. Since the benefit of owning a car is the same, it is a fixed-output problem. The decision can therefore be made solely on the basis of the input, that is, the present worth of the costs, which should be minimized.

Of the given data, the salvage value represents a cash inflow from the investment. It can therefore be subtracted, after accounting for its time value, from the first cost to determine the net cost on the car. If you can visualize the problem without its cashflow diagram, you should be able to follow the analysis below; or else, sketch a diagram to help you.

The present worth of the costs for the three alternatives are

$$\begin{aligned} PW_{\text{costs (Accord)}} &= 20{,}000 + 300(P/A, 9\%, 7) - 3{,}000(P/F, 9\%, 7) \\ &= 20{,}000 + 300 \times 5.033 - 3{,}000 \times 0.5470 \\ &= 20{,}000 + 1{,}510 - 1{,}641 \\ &= \$19{,}869 \end{aligned}$$

$$\begin{aligned} PW_{\text{costs (Prism)}} &= 18{,}000 + 500(P/A, 9\%, 7) - 2{,}000(P/F, 9\%, 7) \\ &= 18{,}000 + 500 \times 5.033 - 2{,}000 \times 0.5470 \\ &= 18{,}000 + 2{,}517 - 1{,}094 \\ &= \$19{,}423 \end{aligned}$$

$$\begin{aligned} PW_{\text{costs (Acclaim)}} &= 17{,}500 + 750(P/A, 9\%, 7) - 1{,}500\,(P/F, 9\%, 7) \\ &= 17{,}500 + 750 \times 5.033 - 1{,}500 \times 0.5470 \\ &= 17{,}500 + 3{,}775 - 820 \\ &= \$20{,}454 \end{aligned}$$

Comparing the three, Hari should buy a Chevy Prism, since its present worth of costs at $19,423 is the lowest.

6.3.2 Fixed Input

The second group of problems involve projects whose inputs are fixed, that is, the alternatives cost the same. The decision maker's task is to select the alternative that maximizes the output (income) from the investment. Such projects are of the *fixed-input-variable-output,* or simply fixed-input, type. The term *variable-output* signifies that the output (benefits) from the various alternatives differ from each other. An example of a fixed-input, variable-output project is hiring someone at a given budgeted salary, where the decision maker selects the best[4] from among the applicants.

[4]Expected to contribute the most toward company goals.

Since the alternative's input is fixed, the inclusion of costs in the analysis will not affect the final decision; it may, however, render the analysis lengthy. Thus, in fixed-input projects we select the alternative whose present worth of benefits is the largest. In other words, we maximize $PW_{benefits}$.

6.3.3 Variable Input and Output

Several engineering economics problems relate to projects whose alternatives' costs and benefits vary. Such problems are of the variable-input, variable-output type. The analyses of such problems are relatively lengthier, since both the costs and the benefits are to be accounted for. The present worth of each alternative is evaluated as illustrated in Example 6.2. Once the alternatives' PWs have been determined, they are compared to select the alternative with the largest positive PW. Example 6.7 offers an illustration.

EXAMPLE 6.7

The downsizing in her company has finally affected Nina. In spite of ten years of dedicated service to the company, she has to go at the end of this year. Fortunately, she has enough savings. After a careful assessment of her financial commitments, Nina decides to venture into business. Her father had left her a parcel of land adjacent to the main street in the town she lives. The business possibilities and the related data are:

Business Type	Initial Investment	Net Annual Income	Terminal Value
Farmer's Market	$120,000	$18,000	$45,000
Gas Station	240,000	36,000	60,000
Grocery Store	140,000	25,000	48,500

Nina plans to sell the business at the end of 25 years for the estimated terminal value. Her savings can earn annually compounded interest of 12% per year in a long-term CD. Which business is the best for Nina on the basis of the PW criterion?

Solution

There are three alternatives here. Note that both the cost of and the benefit from the three businesses are different. It is thus a variable-input, variable-output problem. We therefore evaluate the PWs of the three businesses and compare them to select the most favorable. The value of i is 12%. If Nina did not have her savings to invest, i would have been the interest rate charged by financial institutions on business loans to customers like her.

Again, a cashflow diagram is not necessary if you can comprehend the problem without it. Using Equation (6.1) and the relevant functional notations, we get

$$
\begin{aligned}
\text{PW}_{\text{farmer's market}} &= \text{PW}_{\text{benefits}} - \text{PW}_{\text{costs}} \\
&= [45{,}000(P/F, 12\%, 25) + 18{,}000(P/A, 12\%, 25)] - 120{,}000 \\
&= 45{,}000 \times 0.0588 + 18{,}000 \times 7.843 - 120{,}000 \\
&= 2{,}646 + 141{,}174 - 120{,}000 \\
&= \$23{,}820
\end{aligned}
$$

$$
\begin{aligned}
\text{PW}_{\text{gas station}} &= \text{PW}_{\text{benefits}} - \text{PW}_{\text{costs}} \\
&= [60{,}000(P/F, 12\%, 25) + 36{,}000(P/A, 12\%, 25)] - 240{,}000 \\
&= 60{,}000 \times 0.0588 + 36{,}000 \times 7.843 - 240{,}000 \\
&= 3{,}528 + 282{,}348 - 240{,}000 \\
&= \$45{,}876
\end{aligned}
$$

$$
\begin{aligned}
\text{PW}_{\text{grocery store}} &= \text{PW}_{\text{benefits}} - \text{PW}_{\text{costs}} \\
&= [48{,}500(P/F, 12\%, 25) + 25{,}000(P/A, 12\%, 25)] - 140{,}000 \\
&= 48{,}500 \times 0.0588 + 25{,}000 \times 7.843 - 140{,}000 \\
&= 2{,}852 + 196{,}076 - 140{,}000 \\
&= \$58{,}928
\end{aligned}
$$

A comparison of these present worths suggests that Nina should invest in the grocery store, since its PW of $58,928 is the largest.

6.4 ANALYSIS PERIOD

While analyzing problems under the PW criterion, a situation may arise that needs special attention. This happens when the alternatives' useful lives are different. The relevant question is: What analysis period should be used? In other words, for what time horizon should the alternatives be compared with each other?

Consider a machine that needs replacing, since it has been a production bottleneck. If it is specialized, there may be only one type in the market, that is, there is no choice. Such problems are simple: Collect the cost and benefit data on the new machine covering the period of its useful life and carry out the required analysis. The analysis period is thus the same as the useful life.

Next consider that the machine is general-purpose with one or more alternatives in the market. The PW analysis may involve either of the situations discussed in the following two subsections.

6.4.1 Equal Lives

If the alternative machines A and B have the same useful life, there is little difficulty, since both can be analyzed over this period. Thus, the analysis period is their useful life. Again, gather all the pertinent cost and benefit data, and evaluate and compare their present worths. Select the one whose present worth is more favorable. The same methodology applies to problems involving more than two alternatives, as illustrated in Example 6.8.

EXAMPLE 6.8

The loading and unloading of a machine can be enhanced significantly by using a robotic system. Three systems, whose types and data are given below, are under consideration. If for the investment annual interest rate is 15%, which system should be purchased on the basis of the PW criterion?

Type	Initial Cost	Annual Maintenance	Useful Life in Years
Robota	\$65,000	\$300	10
Automata	70,000	250	10
Robo	75,000	20	10

Solution

Note that the useful life for the three types is the same. Therefore, an analysis period equal to their useful life, that is, 10 years, is decided upon. Based on the input–output concept, the output from each type is the same, that is, each system will be able to carry out the desired loading and unloading tasks. Thus, the decision may be based solely on the present worth of costs, which are

$$\begin{aligned}\text{PW}_{\text{costs (Robota)}} &= 65{,}000 + 300(P/A, 15\%, 10)\\ &= 65{,}000 + 300 \times 5.019\\ &= 65{,}000 + 1{,}506\\ &= \$66{,}506\end{aligned}$$

$$\begin{aligned}\text{PW}_{\text{costs (Automata)}} &= 70{,}000 + 250(P/A, 15\%, 10)\\ &= 70{,}000 + 250 \times 5.019\\ &= 70{,}000 + 1{,}255\\ &= \$71{,}255\end{aligned}$$

$$\begin{aligned}\text{PW}_{\text{costs (Robo)}} &= 75{,}000 + 20(P/A, 15\%, 10)\\ &= 75{,}000 + 20 \times 5.019\\ &= 75{,}000 + 100\\ &= \$75{,}100\end{aligned}$$

Robota is selected, since its present worth of costs at \$66,506 is the minimum.

6.4.2 Unequal Lives

If the useful lives of the alternatives are different, then what analysis period is used for the problem? Let us say that machine A has a useful life of 3 years and B of 5 years. For what time horizon do we compare A with B, 3 years or 5 years?

At the end of 3 years, machine A will need to be replaced again, while B will still be useful. If we replace A and keep it for another 3 years, accounting for the associated costs and benefits, then A will be in use when B will need to be replaced at the end of the fifth year. Thus, we seem to be chasing a time "mirage." This is the difficulty unequal lives create in deciding an appropriate analysis period.

The difficulty is resolved by extending the analysis period to an extent that both A and B terminate together. Such an analysis period[5] corresponds to the *least common multiple (LCM)* of their useful lives. In the case under discussion, it is 15 years (LCM of 3 and 5 years). Thus, the comparison between alternatives A and B should be made by looking through a 15-year time "window." During this period, machine A will be replaced four times ($4 \times 3 = 12$ years of life from replacements, plus 3 years of the first acquisition), and machine B twice. At the end of 15 years both would be ready for simultaneous replacement for further use.

When an analysis period longer than the alternative's useful life is considered in a problem, three major assumptions are implicitly made: (i) the suppliers will continue to be in business, (ii) they will offer the alternative at the same price, and (iii) benefits will remain unchanged throughout the analysis period.

EXAMPLE 6.9

Abraham's personal computer (PC) is slow and lacks features for the current application programs. There are three choices: upgrade the existing system, buy a new PC and software, or buy a one-year-old "fully loaded" PC. For the following data, which one should he opt for if the decision criterion is PW and the applicable yearly interest rate is 9%, compounded annually?

Choice	First Cost	Annual Operating Cost	Useful Life, Years	Salvage Value
Upgrade	$3,000	$200	2	$250
New PC	8,000	100	4	500
Old PC	5,000	150	3	300

Solution

Since PW is the criterion and the useful lives of the three choices are different, the analysis period should be the LCM of their lives of 2, 3, and 4 years, that is, 12 years.

[5]If the LCM-based analysis period becomes too long, then (a) use an analysis period equal to the useful life of one of the alternatives along with its costs and benefits, but use the *prorated* costs and benefits of the other alternatives, or (b) use another analysis method, such as annual worth (Chapter 8) or rate of return (Chapter 9).

Fig. 6.5 shows the cashflow diagrams[6] for the choices. Note that some vectors have not been labeled, since they are repetitive.

The present worths[7] of the choices are found as

$$
\begin{aligned}
PW_{\text{upgrade}} &= -3{,}000 - 200(P/A, 9\%, 12) - (3{,}000 - 250)[(P/F, 9\%, 2) \\
&\quad + (P/F, 9\%, 4) + (P/F, 9\%, 6) + (P/F, 9\%, 8) + (P/F, 9\%, 10)] \\
&\quad + 250(P/F, 9\%, 12) \\
&= -3{,}000 - 200 \times 7.161 - 2{,}750(0.8417 + 0.7084 + 0.5963 \\
&\quad + 0.5019 + 0.4224) + 250 \times 0.3555 \\
&= -3{,}000 - 1{,}432 - 2{,}750 \times 3.0707 + 89 \\
&= -3{,}000 - 1{,}432 - 8{,}444 + 89 \\
&= -\$12{,}788 \\
PW_{\text{new}} &= -8{,}000 - 100(P/A, 9\%, 12) - (8{,}000 - 500)\,[(P/F, 9\%, 4) \\
&\quad + (P/F, 9\%, 8)] + 500(P/F, 9\%, 12) \\
&= -8{,}000 - 100 \times 7.161 - 7{,}500(0.7084 + 0.5019) + 500 \times 0.3555 \\
&= -8{,}000 - 716 - 7{,}500 \times 1.2103 + 178 \\
&= -8{,}716 - 9{,}077 + 178 \\
&= -\$17{,}615 \\
PW_{\text{old}} &= -5{,}000 - 150(P/A, 9\%, 12) - (5{,}000 - 300)[(P/F, 9\%, 3) \\
&\quad + (P/F, 9\%, 6) + (P/F, 9\%, 9)] + 300(P/F, 9\%, 12) \\
&= -5{,}000 - 150 \times 7.161 - 4{,}700(0.7722 + 0.5963 + 0.4604) \\
&\quad + 300 \times 0.3555 \\
&= -5{,}000 - 1{,}074 - 4{,}700 \times 1.8289 + 107 \\
&= -6{,}074 - 8{,}596 + 107 \\
&= -\$14{,}563
\end{aligned}
$$

Comparing the three present worths, which are all negative[8], Abraham decides to upgrade the existing system, since it is most favorable (smallest negative PW, that is, lowest net cost).

6.5 CAPITALIZED COST

Engineers design, develop, and maintain facilities that are expected to provide use or service forever. Highways, bridges, dams, and school buildings are some of the examples. Their useful life is perpetual. Such facilities are usually constructed and maintained by governments with the taxpayers' money.

[6] If you have difficulty comprehending the diagrams, reread Chapter 2.

[7] Following the standard convention that costs (vectors down) are negative, while benefits (vectors up) are positive.

[8] One could have used the *present worth of costs* as the decision variable, since it is a fixed-output problem. In that case, all would have been positive. The decision would nevertheless have been the same.

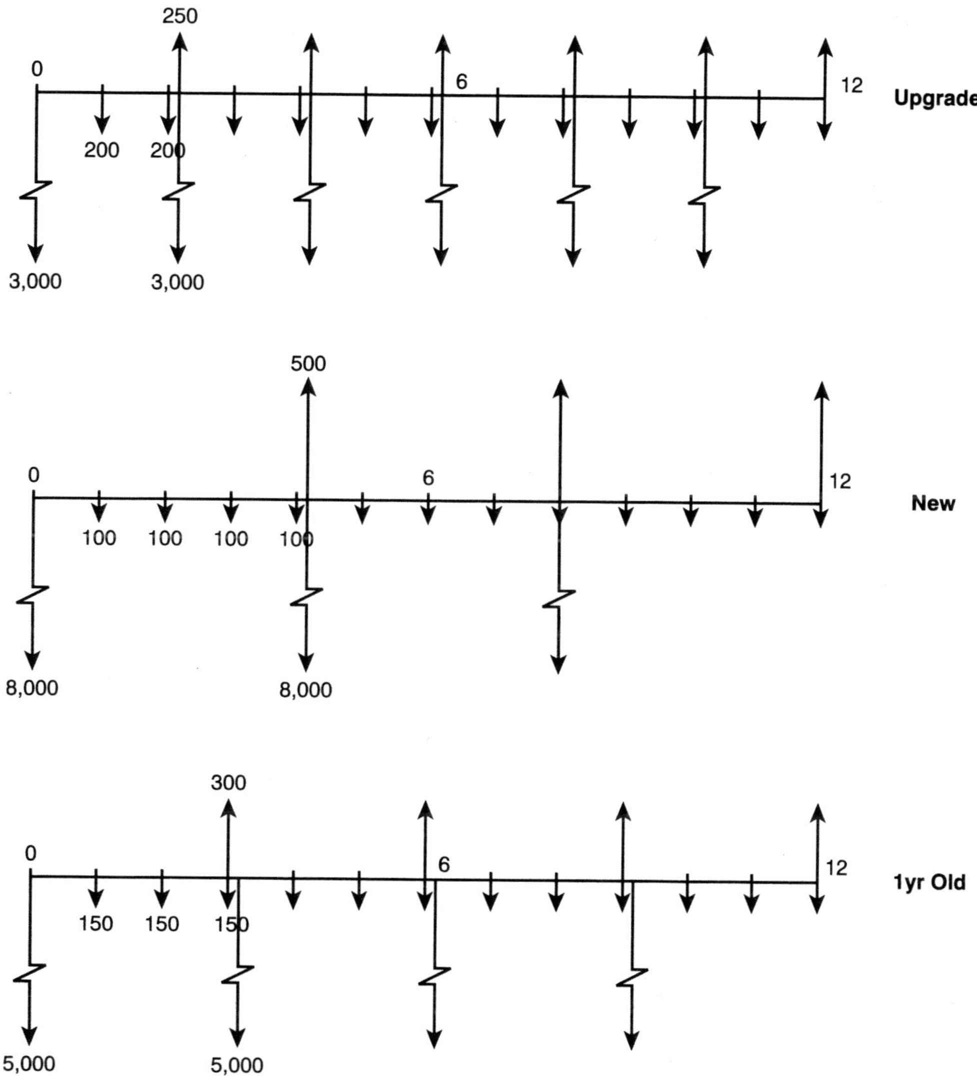

Figure 6.5 Alternatives with Unequal Lives

Since perpetual-life projects last forever, their economic analysis must cover a long period; in fact, the analysis period is infinite. Besides the initial cost, they may require continual maintenance and renovation, as and when necessary. The ongoing costs must be budgeted for, or means to raise them in the future planned for, at the time of initial construction.

If the funds for maintenance are budgeted at the time of initial construction, they are kept in an interest-bearing account and the earned interest used as planned. Such a fund is called *capitalized cost,* since it continues to exist as capital—as *seed corn.* City and other local governments often make provision in their budgets for capitalized costs of various projects. If P is the fund invested as capitalized cost at an annual interest rate i, then the year-end interest is iP. This iP can be disbursed (used up) annually without depleting P.

Thus, a perpetual-life project often has two components: initial cost and capitalized cost. In such cases, no funds will have to be raised in the future for the project. As an alternative to budgeting for the capitalized cost, annual taxation and bonds may be used to raise the maintenance funds as and when necessary.

An example of the capitalized-cost concept is found in higher education. Colleges and universities often raise funds as endowments to offer perpetual scholarships. To offer an annual scholarship A perpetually, the required endowment gift is P so that $iP = A$. In other words,

$$P = A/i$$

where P is the endowment (capitalized cost).

The donor's gift P is invested at an annual interest rate i, generating perpetually an annual income of A. The same logic is applicable to maintaining perpetual projects, as illustrated in Example 6.10.

EXAMPLE 6.10

The City of Hattiesburg finds Hardy Street congested during peak traffic hours and is planing to build a parallel street to reduce congestion. The new street is to be 10 miles long, and the initial cost of construction is \$500,000 per mile. What is the total project cost if a provision of \$10,000 for annual repairs is also made? Assume $i = 5\%$ per year, compounded annually.

Solution

$$\begin{aligned}\text{Initial cost} &= \$500{,}000 \text{ per mile} \times 10 \text{ miles}\\ &= \$5{,}000{,}000\end{aligned}$$

$$\begin{aligned}\text{Capitalized cost for annual repairs, } P &= A/i\\ &= 10{,}000/0.05\\ &= \$200{,}000\end{aligned}$$

Therefore,

$$\begin{aligned}\text{Total project cost} &= \$5{,}000{,}000 + 200{,}000\\ &= \$5.2 \text{ million}\end{aligned}$$

SUMMARY

The present worth (PW) method is commonly practiced by the engineering industry. The application of the PW criterion to engineering projects involves two tasks: determination of the present worth(s), and decision making based on the criterion. A project's PW is the difference between its present worths of benefits and costs. It is usually positive for a project to be funded. In a multialternative project, the alternative with the most favorable PW is selected.

The input–output concept, under which project costs are considered input and benefits the output, streamlines the analysis. Based on this concept problems may be categorized in one of the three types: fixed output, fixed input, or variable input variable output. In fixed-output (same benefit) problems, the alternative with the lowest input (cost) is selected. In fixed-input (same cost) problems, the alternative with the highest output (benefit) is selected. Irrespective of the problem type, the aim is always to maximize the desirable (benefit) and/or minimize the undesirable (cost).

Present worth analysis requires that the alternatives be compared within the same time window (analysis period). This occurs automatically in a project whose alternatives are of equal useful lives. If the alternatives' useful lives are unequal, an analysis period equal to the least common multiple (LCM) of their lives is selected. If the LCM-based analysis period is unreasonably long, there are two choices: (a) Use an analysis period equal to the useful life of one alternative along with its costs and benefits, but use the prorated costs and benefits for the other alternatives, or (b) use another method, such as annual worth (Chapter 8) or rate of return (Chapter 9).

The maintenance and upkeep of projects of perpetual life can be provided for by capitalizing the necessary fund. A project's capitalized cost is the initial lump sum whose interest income pays forever the operation and maintenance cost. The concept of capitalized cost is applicable to nonengineering projects as well, for example in setting up a perpetual scholarship or professorship at a college.

EXERCISES

Discussion Questions

6.1 Give an example from your own experience where the present worth criterion would be appropriate for decision making.

6.2 Explain the meaning of *present* in the present worth method.

6.3 How does the present worth criterion account for the time value of money?

6.4 How does the input-output concept help in analyzing engineering economics problems?

6.5 How useful is the input–output concept in analyzing engineering economics problems?

6.6 Discuss when a project of zero PW can be funded.

Multiple-Choice Questions (Circle the *best* answer.)

6.7 If the present worth of costs exceeds that of the benefits, PW for the project is
a. positive.
b. negative.
c. equal to zero.
d. always an integer.

6.8 Your manager has received approval from the upper management to upgrade your personal computer at a cost of $1,000. This represents a _____ economics problem.
a. variable-input, variable-output
b. fixed-output
c. fixed-input
d. fixed-input, fixed-output

6.9 While comparing the PW of alternatives of unequal useful lives, the following assumptions are made *except* that the
a. suppliers of the resources will continue in business.
b. alternatives will cost the same.
c. benefits will remain unchanged.
d. alternatives are perpetual.

6.10 To attend college you need to purchase a car. This is a project involving
a. fixed input.
b. fixed output.
c. fixed input, fixed output.
d. variable input, variable output.

6.11 John buys a video game for $300. After using it for 2 years, he expects to sell it for $20. If $i = 10\%$, the present worth in dollars is
a. $300(F/P, 10\%, 2) - 20$
b. 280
c. $20(P/F, 10\%, 2) - 300$
d. $300 - 20(P/F, 10\%, 2)$

6.12 There are two choices for replacing a punch press. The basic model has a useful life of 8 years while the deluxe model of 12 years. The most appropriate analysis period for this problem is
a. 20 years
b. 96 years
c. 1.5 years
d. 24 years

6.13 The local city government has an offer of gift from a rich, generous Hollywood actress who grew up in the town and remembers her numerous visits to the city park as a child. She wants to donate toward perpetual upkeep of the newly built children's park. If the upkeep cost is $50,000 per year, and her gift as a long-term investment can earn 10% annually, the gift should be
a. $50,000/12
b. $50,000
c. $500,000
d. $500,000/12

Numerical Problems

6.14 The initial cost of a new photocopier is $12,000. It is expected to generate a net annual income of $4,000 in the first year, decreasing annually by $300. At the end of its 5-year useful life, the photocopier is expected to be sold for $500. Assuming $i = 10\%$ per year with annual compounding, determine the photocopier's present worth.

6.15 Within two years of graduation, Shyam has established a significant market for non-CFC Styrofoam cups. He has decided to purchase a new machine rather than continue to operate with the current one. The maintenance cost of the machine is expected to be $500 in the first year, $1,000 in the second year, $1,500 in the third year, and so on. Assuming an annually compounded interest rate of 12% per year, how much should be included in the current budget to pay for the maintenance if the machine's useful life is 10 years?

6.16 A rental fax machine can be replaced with one of two models. The basic model with a life of 4 years costs $6,000 and is expected to bring in net $2,000 annually. For the deluxe model, the data are 8 years, $10,500, and $2,500 respectively. Assuming no salvage value and an annually compounded interest rate of 8% per year, which model is a better choice on the basis of the PW criterion?

6.17 Two candidate machines are under consideration. For machine A, the initial cost is $100,000, operating benefits are $40,000 per year, and useful life is 6 years. The respective data for machine B are $150,000, $50,000, and 9 years. Neither machine has any salvage value. On the basis of present worth which machine would you select? Assume $i = 8\%$.

6.18 It costs Hattiesburg city $50,000 a year to maintain Hardy street. If this street is resurfaced at a cost of $150,000, the annual maintenance cost will reduce to $20,000 for the first five years, and to $30,000 for the subsequent five years. At the end of ten years, the maintenance cost is expected to revert back to $50,000 a year. Should the city resurface this street if the annually compounded interest rate is 5% per year? Use the PW criterion for the decision.

6.19 Isaac asks his brother-in-law to lend him $16,000 so that he can buy a new car. The car will help Isaac sell more merchandise. Isaac promises to pay back $3,000 at the end of first year, $6,000 at the end of second year, and $9,000 at the end of third year. If his brother-in-law can earn 10% per year on his $16,000 from the local bank, will he do Isaac a favor by lending? Use the present worth criterion.

6.20 Three *mutually exclusive*[9] alternatives X, Y, and Z are being considered as a permanent solution to an assembly-line bottleneck. The data are

[9]The term *mutually exclusive* means that the selection of one of the alternatives precludes the others from being selected. In other words, the alternatives are independent of each other. This is one of the assumptions in multialternative decision making. The term may be explicitly expressed, as in this problem, but in most problems we make an implicit assumption to this effect.

	X	Y	Z
Initial Cost[10]	$10k	$14k	$12k
Annual Benefit	$2,600	$5,500	$1,500
Useful Life, years	5	3	15

Assuming no salvage value, an annually compounded interest rate of 8% per year, and unchanged data at replacement times, which alternative should be selected on the basis of the PW criterion?

6.21 For equipment being considered for acquisition at a cost of $80,000 the following data have been compiled.

Year	Cost	Income
1	$4,000	$20,000
2	5,000	24,000
3	6,000	29,000
4	7,000	32,000
5	8,000	34,000

Assuming 7% annual interest rate, should the equipment be purchased under the PW criterion?

6.22 ZeroDefect Manufacturing Company needs to build additional space for inventory storage. There are three choices:

1. Spend $12 million to build space that will meet the needs for the next 10 years,
2. Spend $8 million now to meet the needs of the first 6 years and another $5 million during the sixth year to meet the needs of the next 4 years, or
3. Rent a nearby space for 10 years at an yearly cost of $2 million.

With reference to the input–output concept, which type of project is this? Which is the best choice on the basis of present worth criterion? Assume no salvage values, and $i = 8\%$ per year.

6.23 Within a few years of graduation from the engineering department of The International University, Tom has amassed enormous wealth from his patent on a microorganism-based solid waste disposal system. He wants to endow a scholarship at his alma mater in his mother's name. The endowment will be placed in an interest-bearing long-term financial instrument earning 12% annual interest. The yearly proceeds should generate a $5,000 income for the annual scholarship. How much should the endowment fund be?

---

[10]k as a prefix in SI units is an acronym for kilo, meaning 1,000.

CHAPTER SEVEN

Future Worth

IN THIS CHAPTER YOU WILL LEARN ABOUT

- The meaning of *future worth* (FW)
- Evaluation of FW
- Determination of a project's FW
- Application of the FW criterion
- When to use the FW criterion

This chapter discusses the future worth (FW) method of analyzing engineering economics problems. It can be considered a proximate extension of the present worth method, whose coverage in the previous chapter is significantly pertinent—the only difference being the reference time. The present worth looks at *now*, while future worth looks at *then*—a certain time in the future. Thus, under the FW criterion, a decision is made on the basis of a project's economic worth at some time in the future. In the FW method, the data are analyzed by referring to the end of useful life or analysis period.

7.1 MEANING

Future worth is the value of something at a certain time in the future. Consider your plan of a car trip lasting several hours. To save the time of stopping for food, you carry two apples with you. The apples have little value for you when you begin the trip just after lunch. But in a few hours when you will be hungry, they will be valuable. Your decision to take the apples with you is based on their future worth. This of course is a rather simplistic explanation of future worth.

FW-based decision making involves evaluation of the project's future worth for each alternative and selection of the most favorable one. A project's FW is the surplus of its benefits' future worth[1] over its costs' future worth. Consider Fig. 7.1, in which the given

[1]In engineering economics we consider only those elements of a project that are expressible in monetary terms. Examples are first cost, maintenance costs, benefits, and salvage value. The *future worth* of a project depends on the values of such elements, that is, the various cashflows pertaining to the project.

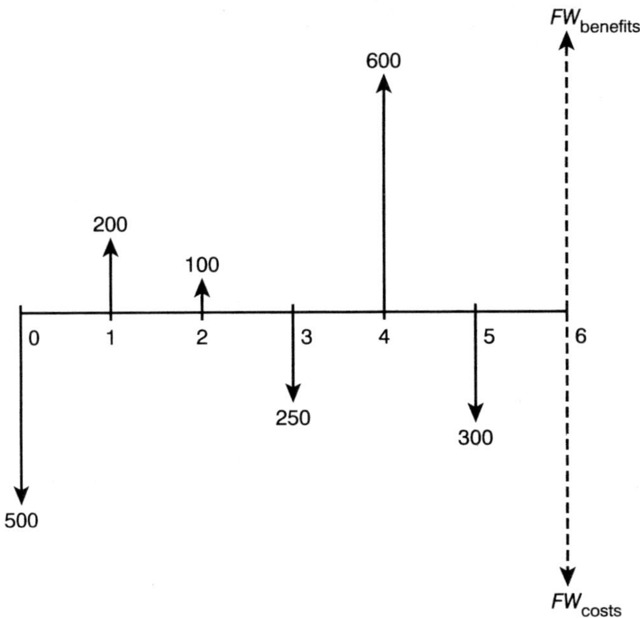

Figure 7.1 Analysis of Future Worth

cashflows are shown by the firm vectors. Following the convention discussed in Chapter 2, a cost of \$500 is incurred now. There are two more costs, \$250 at period 3 and \$300 at period 5. We also see in the diagram three benefits at time periods 1, 2, and 4. For this diagram, the future worth of costs, shown dotted by FW_{costs}, is the *equivalent cost* at period 6 of the three costs. Its determination involves evaluating, and adding together, the F values at period 6 of the \$500 first cost and of the other two costs. In other words, we convert the \$500 cost at period 0, the \$250 cost at period 3, and the \$300 cost at period 5 to their equivalents at period 6, and add these equivalents. For an interest rate i per period, the F values are

$$\begin{aligned} F \text{ value of the \$500 cost} &= 500(F/P, i, 6) \\ F \text{ value of the \$250 cost} &= 250(F/P, i, 3) \\ F \text{ value of the \$300 cost} &= 300(F/P, i, 1) \end{aligned}$$

Adding these F values, the future worth of costs is

$$\text{FW}_{\text{costs}} = 500(F/P, i, 6) + 250(F/P, i, 3) + 300(F/P, i, 1)$$

In the same way we determine the future worth at period 6 of all the benefits, shown dotted in Fig. 7.1 by $\text{FW}_{\text{benefits}}$, as

$$\text{FW}_{\text{benefits}} = 200(F/P, i, 5) + 100(F/P, i, 4) + 600(F/P, i, 2)$$

Thus, in determining the future worth of costs (or benefits), all the costs (or benefits) of the project are transferred as their equivalents to a specific time marker, using the appropriate functional factors. This way the time value of each cashflow is accounted for. Example 7.1 illustrates the application of the FW method.

EXAMPLE 7.1

Khalda begins her four years of college to earn a BS degree in computer science. The annual cost of education is expected to be $15,000. The BS degree is likely to enable her to earn $10,000 more (in relation to if she did not have the degree) per year during her expected professional career of 30 years. What is the worth of additional earnings at career end if $i = 8\%$ per year[2]?

Solution

The annual cost of the BS degree is $15,000 for four years. The financial return from the degree is the $10,000 additional annual income over 30 years. Following the convention that a cashflow is posted at period end, the complete diagram is shown in Fig. 7.2. Note that the question specifically asks about the future worth of only the additional incomes. Thus, the cash outflows at periods 1 through 4 are irrelevant to the solution. The question is also explicit about the time period at which future worth is to be determined. It is at the end of the 30-year career, that is, at period 34 (30 years of professional career following 4 years of education[3]).

The uniformity in the cash inflows is obvious; its exploitation will save time and effort. For the given data, $A = \$10{,}000, n = 30,$ and $i = 8\%,$

$$\begin{aligned} \text{FW}_{\text{benefits}} = F &= A(F/A, i, n) \\ &= 10{,}000(F/A, 8\%, 30) \\ &= 10{,}000 \times 113.283 \\ &= \$1{,}132{,}830 \end{aligned}$$

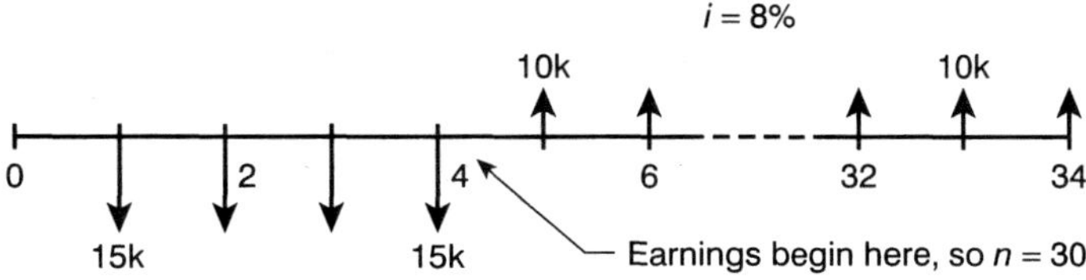

Figure 7.2 Diagram for Example 7.1

---

[2]In the absence of any specific mention, always assume the compounding to be annual. Also, if unspecified, the interest rate should be considered annual.

[3]For the purpose of this example, one can sketch another diagram that does not show periods 1 through 4. Such a diagram will begin at period 4 of Fig. 7.1 as period 0 and end with $n = 30$.

Thus, the future worth of additional earnings due to the BS degree is more than a million dollars.

A project's future worth (FW) accounts for the net effect of benefits over costs and is the basic parameter for decision making under the FW criterion. It is the difference between the future worth of the benefits and the future worth of the costs. Thus,

$$FW = FW_{\text{benefits}} - FW_{\text{costs}} \tag{7.1}$$

Note that we subtract the future worth of the costs from the future worth of the benefits. This renders the project's FW positive in cases where the future worth of the benefits exceeds the future worth of the costs—a desirable aim. Thus, a positive value for FW means that the project is financially attractive and hence normally funded.

EXAMPLE 7.2

Determine the FW of Khalda's higher education in Example 7.1 and comment on its value.

Solution

From Example 7.1, the future worth of the benefits is already known to be \$1,132,830. As a continuation of the discussions therein, we next determine the future worth of the costs. Referring to Fig. 7.2, the four costs—\$15,000 each at periods 1 through 4—need to be transferred to period 34, which we should do as efficiently as possible. One[4] of the most efficient ways is to transfer them first to period 0 as their present value P, and then transfer P to period 34 as the final sum F. Exploiting the uniform-series pattern in these cash outflows,

$$\begin{aligned} P &= A(P/A, i, n) \\ &= 15{,}000(P/A, 8\%, 4) \\ &= 15{,}000 \times 3.312 \\ &= \$49{,}680 \end{aligned}$$

Next,

$$\begin{aligned} FW_{\text{costs}} = F &= P(F/P, i, n) \\ &= 49{,}680(F/P, 8\%, 34) \\ &= 49{,}680 \times 13.690 \\ &= \$680{,}119 \end{aligned}$$

---

[4]The other way is to treat them as A and transfer first to period 4 as F. This F is then treated as P and transferred to period 34 as the F value.

The future worth of the project is thus

$$\begin{aligned} FW &= FW_{benefits} - FW_{costs} \\ &= \$1{,}132{,}830 - \$680{,}119 \\ &= \$452{,}711 \end{aligned}$$

Since the project's FW is positive, the BS degree is a good investment[5] on the basis of the FW criterion.

7.2 METHOD

The FW method of solving engineering economics problems involves determination of the future worth of the project or the future worths of project alternatives. Using the FW criterion, the better of the two alternatives, or the best of all the alternatives, is selected. Thus, the application of this method involves two tasks:

1. Evaluation of the future worth(s)
2. Decision on the basis of the FW criterion

We discussed the first task in Section 7.1. In this section we discuss decision making based on the FW criterion.

From the engineering economics viewpoint, projects can be categorized in two groups: no-alternative projects, or projects with an alternative. In no-alternative projects, there is only one candidate with no choice. The data for the sole candidate are collected and analyzed to determine FW, as discussed in Section 7.1. No-alternative projects do not involve much decision making, beyond the evaluation of FW, as illustrated in Examples 7.2 and 7.3, and their approval if FW is positive.

EXAMPLE 7.3

Determine the future worth of a 10-year project whose first cost is \$12,000, net annual benefit is \$2,500, and salvage value is \$2,000. Assume an annually compounded interest rate of 8% per year.

Solution

If you can visualize the problem without its cashflow diagram, then proceed directly to the calculations. Otherwise, first sketch a diagram to comprehend the problem. Note that the salvage value is recovered at the end of useful life, and as a cash inflow it is a benefit to the project. Since it is not obvious from the problem statement when the

---

[5]Note that we have not accounted for—because it is difficult—other nonmonetary benefits higher education may bring to enhance Khalda's quality of life (directly) and the community she lives in (indirectly).

benefits begin to accrue, we need to make an assumption. A reasonable assumption is that the benefits begin to be realized immediately following the investment.

Using Equation (7.1) and the associated functional factors, the FW is determined[6] as

$$\begin{aligned} \text{FW} &= \text{FW}_{\text{benefits}} - \text{FW}_{\text{costs}} \\ &= [2{,}500(F/A, 8\%, 10) + 2{,}000] - 12{,}000(F/P, 8\%, 10) \\ &= 2{,}500 \times 14.487 + 2{,}000 - 12{,}000 \times 2.159 \\ &= 36{,}218 + 2{,}000 - 25{,}908 \\ &= \$12{,}310 \end{aligned}$$

The preceding example has the same data as Example 6.3, so the present worth answer in Example 6.3 must be equivalent to the future worth answer in Example 7.3. The value of PW (or P) in Example 6.3 was \$5,701, which indeed is equal to the value of FW (or F) in Example 7.3, since

$$\begin{aligned} F &= P(F/P, i, n) \\ &= 5{,}701(F/P, 8\%, 10) \\ &= 5{,}701 \times 2.159 \\ &= \$12{,}308 \end{aligned}$$

The \$2 difference between the \$12,308 based on Example 6.3's PW and the \$12,310 of Example 7.3 is due to the error introduced by the rounded values in the interest tables.

In multialternative projects, there are several feasible choices (candidates) for the same project. For example, as a replacement of an old machine there may be three types of machines—A, B, and C—in the market. In such a case, the process of selecting the best machine is the project[7], while the three candidate machines (A, B, and C) are the alternatives.

The decision making in multialternative projects involves selecting the alternative whose FW is the most favorable, that is, the largest of all the positive FWs. For example, if the FWs of the alternatives are \$300, \$650, \$268, and \$987, then select the one whose FW is \$987. If the FWs of all the alternatives are negative, then none should be selected in a normal situation, since a negative FW means that the investment will incur a loss. If there are compelling business reasons that demand that a selection must be made, then select the alternative whose negative FW is the smallest. For example, if the FWs of four alternatives are −\$450, −\$350, −\$896, and −\$936, then select the one whose FW is −\$350. Being the smallest of the negative FWs, this will minimize the loss in investment and is thus the most attractive of the four alternatives. Where some of the alternatives yield positive FWs and others yield negative FWs, discard those with negative FWs and select the one with the highest positive FW. In short, the decision must eventually be in favor of financially the most *attractive* alternative.

---

[6]If you have difficulty understanding the functional factors used here, reread Chapter 4.

[7]The alternatives themselves may sometimes be called *projects*. In multialternative cases, therefore, the two terms *project* and *alternative* are used synonymously.

EXAMPLE 7.4

World's Best Manufacturer is considering five new projects[8] as part of the next year's R&D efforts. On the basis of future worth, which one should be selected for funding, assuming that the investment must earn an annual return of 12%? All the five projects will take two years to complete and will be profitable during the subsequent five years.

Project	Annual Cost (during development)	Annual Income (during subsequent 5 years)
A	$2,000	$3,500
B	1,800	3,200
C	2,200	2,300
D	1,200	2,000
E	3,500	4,800

Solution

The R&D costs are incurred during the two years of the development. The benefits are realized during the five years following the development. The decision making involves two steps: determination of the FW for each project, and a decision based on the FW criterion.

Let us first evaluate the FWs. Since the projects are similar except the data, their cashflow diagrams will be alike. Referring to the one for project A in Fig. 6.3, its future worth can be determined as

$$\begin{aligned} \mathrm{FW_A} &= \mathrm{FW_{benefits}} - \mathrm{FW_{costs}} \\ &= 3{,}500(F/A, 12\%, 5) - [2{,}000(F/P, 12\%, 5) + 2{,}000(F/P, 12\%, 6)] \\ &= 3{,}500 \times 6.353 - [2{,}000 \times 1.762 + 2{,}000 \times 1.974] \\ &= 3{,}500 \times 6.353 - 2{,}000(1.762 + 1.974) \\ &= 3{,}500 \times 6.353 - 2{,}000 \times 3.736 \\ &= 22{,}236 - 7{,}472 \\ &= \$14{,}764 \end{aligned}$$

The calculations for the other projects will be similar. Thus,

$$\begin{aligned} \mathrm{FW_B} &= 3{,}200 \times 6.353 - 1{,}800 \times 3.736 \\ &= 20{,}330 - 6{,}725 \\ &= \$13{,}605 \end{aligned}$$

$$\begin{aligned} \mathrm{FW_C} &= 2{,}300 \times 6.353 - 2{,}200 \times 3.736 \\ &= 14{,}612 - 8{,}219 \\ &= \$6{,}393 \end{aligned}$$

[8]Here, the term *projects* is more appropriate than *alternatives*, since each may be technically different from the others. From the analysis viewpoint, however, they are in fact alternatives.

$$
\begin{aligned}
FW_D &= 2{,}000 \times 6.353 - 1{,}200 \times 3.736 \\
&= 12{,}706 - 4{,}483 \\
&= \$8{,}223
\end{aligned}
$$

$$
\begin{aligned}
FW_E &= 4{,}800 \times 6.353 - 3{,}500 \times 3.736 \\
&= 30{,}494 - 13{,}076 \\
&= \$17{,}418
\end{aligned}
$$

The second step involves selecting the project[9]. As discussed earlier, the most attractive project should be selected. In this case, the future worths of all five projects are positive, so select the one with the largest positive FW[10], that is, project E.

7.3 WHEN TO USE

The FW criterion, and its application to engineering economics problems, is very similar to the PW criterion. An obvious question is when to use one, and not the other. The simple rule is this: If the *present* is deemed to be important in decision making, use the PW method; by the same token, if the *future* is important, then use the FW criterion. The time period to which *future* refers should be the one of *relevance.*

Consider financial planning by an investor who desires to build retirement equity. In here, *future*—rather than the present—is relevant. And the future occurs at the time the investor would like to retire, which becomes the pertinent time period. The obvious criterion in this case is future worth. In contrast, for capital investment projects the appropriate criterion is usually the present worth. Here, the aim is to invest in a resource now for generating profits later. In general, engineering companies are not in the business of financial investment, but in engineering investment. So, the PW criterion is more suited to the engineering industry, the FW criterion to the financial industry.

---

[9]Being the same as Example 6.4 except the criterion, this example offers an opportunity to compare its FW result with the corresponding PW result in Example 6.4. For instance, FW_B in Example 7.4 can be compared with the FW_B of Example 6.4 as obtained from PW_B,

$$
\begin{aligned}
FW_B &= PW_B(F/P, 12\%, 7) \\
&= 6{,}155 \times 2.211 \\
&= \$13{,}608
\end{aligned}
$$

The $3 difference in this value ($13,608) of FW_B and that ($13,605) in Example 7.4 is due to the rounding-off error in the functional factor's tabular value.

[10]As an alternative approach, the FWs in Example 7.4 could have been determined directly from the corresponding PWs in Example 6.4.

SUMMARY

The application of the future worth (FW) criterion consists of two steps: evaluation of FW, and decision making to select the best alternative. A project's FW is defined as the surplus of its benefits' future worth over its costs' future worth. It is evaluated by analyzing the given cashflows through first principles or using Equation (7.1). Alternatively, it can be evaluated by transferring the project's PW, if known, to the specific future time. In multialternative projects, the criterion of decision making is the largest positive FW. If all the FWs are negative, the best choice is the smallest negative FW. The FW criterion is most suited for projects where savings accumulate for disbursement in the future. Compared to the PW criterion, the FW criterion finds fewer applications in engineering investments. It is more appealing to financial investors.

EXERCISES

Discussion Questions

7.1 Give an example from your own experience where the future worth criterion would be most appropriate for decision making.

7.2 How does the future worth criterion account for the time value of money?

7.3 Explain the significance of *future* in the future worth method.

7.4 Can the FW criterion lead to a decision different from that based on the PW criterion? If yes, under what circumstances? If no, why not?

7.5 Of the two steps involved in FW analysis, which one is more difficult and why?

Multiple-Choice Questions (Circle the *best* answer.)

7.6 If the future worth of costs exceeds the future worth of benefits, the project FW will always be
a. positive.
b. negative.
c. a fraction.
d. an integer.

7.7 John buys a video game for \$300. After using it for 2 years, he expects to sell it for \$20. If $i = 10\%$ per year, the future worth in dollars is
a. $300(F/P, 10\%, 2) - 20$
b. $280 + 10\%$ of 280
c. $20 - 300(F/P, 10\%, 2)$
d. $300 - 20(P/F, 10\%, 2)$

7.8 For most projects, FW is likely to be
a. smaller than PW.
b. equal to PW.
c. greater than PW.
d. any of the above depending on the cashflow pattern

7.9 Bob is worried about the harm he might be doing his lungs by smoking. Besides, he has begun to see the benefit of quitting since he enrolled in engineering economics. He smokes \$60 worth of cigarettes a month. If he quits and saves this money each month in a 20-year CD that pays 12% annual interest, compounded monthly, the future worth of the economic benefit from quitting is

a. $60(F/A, 12\%, 240)$
b. $60(F/A, 12\%, 20)$
c. $60(F/A, 1\%, 240)$
d. $60(F/P, 1\%, 240)$

7.10 With help from her financial advisor, Gita is planning to retire in ten years. The criterion most suitable in this planning is

a. payback period.
b. present worth.
c. future worth.
d. benefit–cost ratio.

Numerical Problems

7.11 The initial cost of a new photocopier is \$6,000. It is expected to generate a net annual income of \$2,000 in the first year, decreasing thereafter by \$300 each year. At the end of its 7-year useful life the photocopier is expected to be sold for \$750. Assuming $i = 10\%$ with annual compounding, determine the photocopier's future worth.

7.12 Determine the FW of the project whose cashflow diagram is depicted in Fig. 7.3. Assume $i = 10\%$.

7.13 There are two models of metrology equipment in the market. The basic model with a life of 4 years costs \$6,000 and is expected to bring in \$2,000 net annually. For the deluxe model, these data are 8 years, \$10,500, and \$2,500 respectively. Assuming no salvage value and an annually compounded interest rate of 8% per year, which model is a better choice on the basis of the FW criterion?

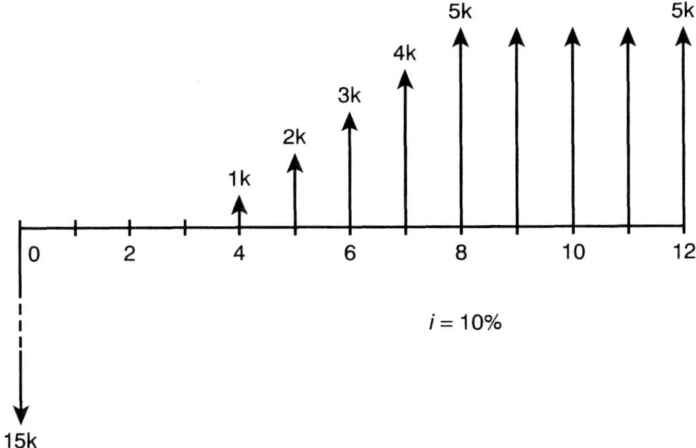

Figure 7.3 Diagram for Problem 7.12

7.14 Saira Banu turns 15 today. Her elder sister, who is enrolled in engineering economics, has helped her appreciate the power of compounding in building up equity. Saira decides to set aside the cash gift from her grandparents' trust at each birthday, beginning the 16th and ending with the 55th. The trust provides a dollar gift ten times her age. If her savings can earn an annually compounded interest rate of 10% per year, how much will have accumulated on her 60th birthday? Though she does not save beyond the 55th birthday, the savings continue to earn compounded interest.

7.15 Two candidate machines are under consideration. Machine A's initial cost is \$100,000, operating cost is \$4,000 per year, and useful life is 6 years. The respective data for machine B are \$150,000, \$5,000, and 9 years. Neither machine has any salvage value. On the basis of future worth, which machine would you select if $i = 7\%$?

7.16 Madhuri begins her career as an electronics engineer at an annual salary of \$40,000. She expects a yearly increase of \$2,000 in her salary during a 30-year professional career. If she invests 5% of her salary each year in an IRA (individual retirement account) at 10% annually compounded interest, how much will she get at the end of her career?

7.17 Isaac asks his sister-in-law to lend him \$16,000 so that he can buy a car. The car will help him sell more merchandise. Isaac promises to pay back \$3,000 at the end of first year, \$6,000 at the end of second year, and \$9,000 at the end of third year. If his sister-in-law's \$16,000 can earn 10% annual interest from the local bank, is she doing him a favor by lending? Use the FW criterion for your decision.

7.18 ZeroDefect Manufacturing Company needs to build additional space for inventory storage. There are three choices:

1. Spend \$12 million to build space that will meet the needs for the next 10 years
2. Spend \$8 million now to meet the needs of the first 6 years and another \$5 million during the sixth year to meet the needs of the next 4 years
3. Rent a nearby space for 10 years at an yearly cost of \$2 million

Which is the best choice on the basis of future worth criterion? Assume zero salvage value and $i = 8\%$.

7.19 For equipment being considered for acquisition the following data have been compiled.

Year	Cost	Income
0	\$70,000	None
1	4,000	\$20,000
2	5,000	24,000
3	6,000	29,000
4	7,000	32,000
5	8,000	34,000

Assuming that money is worth 7%, should the equipment be acquired if decision criterion is future worth?

7.20 For the following cashflows, determine the FW for annually compounded interest rate of 9% per year.

Year	Cashflow
0	−$6,000
1	3,000
2	5,000
3	−1,500
4	500
5	2,500

What will the percentage change be in FW if compounding is monthly?

CHAPTER EIGHT

Annual Worth

IN THIS CHAPTER YOU WILL LEARN ABOUT

- The meaning of *annual worth* (AW)
- How to determine a project's AW
- Application of the AW criterion
- How to account for salvage value
- Application of the input–output concept in the AW method
- Irrelevancy of the analysis period

In Chapter 6, we learned how to apply the present worth (PW) criterion to engineering projects. In Chapter 7, we learned about the future worth (FW) method. In this chapter, we discuss another important criterion, called *annual worth* (AW), and its application to engineering projects. Like the PW or FW criteria, the AW criterion also accounts for the time value of money. Annual worth's greatest attraction is its compatability with the commonly practiced yearly accounting for profit, taxes, and other similar purposes.

AW-based decisions rely on spreading the costs and benefits uniformly over the resource's useful life. The given cashflow data are transformed into their yearly equivalent, whose value dictates the investment worthiness of the project. In the AW method we look through the "year[1] as the window."

8.1 MEANING

Annual worth is the equivalent net cash inflow each year over the useful life. Consider a project whose cashflow diagram is Fig. 8.1(a). There are two cash outflows (costs), M

[1]The time period in the AW method is usually a year. There is no reason why another time period can't be used in the analysis. If month is used, the decision will be based on *monthly worth*. All the discussions of this chapter remain pertinent to monthly worth analysis, or for that matter to analyses based on any time period.

and N, and four inflows (benefits), Q, R, S, and T. Let us comprehend the meaning of AW with reference to this diagram.

In the AW method, we determine the annual equivalents of project costs and benefits, as in Fig. 8.1(a), in terms of uniform-series costs and benefits, as in Fig. 8.1(b). The two costs M and N are shown in Fig. 8.1(b) by their annual equivalent D, while the four benefits Q, R, S, and T are shown by their annual equivalent E. These annual equivalents, D and E, are respectively called *equivalent uniform annual cost* (EUAC) and *equivalent uniform annual benefit* (EUAB). In this text, we prefer to use AW_{costs} for EUAC and $AW_{benefits}$ for EUAB. Since we have used PW_{costs} and $PW_{benefits}$ in Chapter 6, and FW_{costs} and $FW_{benefits}$ in Chapter 7, it will be reader friendly to use AW_{costs} and $AW_{benefits}$. The excess of $AW_{benefits}$ over AW_{costs}, representing the net annual cash inflow, is the annual worth (AW) of a project. In other words,

$$AW = AW_{benefits} - AW_{costs} \quad (8.1)$$

From this equation, AW will be positive when $AW_{benefits}$ is greater than AW_{costs}. So a positive AW means that the project is investment worthy. Also, greater the value of positive AW, more attractive the project.

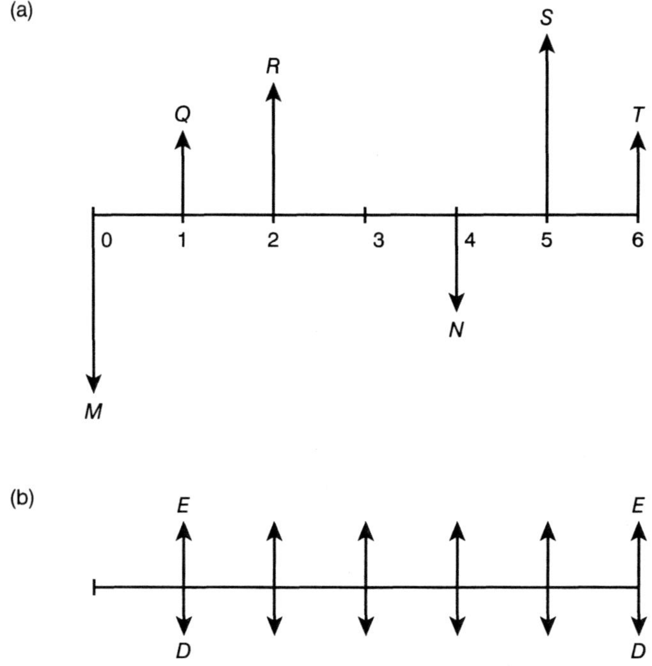

Figure 8.1 Concept of Annual Worth

It is obvious from this discussion, as well as from Equation (8.1), that AW analysis involves determining[2] $AW_{benefits}$ and AW_{costs}. This means converting the given cashflows into equivalent annual cashflows. How can this conversion be done efficiently depends primarily on the cashflows' distribution. We have already learned in Chapters 3 and 4 how to carry out conversions.

In the absence of any pattern in the cashflows, we follow the straightforward approach, treating each cashflow separately. In case some pattern exists, we utilize any of the equations derived in Chapter 4 for series cashflows. In many cases the given cashflows may have to be converted to their P value at period 0 before being converted to the equivalent annual cashflow. Examples 8.1 through 8.3 illustrate the procedure.

EXAMPLE 8.1

Judy has just won \$400,000 in a lottery. Compare this windfall with her husband's annual salary of \$40,000, if he expects to continue in his job for the next 25 years. Assume $i = 6\%$ per year.

Solution

Judy's windfall of \$400,000 is a benefit which she receives now. Her husband, on the other hand, expects to earn \$40,000 annually for the next 25 years. Their cash inflows can be compared by looking at the data through the same time "window." There are two ways to compare. We can determine the present worth (PW) of her husband's future salaries and compare it with her \$400,000 windfall. Alternatively, we can determine her windfall's equivalent uniform-series annual worth AW and compare it with her husband's \$40,000 annual salary. Following the latter, AW of Judy's windfall for $n = 25$ is

$$\begin{aligned} AW &= 400{,}000(A/P, 6\%, 25) \\ &= 400{,}000 \times 0.0782 \\ &= \$31{,}280 \end{aligned}$$

Thus, Judy's windfall is equivalent to an annual salary of \$31,280 over the next 25 years. Her husband will thus earn \$40,000 − \$31,280 = \$8,720 more per year during the period.

EXAMPLE 8.2

What is the annual worth of the cashflows in Fig. 8.1(a) if $M = \$1{,}200, N = \$400, Q = \$300, R = \$600, S = M$, and $T = Q$? Assume $i = 6\%$ per year.

---

[2] $AW_{benefits}$ and AW_{costs} are different notations for uniform series vectors A discussed in Chapter 4. $AW_{benefits}$ is the uniform cash inflow vector, while AW_{costs} is the uniform cash outflow vector.

Solution

The annual worth AW is the difference between $AW_{benefits}$ and AW_{costs}. To determine it, evaluate $AW_{benefits}$ and AW_{costs}, which, referring to Fig. 8.1(b), are E and D.

Since there exists no "strong" pattern in the given cashflows, we will follow the straightforward approach. But how do we transfer the given cashflows into E and D? There is no direct way to do this. Instead, the given irregular benefits and costs are converted first to their P values at period 0. These P values are then transformed into their A values (E and D).

Note that $S = M = \$1{,}200$, and $T = Q = \$300$. For the P value of the cash inflows (benefits) we have

$$\begin{aligned} PW_{benefits} &= Q(P/F, 6\%, 1) + R(P/F, 6\%, 2) + S(P/F, 6\%, 5) + T(P/F, 6\%, 6) \\ &= 300 \times 0.9434 + 600 \times 0.8900 + 1{,}200 \times 0.7473 + 300 \times 0.7050 \\ &= 283.0 + 534.0 + 896.8 + 211.5 \\ &= \$1{,}925.3 \end{aligned}$$

For the P value of the cash outflows (costs) we have

$$\begin{aligned} PW_{costs} &= M + N(P/F, 6\%, 4) \\ &= 1{,}200 + 400 \times 0.7921 \\ &= 1{,}200 + 316.8 \\ &= \$1{,}516.8 \end{aligned}$$

We next convert these P values to their annual equivalents yielding us $AW_{benefits}$ and AW_{costs}. Thus,

$$\begin{aligned} AW_{benefits} &= PW_{benefits}\,(A/P, 6\%, 6) \\ &= 1{,}925.3 \times 0.2034 \\ &= \$391.6 \end{aligned}$$

$$\begin{aligned} AW_{costs} &= PW_{costs}\,(A/P, 6\%, 6) \\ &= 1{,}516.8 \times 0.2034 \\ &= \$308.5 \end{aligned}$$

As a difference between $AW_{benefits}$ and AW_{costs} (the same as using Equation 8.1), the annual worth is

$$\begin{aligned} AW &= AW_{benefits} - AW_{costs} \\ &= 391.6 - 308.5 \\ &= \$83.1 \end{aligned}$$

EXAMPLE 8.3

For the project whose cashflow diagram is Fig. 8.2, what is the value of $AW_{benefits}$? Assume $i = 8\%$ per year.

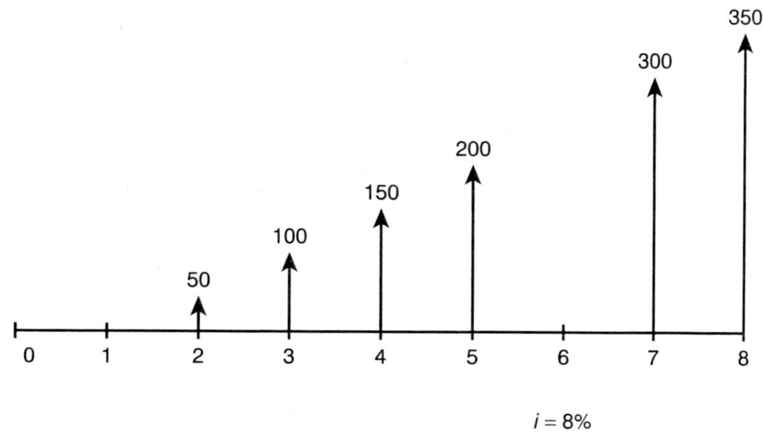

Figure 8.2 Application of Tailoring

Solution

The value of $AW_{benefits}$ depends on the six cash inflows which follow an arithmetic-series pattern, except the absence of cashflow at period 6. We can tailor the cash inflows by inserting a $250 upward vector at period 6. This will yield an arithmetic series with $G = 50$. To compensate for the inserted $250 upward vector a downward vector of the same magnitude is set at period 6.

Now, how do we convert the arithmetic series data to their equivalent uniform annual cash inflow? Have we discussed any relationship earlier that can do it? Yes indeed, we have, and it is Equation 4.6(b) in Chapter 4. Thus, the equivalent uniform annual cash inflow R_1 for the tailored arithmetic series is

$$\begin{aligned} R_1 &= G(A/G, i, n) \\ &= 50(A/G, 8\%, 8) \\ &= 50 \times 3.099 \\ &= \$154.95 \end{aligned}$$

We must include the effect of the compensating downward vector at period 6 on $AW_{benefits}$. This requires that we first transfer the $250 vector from period 6 to its P value at period 0, and then convert this P value to its equivalent A value, say R_2, as follows.

$$\begin{aligned} R_2 &= 250(P/F, 8\%, 6)(A/P, 8\%, 8) \\ &= 250 \times 0.6320 \times 0.1740 \\ &= \$27.49 \end{aligned}$$

Since R_1 is upward and R_2 is downward, their proper difference yields the value of $AW_{benefits}$. Thus,

$$\begin{aligned} AW_{benefits} &= R_1 - R_2 \\ &= \$154.95 - \$27.49 \\ &= \$127.46 \end{aligned}$$

8.2 METHOD

The application of the AW method involves two tasks:

1. Evaluation of the annual worth(s)
2. Decision making under the AW criterion (selection of the most favorable project)

We discussed the first task in Section 8.1. In this section we discuss the AW criterion and how to apply it.

Engineering economics projects can be categorized into one of two groups: no-alternative projects, or multialternative projects. In no-alternative projects, there is only one candidate, with no choice. The data for the sole project is collected and AW determined, as discussed in Section 8.1. If the AW is positive, the project is funded. No-alternative projects therefore do not involve much of the decision making, beyond the evaluation of AW, as illustrated earlier in Example 8.2 and next in Example 8.4.

In multialternative projects, there are several choices. Decision making involves selecting the alternative whose AW is the most favorable, that is, the largest of all the positive AWs. For example, if the alternatives' AWs are \$300, \$650, \$268, and \$987, then select the one whose AW is \$987. If the AWs of all the alternative[3] are negative, then none should be selected in a normal situation, since negative AW means that the annual cost is greater than the annual benefit. However, if there is a business compulsion that demands that one of the alternatives must be selected (or there are significant intangible benefits), then select the one whose negative AW is the smallest. For example, if the AWs of four project alternatives are −\$450, −\$350, −\$896, and −\$936, then select the one whose AW is −\$350. Being the smallest of the negative AWs, this minimizes the loss, and therefore is the most favorable. In case some of the alternatives yield positive AWs while others yield negative AWs, then discard the alternatives with negative AWs and select from the remaining the one with the largest positive AW. In short, the decision must be in favor of financially the most *attractive* alternative. Example 8.5 illustrates the application of the AW method to a multialternative project.

8.3 SALVAGE VALUE

The salvage value is recovered at the end of the useful life. It should be converted, by treating it as a future sum, to its A value. This cash inflow can be thought of as a benefit and is generally added to $AW_{benefits}$, as in Example 8.4. It can, on the other hand, be thought of as partial recovery of the first (initial) cost and therefore subtracted from the project's AW_{costs}. In AW analysis, the final decision in a multialternative project is not affected by how we treat the salvage value—as part of the benefit or as a recovery of the first cost.

[3]The terms *project* and *alternative* are used synonymously.

EXAMPLE 8.4

Determine the annual worth of a 10-year project whose first cost is \$12,000, annual benefit is \$2,500, and salvage value is \$2,000. Assume an annually compounded interest rate of 8% per year.

Solution

If you can visualize the problem without its cashflow diagram, then proceed directly to the calculations. Otherwise, sketch a diagram to comprehend the problem. Note that the salvage value is recovered at the end of useful life as a benefit (inflow to the project).

Using Equation (8.1) and the associated functional notations, AW is given[4] by

$$\begin{aligned} AW &= AW_{\text{benefits}} - AW_{\text{costs}} \\ &= 2{,}500 + 2{,}000(A/F, 8\%, 10) - 12{,}000(A/P, 8\%, 10) \\ &= 2{,}500 + 2{,}000 \times 0.0690 - 12{,}000 \times 0.1490 \\ &= 2{,}500 + 138 - 1{,}788 \\ &= \$850 \end{aligned}$$

EXAMPLE 8.5

Best Manufacturers is considering the following five new projects as next year's R&D initiatives. On the basis of annual worth which one should be selected for funding, assuming that the investment must earn an annual return of 12%? All five projects will take two years to develop and will be profitable during the subsequent five years.

Project	Annual Cost (during development)	Annual Income (during 5-year life)
A	\$2,000	\$3,500
B	1,800	3,200
C	2,200	2,300
D	1,200	2,000
E	3,500	4,800

Solution

The R&D costs are incurred during the first two years of the development. The benefits are realized during the five years following development. The decision making involves twin steps: determination of each project's AW, and selection based on the AW criterion.

Let us go through the first step and evaluate the AW of each project. Since the cashflow patterns of the projects are similar, their cashflow diagrams will be alike. Consider project A first, whose cashflow diagram is shown in Fig. 6.3. To determine its annual

[4]If you have difficulty understanding the functional notations used here, reread Chapter 4.

worth, some tailoring of the cash inflows is required to exploit the uniform-series pattern. The tailoring involves adding two \$3,500 vectors *upward*, one at period 1 and the other at period 2, so that the cash inflows of \$3,500 can become $AW_{benefits}$. The added vectors are compensated for by having two \$3,500 vectors *downward* at the same time periods, that is, at periods 1 and 2. This compensation increases the two cash outflows at these periods to \$5,500 from the original \$2,000.

The tailored diagram is next analyzed to determine AW_{costs}. The two downward \$5,500 vectors are transferred to period 0 as the P value, which is then converted to the A value spread over 7 years, that is, to AW_{costs}. This yields

$$\begin{aligned} AW_A &= AW_{benefits} - AW_{costs} \\ &= 3{,}500 - 5{,}500(P/A, 12\%, 2)(A/P, 12\%, 7) \\ &= 3{,}500 - 5{,}500 \times 1.690 \times 0.2191 \\ &= 3{,}500 - 5{,}500 \times 0.3703 \\ &= 3{,}500 - 2{,}037 \\ &= \$1{,}463 \end{aligned}$$

In a similar way,

$$\begin{aligned} AW_B &= 3{,}200 - 5{,}000 \times 0.3703 \\ &= 3{,}200 - 1{,}852 \\ &= \$1{,}348 \end{aligned}$$

$$\begin{aligned} AW_C &= 2{,}300 - 4{,}500 \times 0.3703 \\ &= 2{,}300 - 1{,}666 \\ &= \$634 \end{aligned}$$

$$\begin{aligned} AW_D &= 2{,}000 - 3{,}200 \times 0.3703 \\ &= 2{,}000 - 1{,}185 \\ &= \$815 \end{aligned}$$

$$\begin{aligned} AW_E &= 4{,}800 - 8{,}300 \times 0.3703 \\ &= 4{,}800 - 3{,}073 \\ &= \$1{,}727 \end{aligned}$$

The second step involves deciding which project to select. Since the AWs of all the five projects are positive, the most favorable project is the one whose AW is the largest. So project E is selected.

8.4 INPUT–OUTPUT CONCEPT

The input–output concept discussed in Chapter 6 is equally useful in annual worth analysis. On the basis of input–output, engineering economics projects can be grouped into three categories:

1. *Fixed Output.* In many projects, the output is fixed, that is, the benefit from each alternative is the same. Such multialternative projects are solved by comparing the alternatives' annual worths of costs, since the annual worth of benefit(s) is the same (fixed) for each alternative. Excluding benefits from the analysis keeps the calculations simpler. Obviously, the criterion of decision making in such problems[5] is to *minimize the annual worth of the costs*, as illustrated in Example 8.6.

EXAMPLE 8.6

Hari has been using an old car since his first year at college. During the last three years as an industrial technologist he has saved enough money to buy a new car. He is considering a Honda Accord, a Chevy Prism, and a Plymouth Acclaim. Given the following data, which one should he buy if the decision criterion is annual worth? Assume each car's useful life to be seven years and an annually compounded interest rate of 9% per year.

Car	First Cost	Annual Cost	Salvage Value
Honda Accord	$20,000	$300	$3,000
Chevy Prism	18,000	500	2,000
Plymouth Acclaim	17,500	750	1,500

Solution

In this problem, the output, which is the convenience derived from owning a car, is fixed. The convenience translates into indirect benefits that are difficult to quantify in monetary terms. Of the given data, the salvage value represents a cash inflow and is treated as a benefit. Since it is a fixed-output problem, the selection could be based on the annual worth of the costs. If you can visualize the problem without its cashflow diagram, you should be able to digest the following analysis; otherwise, sketch a diagram to help you. The annual worths[6] for the three choices, from Equation (8.1), are

$$\begin{aligned} AW_{\text{Accord}} &= 3{,}000(A/F, 9\%, 7) - [20{,}000(A/P, 9\%, 7) + 300] \\ &= 3{,}000 \times 0.1087 - (20{,}000 \times 0.1987 + 300) \\ &= 326 - (3{,}974 + 300) \\ &= -\$3{,}948 \end{aligned}$$

[5]The terms *problem* and *project* have been used synonymously.

[6]This problem can be treated differently by subtracting the salvage value's equivalent annual worth from that of the initial cost's. This, for example, will yield for Honda Accord

$$AW_{\text{costs (Accord)}} = 300 + [20{,}000(A/P, 9\%, 7) - 3{,}000(A/F, 9\%, 7)]$$

This approach too will lead to the same decision. Try doing it.

$$\begin{aligned}
AW_{\text{Prism}} &= 2{,}000(A/F, 9\%, 7) - [18{,}000(A/P, 9\%, 7) + 500] \\
&= 2{,}000 \times 0.1087 - (18{,}000 \times 0.1987 + 500) \\
&= 217 - (3{,}577 + 500) \\
&= -\$3{,}860
\end{aligned}$$

$$\begin{aligned}
AW_{\text{Acclaim}} &= 1{,}500(A/F, 9\%, 7) - [17{,}500(A/P, 9\%, 7) + 750] \\
&= 1{,}500 \times 0.1087 - (17{,}500 \times 0.1987 + 750) \\
&= 163 - (3{,}477 + 750) \\
&= -\$4{,}064
\end{aligned}$$

All three AWs[7] are negative. Comparing them, Hari should buy a Chevy Prism, since, with its lowest negative annual worth, it is the most favorable. In other words, Prism is selected because the equivalent annual cost of purchasing and operating it is the least of all.

---

2. *Fixed Input.* In the second group of projects, the input is fixed, that is, the alternatives cost the same. Such projects are of the *fixed-input, variable-output* type. By variable-output we mean that the benefits (output) from the alternatives differ from each other. The decision maker's task is to select that alternative which maximizes the output (benefit). The decision criterion in such projects is the largest positive annual worth of benefits. In other words, maximize AW_{benefits}. Since the input is fixed, its inclusion in calculations for alternatives' annual worths would not affect the decision but would unnecessarily complicate the analysis. An example of a fixed-input project is hiring an employee at a fixed budgeted salary, where the decision maker's task is to select the best from among the applicants. Another example is lump-sum investment in a CD after "shopping" around for the highest effective interest rate to maximize the return.
3. *Input and Output Vary.* Several engineering projects involve alternatives in which neither input nor output is fixed. The economic analyses of such variable-input, variable-output projects are relatively lengthier. The annual worth of each alternative is evaluated using Equation (8.1), as illustrated in Example 8.5 and elsewhere. Once AWs are known, selection is made in favor of the most attractive project, usually the one with the largest positive AW (smallest negative AW if alternatives' AWs are all negative).

8.5 ANALYSIS PERIOD

In Chapter 5 we discussed the situation where alternatives' useful lives are different. There we answered the question, What analysis period should be used? This question does not arise at all in AW analysis, because AW translates the given cashflows in an

---

[7]Note that each car's annual worth is negative. Then, why invest in a car? The explanation is that we did not, perhaps could not, account for the benefits of owning a car; only the salvage value contributed toward the benefit. Such a situation arises in many projects where the benefits are difficult to quantify. If we account for all the benefits of owning a car, for example, savings on bus or taxi fares, cost of time saved from driving rather than walking to work, and so forth, we are likely to find that a car's AW is actually positive. Moreover, there is the intangible beneficial feeling of prestige in owning a car, which can at times be egoistic, especially with expensive cars.

equivalent uniform net annual cashflow. As long as the alternatives of different useful lives can be replaced for the same cost and yield on replacement the same benefits—usually a sensible assumption—AW is unaffected by the analysis period. Thus, in AW analysis of alternatives with unequal lives, each alternative's analysis period is its own useful life[8]; there is no need for a LCM-based analysis period as in PW analysis. This renders the AW analysis simpler, as illustrated in Example 8.7.

EXAMPLE 8.7

Abraham's personal computer (PC) is slow and lacks features essential for running current application programs. There are three choices: Upgrade the existing system, buy a new PC, or buy a one-year-old PC. For the following data, which one should he select if the yearly interest rate is 9%, compounded annually? Use the AW criterion for decision making.

Choice	First Cost	Annual Operating Cost	Useful Life, Years	Salvage Value
Upgrade	$3,000	$200	2	$250
New PC	8,000	100	4	500
Old PC	5,000	150	3	300

Solution

Since the AW criterion is the basis of decision making, it does not matter that the useful lives of the three choices are different, especially if we assume that the replacement cost and the resulting benefit data will remain unchanged.

We need to evaluate the annual worths of the three choices and select the most favorable one. Considering the salvage value as a benefit, their annual worths[9] are

$$\begin{aligned} \text{AW}_{\text{upgrade}}{}^{10} &= \text{AW}_{\text{benefits}} - \text{AW}_{\text{costs}} \\ &= 250\,(A/F, 9\%, 2) - [3{,}000(A/P, 9\%, 2) + 200] \\ &= 250 \times 0.4785 - (3{,}000 \times 0.5685 + 200) \\ &= 120 - (1{,}706 + 200) \\ &= -\$1{,}786 \end{aligned}$$

---

[8]Thus, the alternatives are analyzed over different analysis periods.

[9]Follow the standard convention that costs (vectors down) are negative, while benefits (vectors up) are positive.

[10]The convention for subscripts followed in this text is the same everywhere. If there is no subscript the notation AW (or PW) relates to the project as a whole, used normally for no-alternative projects. With *benefits* or *costs* as subscript, the AW (or PW) is of the benefits or costs of the project. In multialternative projects, a key term is used as a subscript, as in $\text{AW}_{\text{upgrade}}$, to represent the AW (or PW) of the alternative. An additional subscript is used to denote benefits or costs, for example $\text{AW}_{\text{costs (Accord)}}$ in the footnote under Example 8.6.

$$\begin{aligned} AW_{new} &= AW_{benefits} - AW_{costs} \\ &= 500(A/F, 9\%, 4) - [8{,}000(A/P, 9\%, 4) + 100] \\ &= 500 \times 0.2187 - (8{,}000 \times 0.3087 + 100) \\ &= 109 - (2{,}470 + 100) \\ &= -\$2{,}461 \end{aligned}$$

$$\begin{aligned} AW_{old} &= AW_{benefits} - AW_{costs} \\ &= 300(A/F, 9\%, 3) - [5{,}000(A/P, 9\%, 3) + 150] \\ &= 300 \times 0.3051 - (5{,}000 \times 0.3951 + 150) \\ &= 92 - (1{,}976 + 150) \\ &= -\$2{,}034 \end{aligned}$$

Comparing[11] the three annual worths, which are all negative, Abraham decides to upgrade the existing system, since, with its smallest negative AW, it is the most favorable.

SUMMARY

A project's annual worth (AW) is the surplus of its equivalent uniform annual benefit, $AW_{benefits}$, over its equivalent uniform annual cost, AW_{costs}. Its determination therefore involves evaluation of $AW_{benefits}$ and AW_{costs} for the project. The application of the AW criterion to engineering projects is a two-step process: determination of AW(s), and selection of the financially most attractive alternative. The input–output concept categorizes problems into fixed output, fixed input, and variable-input variable-output types. By input we mean cost(s) and by output we mean benefit(s). In fixed-output problems, the alternatives yield the same benefit; therefore, the decision is based on minimization of AW_{costs}. In fixed-input problems, on the other hand, maximum $AW_{benefits}$ is the decision criterion. In variable-input, variable-output problems, the alternative with the largest positive AW is usually selected. If the alternatives' AWs are all negative, then select the alternative whose negative AW is the smallest. The AW method is better suited for analyzing projects with alternatives of different useful lives than the PW or FW methods. It takes care of unequal lives by itself, since AW normalizes the cashflows on yearly basis.

EXERCISES

Discussion Questions

8.1 Explain the significance of *annual* in the annual worth method.

8.2 Give an example from personal financing where the annual worth criterion would be most appropriate for making a decision.

8.3 How does the annual worth criterion account for time value of money?

---

[11]Note the brevity of analysis in Example 8.7 based on the AW method in comparison to that in Example 6.9 based on the PW method. This is because PW analysis is conducted over a long analysis period based on LCM. The AW method does not require a common analysis period; instead the alternatives are analyzed over their own useful lives.

8.4 How is the input–output concept helpful in AW analysis?

8.5 Why do engineering economists not worry about alternatives' unequal useful lives while analyzing problems by the AW method.

Multiple-Choice Questions (Circle the *best* answer.)

8.6 AW analysis of engineering economics problems may be preferred over other methods because
a. it is the simplest.
b. the LCM-based analysis period becomes irrelevant.
c. financial institutions require it.
d. it is easier to budget for.

8.7 In fixed-output problems the objective should be to
a. maximize AW.
b. maximize $AW_{benefits}$.
c. minimize AW_{costs}.
d. minimize AW.

8.8 In fixed-input problems the objective should be to
a. maximize AW.
b. maximize $AW_{benefits}$.
c. minimize AW_{costs}.
d. minimize AW.

8.9 John buys a video game for \$300. After using it for 2 years, he expects to sell it for \$20. If $i = 10\%$ per year, the annual worth in dollars is
a. $20(A/F, 10\%, 2) - 300(A/P, 10\%, 2)$
b. $280 + 10\%$ of 300
c. $20(F/A, 10\%, 2) - 300(P/A, 10\%, 2)$
d. $300(P/A, 10\%, 2) - 20(F/A, 10\%, 2)$

8.10 There are two choices for replacing a punch press. The basic model has a useful life of 8 years, the deluxe model of 12 years. The analysis period for this problem under the AW criterion is
a. 24 years for both.
b. 4 years for basic and 6 years for deluxe.
c. 8 years for basic and 12 years for deluxe.
d. 20 years.

8.11 In analyzing a variable-input, variable-output problem we need to determine
a. $AW_{benefits}$ only.
b. AW_{costs} only.
c. a or b depending on which one is positive.
d. both a and b

8.12 In analyzing a project whose alternatives have unequal lives, the best method is
a. present worth.
b. future worth.
c. annual worth.
d. a, b, or c, depending on the number of alternatives

Numerical Problems

8.13 Amanda has bought a second-hand car for $4,000. She plans to keep it for four years, at the end of which it is likely to fetch $500. If the maintenance cost is $350 per year, and $i = 8\%$ per year, what is the car's net annual cost?

8.14 Urmila is interested in a mink coat she is likely to wear thrice a year while partying. The choices are to buy for $5,000, or rent each time when needed for $120. If she owns the coat, she insures it for an annual premium of $50. If she rents it she spends $20 on dry cleaning each time she uses it. Based on the AW criterion should she buy or rent? Assume $i = 5\%$ per year, $n = 10$ years, and zero salvage value if the coat is bought.

8.15 A fax machine needs replacing. There are two models on the market. The basic model, with a life of 4 years, costs $6,000 and is expected to generate annually a net income of $2,000. For the deluxe model, these data are 8 years, $10,500, and $2,500 respectively. Assuming zero salvage value and an annually compounded interest rate of 8% per year, which model is a better choice on the basis of the AW criterion?

8.16 Two candidate machines are under consideration. Machine A has an initial cost of $100,000, operating costs of $4,000 per year, and a useful life of 6 years. The respective data for machine B are $150,000, $5,000, and 9 years. Neither machine has any salvage value. On the basis of annual worth which machine would you select? Assume $i = 8\%$ per year.

8.17 A CNC lathe costs $50,000 to purchase. Its expected useful life is 10 years, and its salvage value is likely to be 10% of the initial cost. The annual operating cost will be $800. The first-year maintenance cost is estimated to be $700, increasing each year by $300. Determine AW_{costs} for an annual interest rate of 12% compounded twice a year.

8.18 Three *mutually exclusive*[12] alternatives X, Y, and Z are being considered as a permanent solution to an assembly-line bottleneck. The data are

	X	Y	Z
Initial cost[13]	$10k	$15k	$20k
Annual benefit	1.6k	1.5k	1.9k
Useful life, years	5	10	15

Assuming zero salvage value and an annually compounded interest rate of 8% per year, which alternative should be selected based on the AW criterion?

[12]The term *mutually exclusive* means that the selection of an alternative precludes the others from being selected. In other words, the alternatives (choices) are independent of each other. This is a basic assumption in multialternative projects and may be explicitly expressed, as in this problem. In most problems, however, such an assumption is made implicitly.

[13]As a prefix in SI units, k is an acronym for kilo, meaning 1,000.

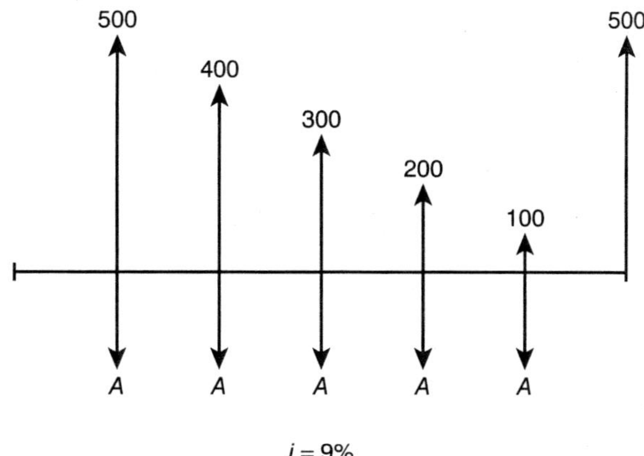

Figure 8.3 Diagram for Problem 8.21

8.19 For equipment being considered for acquisition the following data have been compiled. Assuming a yearly interest rate of 7%, should the equipment be procured if the decision criterion is annual worth?

Year	Cost	Income
0	\$70,000	None
1	4,000	\$20,000
2	5,000	24,000
3	6,000	29,000
4	7,000	32,000
5	8,000	34,000

8.20 An automation system comprising two robots can do the job of an operator who annually costs \$30,000. The robots cost \$40,000 each, are expected to remain useful for 10 years, and will have no salvage value. The annual maintenance and operating cost for the system is estimated to be \$3,000. If $i = 10\%$ per year, what is the AW?

8.21 Evaluate A in Fig. 8.3.

8.22 Tony is considering setting up a faxing service business. The equipment and accessories will cost \$8,000 and last for 4 years with no resale value. Rent, labor, insurance, and maintenance costs add up to \$10,000 a year. If he expects an annual benefit of \$25,000 should he set up the business on the basis of the AW criterion? Assume $i = 15\%$ per year.

Compound Interest Tables

	Single Payment		Uniform Payment Series				Arithmetic Gradient		
	Compound Amount Factor	Present Worth Factor	Sinking Fund Factor	Capital Recovery Factor	Compound Amount Factor	Present Worth Factor	Gradient Uniform Series	Gradient Present Worth	
n	Find F Given P F/P	Find P Given F P/F	Find A Given F A/F	Find A Given P A/P	Find F Given A F/A	Find P Given A P/A	Find A Given G A/G	Find P Given G P/G	n
1	1.003	.9975	1.0000	1.0025	1.000	0.998	0	0	1
2	1.005	.9950	.4994	.5019	2.003	1.993	0.504	1.005	2
3	1.008	.9925	.3325	.3350	3.008	2.985	1.005	2.999	3
4	1.010	.9901	.2491	.2516	4.015	3.975	1.501	5.966	4
5	1.013	.9876	.1990	.2015	5.025	4.963	1.998	9.916	5
6	1.015	.9851	.1656	.1681	6.038	5.948	2.498	14.861	6
7	1.018	.9827	.1418	.1443	7.053	6.931	2.995	20.755	7
8	1.020	.9802	.1239	.1264	8.070	7.911	3.490	27.611	8
9	1.023	.9778	.1100	.1125	9.091	8.889	3.987	35.440	9
10	1.025	.9753	.0989	.1014	10.113	9.864	4.483	44.216	10
11	1.028	.9729	.0898	.0923	11.139	10.837	4.978	53.950	11
12	1.030	.9705	.0822	.0847	12.167	11.807	5.474	64.634	12
13	1.033	.9681	.0758	.0783	13.197	12.775	5.968	76.244	13
14	1.036	.9656	.0703	.0728	14.230	13.741	6.464	88.826	14
15	1.038	.9632	.0655	.0680	15.266	14.704	6.957	102.301	15
16	1.041	.9608	.0613	.0638	16.304	15.665	7.451	116.716	16
17	1.043	.9584	.0577	.0602	17.344	16.624	7.944	132.063	17
18	1.046	.9561	.0544	.0569	18.388	17.580	8.437	148.319	18
19	1.049	.9537	.0515	.0540	19.434	18.533	8.929	165.492	19
20	1.051	.9513	.0488	.0513	20.482	19.485	9.421	183.559	20
21	1.054	.9489	.0464	.0489	21.534	20.434	9.912	202.531	21
22	1.056	.9465	.0443	.0468	22.587	21.380	10.404	222.435	22
23	1.059	.9442	.0423	.0448	23.644	22.324	10.894	243.212	23
24	1.062	.9418	.0405	.0430	24.703	23.266	11.384	264.854	24
25	1.064	.9395	.0388	.0413	25.765	24.206	11.874	287.407	25
26	1.067	.9371	.0373	.0398	26.829	25.143	12.363	310.848	26
27	1.070	.9348	.0358	.0383	27.896	26.078	12.852	335.150	27
28	1.072	.9325	.0345	.0370	28.966	27.010	13.341	360.343	28
29	1.075	.9301	.0333	.0358	30.038	27.940	13.828	386.366	29
30	1.078	.9278	.0321	.0346	31.114	28.868	14.317	413.302	30
36	1.094	.9140	.0266	.0291	37.621	34.387	17.234	592.632	36
40	1.105	.9049	.0238	.0263	42.014	38.020	19.171	728.882	40
48	1.127	.8871	.0196	.0221	50.932	45.179	23.025	1 040.22	48
50	1.133	.8826	.0188	.0213	53.189	46.947	23.984	1 125.96	50
52	1.139	.8782	.0180	.0205	55.458	48.705	24.941	1 214.76	52
60	1.162	.8609	.0155	.0180	64.647	55.653	28.755	1 600.31	60
70	1.191	.8396	.0131	.0156	76.395	64.144	33.485	2 147.87	70
72	1.197	.8355	.0127	.0152	78.780	65.817	34.426	2 265.81	72
80	1.221	.8189	.0113	.0138	88.440	72.427	38.173	2 764.74	80
84	1.233	.8108	.0107	.0132	93.343	75.682	40.037	3 030.06	84
90	1.252	.7987	.00992	.0124	100.789	80.504	42.820	3 447.19	90
96	1.271	.7869	.00923	.0117	108.349	85.255	45.588	3 886.62	96
100	1.284	.7790	.00881	.0113	113.451	88.383	47.425	4 191.60	100
104	1.297	.7713	.00843	.0109	118.605	91.480	49.256	4 505.93	104
120	1.349	.7411	.00716	.00966	139.743	103.563	56.512	5 852.52	120
240	1.821	.5492	.00305	.00555	328.306	180.312	107.590	19 399.75	240
360	2.457	.4070	.00172	.00422	582.745	237.191	152.894	36 264.96	360
480	3.315	.3016	.00108	.00358	926.074	279.343	192.673	53 821.93	480

	Single Payment		Uniform Payment Series				Arithmetic Gradient		
	Compound Amount Factor	Present Worth Factor	Sinking Fund Factor	Capital Recovery Factor	Compound Amount Factor	Present Worth Factor	Gradient Uniform Series	Gradient Present Worth	
	Find *F* Given *P*	Find *P* Given *F*	Find *A* Given *F*	Find *A* Given *P*	Find *F* Given *A*	Find *P* Given *A*	Find *A* Given *G*	Find *P* Given *G*	
n	*F/P*	*P/F*	*A/F*	*A/P*	*F/A*	*P/A*	*A/G*	*P/G*	*n*
1	1.005	.9950	1.0000	1.0050	1.000	0.995	0	0	1
2	1.010	.9901	.4988	.5038	2.005	1.985	0.499	0.991	2
3	1.015	.9851	.3317	.3367	3.015	2.970	0.996	2.959	3
4	1.020	.9802	.2481	.2531	4.030	3.951	1.494	5.903	4
5	1.025	.9754	.1980	.2030	5.050	4.926	1.990	9.803	5
6	1.030	.9705	.1646	.1696	6.076	5.896	2.486	14.660	6
7	1.036	.9657	.1407	.1457	7.106	6.862	2.980	20.448	7
8	1.041	.9609	.1228	.1278	8.141	7.823	3.474	27.178	8
9	1.046	.9561	.1089	.1139	9.182	8.779	3.967	34.825	9
10	1.051	.9513	.0978	.1028	10.228	9.730	4.459	43.389	10
11	1.056	.9466	.0887	.0937	11.279	10.677	4.950	52.855	11
12	1.062	.9419	.0811	.0861	12.336	11.619	5.441	63.218	12
13	1.067	.9372	.0746	.0796	13.397	12.556	5.931	74.465	13
14	1.072	.9326	.0691	.0741	14.464	13.489	6.419	86.590	14
15	1.078	.9279	.0644	.0694	15.537	14.417	6.907	99.574	15
16	1.083	.9233	.0602	.0652	16.614	15.340	7.394	113.427	16
17	1.088	.9187	.0565	.0615	17.697	16.259	7.880	128.125	17
18	1.094	.9141	.0532	.0582	18.786	17.173	8.366	143.668	18
19	1.099	.9096	.0503	.0553	19.880	18.082	8.850	160.037	19
20	1.105	.9051	.0477	.0527	20.979	18.987	9.334	177.237	20
21	1.110	.9006	.0453	.0503	22.084	19.888	9.817	195.245	21
22	1.116	.8961	.0431	.0481	23.194	20.784	10.300	214.070	22
23	1.122	.8916	.0411	.0461	24.310	21.676	10.781	233.680	23
24	1.127	.8872	.0393	.0443	25.432	22.563	11.261	254.088	24
25	1.133	.8828	.0377	.0427	26.559	23.446	11.741	275.273	25
26	1.138	.8784	.0361	.0411	27.692	24.324	12.220	297.233	26
27	1.144	.8740	.0347	.0397	28.830	25.198	12.698	319.955	27
28	1.150	.8697	.0334	.0384	29.975	26.068	13.175	343.439	28
29	1.156	.8653	.0321	.0371	31.124	26.933	13.651	367.672	29
30	1.161	.8610	.0310	.0360	32.280	27.794	14.127	392.640	30
36	1.197	.8356	.0254	.0304	39.336	32.871	16.962	557.564	36
40	1.221	.8191	.0226	.0276	44.159	36.172	18.836	681.341	40
48	1.270	.7871	.0185	.0235	54.098	42.580	22.544	959.928	48
50	1.283	.7793	.0177	.0227	56.645	44.143	23.463	1 035.70	50
52	1.296	.7716	.0169	.0219	59.218	45.690	24.378	1 113.82	52
60	1.349	.7414	.0143	.0193	69.770	51.726	28.007	1 448.65	60
70	1.418	.7053	.0120	.0170	83.566	58.939	32.468	1 913.65	70
72	1.432	.6983	.0116	.0166	86.409	60.340	33.351	2 012.35	72
80	1.490	.6710	.0102	.0152	98.068	65.802	36.848	2 424.65	80
84	1.520	.6577	.00961	.0146	104.074	68.453	38.576	2 640.67	84
90	1.567	.6383	.00883	.0138	113.311	72.331	41.145	2 976.08	90
96	1.614	.6195	.00814	.0131	122.829	76.095	43.685	3 324.19	96
100	1.647	.6073	.00773	.0127	129.334	78.543	45.361	3 562.80	100
104	1.680	.5953	.00735	.0124	135.970	80.942	47.025	3 806.29	104
120	1.819	.5496	.00610	.0111	163.880	90.074	53.551	4 823.52	120
240	3.310	.3021	.00216	.00716	462.041	139.581	96.113	13 415.56	240
360	6.023	.1660	.00100	.00600	1 004.500	166.792	128.324	21 403.32	360
480	10.957	.0913	.00050	.00550	1 991.500	181.748	151.795	27 588.37	480

	Single Payment		Uniform Payment Series				Arithmetic Gradient		
	Compound Amount Factor	Present Worth Factor	Sinking Fund Factor	Capital Recovery Factor	Compound Amount Factor	Present Worth Factor	Gradient Uniform Series	Gradient Present Worth	
n	Find *F* Given *P* *F/P*	Find *P* Given *F* *P/F*	Find *A* Given *F* *A/F*	Find *A* Given *P* *A/P*	Find *F* Given *A* *F/A*	Find *P* Given *A* *P/A*	Find *A* Given *G* *A/G*	Find *P* Given *G* *P/G*	*n*
1	1.008	.9926	1.0000	1.0075	1.000	0.993	0	0	1
2	1.015	.9852	.4981	.5056	2.008	1.978	0.499	0.987	2
3	1.023	.9778	.3308	.3383	3.023	2.956	0.996	2.943	3
4	1.030	.9706	.2472	.2547	4.045	3.926	1.492	5.857	4
5	1.038	.9633	.1970	.2045	5.076	4.889	1.986	9.712	5
6	1.046	.9562	.1636	.1711	6.114	5.846	2.479	14.494	6
7	1.054	.9490	.1397	.1472	7.160	6.795	2.971	20.187	7
8	1.062	.9420	.1218	.1293	8.213	7.737	3.462	26.785	8
9	1.070	.9350	.1078	.1153	9.275	8.672	3.951	34.265	9
10	1.078	.9280	.0967	.1042	10.344	9.600	4.440	42.619	10
11	1.086	.9211	.0876	.0951	11.422	10.521	4.927	51.831	11
12	1.094	.9142	.0800	.0875	12.508	11.435	5.412	61.889	12
13	1.102	.9074	.0735	.0810	13.602	12.342	5.897	72.779	13
14	1.110	.9007	.0680	.0755	14.704	13.243	6.380	84.491	14
15	1.119	.8940	.0632	.0707	15.814	14.137	6.862	97.005	15
16	1.127	.8873	.0591	.0666	16.932	15.024	7.343	110.318	16
17	1.135	.8807	.0554	.0629	18.059	15.905	7.822	124.410	17
18	1.144	.8742	.0521	.0596	19.195	16.779	8.300	139.273	18
19	1.153	.8676	.0492	.0567	20.339	17.647	8.777	154.891	19
20	1.161	.8612	.0465	.0540	21.491	18.508	9.253	171.254	20
21	1.170	.8548	.0441	.0516	22.653	19.363	9.727	188.352	21
22	1.179	.8484	.0420	.0495	23.823	20.211	10.201	206.170	22
23	1.188	.8421	.0400	.0475	25.001	21.053	10.673	224.695	23
24	1.196	.8358	.0382	.0457	26.189	21.889	11.143	243.924	24
25	1.205	.8296	.0365	.0440	27.385	22.719	11.613	263.834	25
26	1.214	.8234	.0350	.0425	28.591	23.542	12.081	284.421	26
27	1.224	.8173	.0336	.0411	29.805	24.360	12.548	305.672	27
28	1.233	.8112	.0322	.0397	31.029	25.171	13.014	327.576	28
29	1.242	.8052	.0310	.0385	32.261	25.976	13.479	350.122	29
30	1.251	.7992	.0298	.0373	33.503	26.775	13.942	373.302	30
36	1.309	.7641	.0243	.0318	41.153	31.447	16.696	525.038	36
40	1.348	.7416	.0215	.0290	46.447	34.447	18.507	637.519	40
48	1.431	.6986	.0174	.0249	57.521	40.185	22.070	886.899	48
50	1.453	.6882	.0166	.0241	60.395	41.567	22.949	953.911	50
52	1.475	.6780	.0158	.0233	63.312	42.928	23.822	1 022.64	52
60	1.566	.6387	.0133	.0208	75.425	48.174	27.268	1 313.59	60
70	1.687	.5927	.0109	.0184	91.621	54.305	31.465	1 708.68	70
72	1.713	.5839	.0105	.0180	95.008	55.477	32.289	1 791.33	72
80	1.818	.5500	.00917	.0167	109.074	59.995	35.540	2 132.23	80
84	1.873	.5338	.00859	.0161	116.428	62.154	37.137	2 308.22	84
90	1.959	.5104	.00782	.0153	127.881	65.275	39.496	2 578.09	90
96	2.049	.4881	.00715	.0147	139.858	68.259	41.812	2 854.04	96
100	2.111	.4737	.00675	.0143	148.147	70.175	43.332	3 040.85	100
104	2.175	.4597	.00638	.0139	156.687	72.035	44.834	3 229.60	104
120	2.451	.4079	.00517	.0127	193.517	78.942	50.653	3 998.68	120
240	6.009	.1664	.00150	.00900	667.901	111.145	85.422	9 494.26	240
360	14.731	.0679	.00055	.00805	1 830.800	124.282	107.115	13 312.50	360
480	36.111	.0277	.00021	.00771	4 681.500	129.641	119.662	15 513.16	480

	Single Payment		Uniform Payment Series				Arithmetic Gradient		
	Compound Amount Factor	Present Worth Factor	Sinking Fund Factor	Capital Recovery Factor	Compound Amount Factor	Present Worth Factor	Gradient Uniform Series	Gradient Present Worth	
n	Find *F* Given *P* *F/P*	Find *P* Given *F* *P/F*	Find *A* Given *F* *A/F*	Find *A* Given *P* *A/P*	Find *F* Given *A* *F/A*	Find *P* Given *A* *P/A*	Find *A* Given *G* *A/G*	Find *P* Given *G* *P/G*	*n*
1	1.010	.9901	1.0000	1.0100	1.000	0.990	0	0	1
2	1.020	.9803	.4975	.5075	2.010	1.970	0.498	0.980	2
3	1.030	.9706	.3300	.3400	3.030	2.941	0.993	2.921	3
4	1.041	.9610	.2463	.2563	4.060	3.902	1.488	5.804	4
5	1.051	.9515	.1960	.2060	5.101	4.853	1.980	9.610	5
6	1.062	.9420	.1625	.1725	6.152	5.795	2.471	14.320	6
7	1.072	.9327	.1386	.1486	7.214	6.728	2.960	19.917	7
8	1.083	.9235	.1207	.1307	8.286	7.652	3.448	26.381	8
9	1.094	.9143	.1067	.1167	9.369	8.566	3.934	33.695	9
10	1.105	.9053	.0956	.1056	10.462	9.471	4.418	41.843	10
11	1.116	.8963	.0865	.0965	11.567	10.368	4.900	50.806	11
12	1.127	.8874	.0788	.0888	12.682	11.255	5.381	60.568	12
13	1.138	.8787	.0724	.0824	13.809	12.134	5.861	71.112	13
14	1.149	.8700	.0669	.0769	14.947	13.004	6.338	82.422	14
15	1.161	.8613	.0621	.0721	16.097	13.865	6.814	94.481	15
16	1.173	.8528	.0579	.0679	17.258	14.718	7.289	107.273	16
17	1.184	.8444	.0543	.0643	18.430	15.562	7.761	120.783	17
18	1.196	.8360	.0510	.0610	19.615	16.398	8.232	134.995	18
19	1.208	.8277	.0481	.0581	20.811	17.226	8.702	149.895	19
20	1.220	.8195	.0454	.0554	22.019	18.046	9.169	165.465	20
21	1.232	.8114	.0430	.0530	23.239	18.857	9.635	181.694	21
22	1.245	.8034	.0409	.0509	24.472	19.660	10.100	198.565	22
23	1.257	.7954	.0389	.0489	25.716	20.456	10.563	216.065	23
24	1.270	.7876	.0371	.0471	26.973	21.243	11.024	234.179	24
25	1.282	.7798	.0354	.0454	28.243	22.023	11.483	252.892	25
26	1.295	.7720	.0339	.0439	29.526	22.795	11.941	272.195	26
27	1.308	.7644	.0324	.0424	30.821	23.560	12.397	292.069	27
28	1.321	.7568	.0311	.0411	32.129	24.316	12.852	312.504	28
29	1.335	.7493	.0299	.0399	33.450	25.066	13.304	333.486	29
30	1.348	.7419	.0287	.0387	34.785	25.808	13.756	355.001	30
36	1.431	.6989	.0232	.0332	43.077	30.107	16.428	494.620	36
40	1.489	.6717	.0205	.0305	48.886	32.835	18.178	596.854	40
48	1.612	.6203	.0163	.0263	61.223	37.974	21.598	820.144	48
50	1.645	.6080	.0155	.0255	64.463	39.196	22.436	879.417	50
52	1.678	.5961	.0148	.0248	67.769	40.394	23.269	939.916	52
60	1.817	.5504	.0122	.0222	81.670	44.955	26.533	1 192.80	60
70	2.007	.4983	.00993	.0199	100.676	50.168	30.470	1 528.64	70
72	2.047	.4885	.00955	.0196	104.710	51.150	31.239	1 597.86	72
80	2.217	.4511	.00822	.0182	121.671	54.888	34.249	1 879.87	80
84	2.307	.4335	.00765	.0177	130.672	56.648	35.717	2 023.31	84
90	2.449	.4084	.00690	.0169	144.863	59.161	37.872	2 240.56	90
96	2.599	.3847	.00625	.0163	159.927	61.528	39.973	2 459.42	96
100	2.705	.3697	.00587	.0159	170.481	63.029	41.343	2 605.77	100
104	2.815	.3553	.00551	.0155	181.464	64.471	42.688	2 752.17	104
120	3.300	.3030	.00435	.0143	230.039	69.701	47.835	3 334.11	120
240	10.893	.0918	.00101	.0110	989.254	90.819	75.739	6 878.59	240
360	35.950	.0278	.00029	.0103	3 495.000	97.218	89.699	8 720.43	360
480	118.648	.00843	.00008	.0101	11 764.800	99.157	95.920	9 511.15	480

	Single Payment		Uniform Payment Series				Arithmetic Gradient		
n	Compound Amount Factor Find F Given P F/P	Present Worth Factor Find P Given F P/F	Sinking Fund Factor Find A Given F A/F	Capital Recovery Factor Find A Given P A/P	Compound Amount Factor Find F Given A F/A	Present Worth Factor Find P Given A P/A	Gradient Uniform Series Find A Given G A/G	Gradient Present Worth Find P Given G P/G	n
1	1.013	.9877	1.0000	1.0125	1.000	0.988	0	0	1
2	1.025	.9755	.4969	.5094	2.013	1.963	0.497	0.976	2
3	1.038	.9634	.3292	.3417	3.038	2.927	0.992	2.904	3
4	1.051	.9515	.2454	.2579	4.076	3.878	1.485	5.759	4
5	1.064	.9398	.1951	.2076	5.127	4.818	1.976	9.518	5
6	1.077	.9282	.1615	.1740	6.191	5.746	2.464	14.160	6
7	1.091	.9167	.1376	.1501	7.268	6.663	2.951	19.660	7
8	1.104	.9054	.1196	.1321	8.359	7.568	3.435	25.998	8
9	1.118	.8942	.1057	.1182	9.463	8.462	3.918	33.152	9
10	1.132	.8832	.0945	.1070	10.582	9.346	4.398	41.101	10
11	1.146	.8723	.0854	.0979	11.714	10.218	4.876	49.825	11
12	1.161	.8615	.0778	.0903	12.860	11.079	5.352	59.302	12
13	1.175	.8509	.0713	.0838	14.021	11.930	5.827	69.513	13
14	1.190	.8404	.0658	.0783	15.196	12.771	6.299	80.438	14
15	1.205	.8300	.0610	.0735	16.386	13.601	6.769	92.058	15
16	1.220	.8197	.0568	.0693	17.591	14.420	7.237	104.355	16
17	1.235	.8096	.0532	.0657	18.811	15.230	7.702	117.309	17
18	1.251	.7996	.0499	.0624	20.046	16.030	8.166	130.903	18
19	1.266	.7898	.0470	.0595	21.297	16.819	8.628	145.119	19
20	1.282	.7800	.0443	.0568	22.563	17.599	9.088	159.940	20
21	1.298	.7704	.0419	.0544	23.845	18.370	9.545	175.348	21
22	1.314	.7609	.0398	.0523	25.143	19.131	10.001	191.327	22
23	1.331	.7515	.0378	.0503	26.458	19.882	10.455	207.859	23
24	1.347	.7422	.0360	.0485	27.788	20.624	10.906	224.930	24
25	1.364	.7330	.0343	.0468	29.136	21.357	11.355	242.523	25
26	1.381	.7240	.0328	.0453	30.500	22.081	11.803	260.623	26
27	1.399	.7150	.0314	.0439	31.881	22.796	12.248	279.215	27
28	1.416	.7062	.0300	.0425	33.280	23.503	12.691	298.284	28
29	1.434	.6975	.0288	.0413	34.696	24.200	13.133	317.814	29
30	1.452	.6889	.0277	.0402	36.129	24.889	13.572	337.792	30
36	1.564	.6394	.0222	.0347	45.116	28.847	16.164	466.297	36
40	1.644	.6084	.0194	.0319	51.490	31.327	17.852	559.247	40
48	1.815	.5509	.0153	.0278	65.229	35.932	21.130	759.248	48
50	1.861	.5373	.0145	.0270	68.882	37.013	21.930	811.692	50
52	1.908	.5242	.0138	.0263	72.628	38.068	22.722	864.960	52
60	2.107	.4746	.0113	.0238	88.575	42.035	25.809	1 084.86	60
70	2.386	.4191	.00902	.0215	110.873	46.470	29.492	1 370.47	70
72	2.446	.4088	.00864	.0211	115.675	47.293	30.205	1 428.48	72
80	2.701	.3702	.00735	.0198	136.120	50.387	32.983	1 661.89	80
84	2.839	.3522	.00680	.0193	147.130	51.822	34.326	1 778.86	84
90	3.059	.3269	.00607	.0186	164.706	53.846	36.286	1 953.85	90
96	3.296	.3034	.00545	.0179	183.643	55.725	38.180	2 127.55	96
100	3.463	.2887	.00507	.0176	197.074	56.901	39.406	2 242.26	100
104	3.640	.2747	.00474	.0172	211.190	58.021	40.604	2 355.90	104
120	4.440	.2252	.00363	.0161	275.220	61.983	45.119	2 796.59	120
240	19.716	.0507	.00067	.0132	1 497.3	75.942	67.177	5 101.55	240
360	87.543	.0114	.00014	.0126	6 923.4	79.086	75.840	5 997.91	360
480	388.713	.00257	.00003	.0125	31 017.1	79.794	78.762	6 284.74	480

	Single Payment		Uniform Payment Series				Arithmetic Gradient		
	Compound Amount Factor	Present Worth Factor	Sinking Fund Factor	Capital Recovery Factor	Compound Amount Factor	Present Worth Factor	Gradient Uniform Series	Gradient Present Worth	
n	Find *F* Given *P* *F/P*	Find *P* Given *F* *P/F*	Find *A* Given *F* *A/F*	Find *A* Given *P* *A/P*	Find *F* Given *A* *F/A*	Find *P* Given *A* *P/A*	Find *A* Given *G* *A/G*	Find *P* Given *G* *P/G*	*n*
1	1.015	.9852	1.0000	1.0150	1.000	0.985	0	0	1
2	1.030	.9707	.4963	.5113	2.015	1.956	0.496	0.970	2
3	1.046	.9563	.3284	.3434	3.045	2.912	0.990	2.883	3
4	1.061	.9422	.2444	.2594	4.091	3.854	1.481	5.709	4
5	1.077	.9283	.1941	.2091	5.152	4.783	1.970	9.422	5
6	1.093	.9145	.1605	.1755	6.230	5.697	2.456	13.994	6
7	1.110	.9010	.1366	.1516	7.323	6.598	2.940	19.400	7
8	1.126	.8877	.1186	.1336	8.433	7.486	3.422	25.614	8
9	1.143	.8746	.1046	.1196	9.559	8.360	3.901	32.610	9
10	1.161	.8617	.0934	.1084	10.703	9.222	4.377	40.365	10
11	1.178	.8489	.0843	.0993	11.863	10.071	4.851	48.855	11
12	1.196	.8364	.0767	.0917	13.041	10.907	5.322	58.054	12
13	1.214	.8240	.0702	.0852	14.237	11.731	5.791	67.943	13
14	1.232	.8118	.0647	.0797	15.450	12.543	6.258	78.496	14
15	1.250	.7999	.0599	.0749	16.682	13.343	6.722	89.694	15
16	1.269	.7880	.0558	.0708	17.932	14.131	7.184	101.514	16
17	1.288	.7764	.0521	.0671	19.201	14.908	7.643	113.937	17
18	1.307	.7649	.0488	.0638	20.489	15.673	8.100	126.940	18
19	1.327	.7536	.0459	.0609	21.797	16.426	8.554	140.505	19
20	1.347	.7425	.0432	.0582	23.124	17.169	9.005	154.611	20
21	1.367	.7315	.0409	.0559	24.470	17.900	9.455	169.241	21
22	1.388	.7207	.0387	.0537	25.837	18.621	9.902	184.375	22
23	1.408	.7100	.0367	.0517	27.225	19.331	10.346	199.996	23
24	1.430	.6995	.0349	.0499	28.633	20.030	10.788	216.085	24
25	1.451	.6892	.0333	.0483	30.063	20.720	11.227	232.626	25
26	1.473	.6790	.0317	.0467	31.514	21.399	11.664	249.601	26
27	1.495	.6690	.0303	.0453	32.987	22.068	12.099	266.995	27
28	1.517	.6591	.0290	.0440	34.481	22.727	12.531	284.790	28
29	1.540	.6494	.0278	.0428	35.999	23.376	12.961	302.972	29
30	1.563	.6398	.0266	.0416	37.539	24.016	13.388	321.525	30
36	1.709	.5851	.0212	.0362	47.276	27.661	15.901	439.823	36
40	1.814	.5513	.0184	.0334	54.268	29.916	17.528	524.349	40
48	2.043	.4894	.0144	.0294	69.565	34.042	20.666	703.537	48
50	2.105	.4750	.0136	.0286	73.682	35.000	21.428	749.955	50
52	2.169	.4611	.0128	.0278	77.925	35.929	22.179	796.868	52
60	2.443	.4093	.0104	.0254	96.214	39.380	25.093	988.157	60
70	2.835	.3527	.00817	.0232	122.363	43.155	28.529	1 231.15	70
72	2.921	.3423	.00781	.0228	128.076	43.845	29.189	1 279.78	72
80	3.291	.3039	.00655	.0215	152.710	46.407	31.742	1 473.06	80
84	3.493	.2863	.00602	.0210	166.172	47.579	32.967	1 568.50	84
90	3.819	.2619	.00532	.0203	187.929	49.210	34.740	1 709.53	90
96	4.176	.2395	.00472	.0197	211.719	50.702	36.438	1 847.46	96
100	4.432	.2256	.00437	.0194	228.802	51.625	37.529	1 937.43	100
104	4.704	.2126	.00405	.0190	246.932	52.494	38.589	2 025.69	104
120	5.969	.1675	.00302	.0180	331.286	55.498	42.518	2 359.69	120
240	35.632	.0281	.00043	.0154	2 308.8	64.796	59.737	3 870.68	240
360	212.700	.00470	.00007	.0151	14 113.3	66.353	64.966	4 310.71	360
480	1 269.700	.00079	.00001	.0150	84 577.8	66.614	66.288	4 415.74	480

	Single Payment		Uniform Payment Series				Arithmetic Gradient		
n	Compound Amount Factor Find F Given P F/P	Present Worth Factor Find P Given F P/F	Sinking Fund Factor Find A Given F A/F	Capital Recovery Factor Find A Given P A/P	Compound Amount Factor Find F Given A F/A	Present Worth Factor Find P Given A P/A	Gradient Uniform Series Find A Given G A/G	Gradient Present Worth Find P Given G P/G	n
1	1.018	.9828	1.0000	1.0175	1.000	0.983	0	0	1
2	1.035	.9659	.4957	.5132	2.018	1.949	0.496	0.966	2
3	1.053	.9493	.3276	.3451	3.053	2.898	0.989	2.865	3
4	1.072	.9330	.2435	.2610	4.106	3.831	1.478	5.664	4
5	1.091	.9169	.1931	.2106	5.178	4.748	1.965	9.332	5
6	1.110	.9011	.1595	.1770	6.269	5.649	2.450	13.837	6
7	1.129	.8856	.1355	.1530	7.378	6.535	2.931	19.152	7
8	1.149	.8704	.1175	.1350	8.508	7.405	3.409	25.245	8
9	1.169	.8554	.1036	.1211	9.656	8.261	3.885	32.088	9
10	1.189	.8407	.0924	.1099	10.825	9.101	4.357	39.655	10
11	1.210	.8263	.0832	.1007	12.015	9.928	4.827	47.918	11
12	1.231	.8121	.0756	.0931	13.225	10.740	5.294	56.851	12
13	1.253	.7981	.0692	.0867	14.457	11.538	5.758	66.428	13
14	1.275	.7844	.0637	.0812	15.710	12.322	6.219	76.625	14
15	1.297	.7709	.0589	.0764	16.985	13.093	6.677	87.417	15
16	1.320	.7576	.0547	.0722	18.282	13.851	7.132	98.782	16
17	1.343	.7446	.0510	.0685	19.602	14.595	7.584	110.695	17
18	1.367	.7318	.0477	.0652	20.945	15.327	8.034	123.136	18
19	1.390	.7192	.0448	.0623	22.311	16.046	8.481	136.081	19
20	1.415	.7068	.0422	.0597	23.702	16.753	8.924	149.511	20
21	1.440	.6947	.0398	.0573	25.116	17.448	9.365	163.405	21
22	1.465	.6827	.0377	.0552	26.556	18.130	9.804	177.742	22
23	1.490	.6710	.0357	.0532	28.021	18.801	10.239	192.503	23
24	1.516	.6594	.0339	.0514	29.511	19.461	10.671	207.671	24
25	1.543	.6481	.0322	.0497	31.028	20.109	11.101	223.225	25
26	1.570	.6369	.0307	.0482	32.571	20.746	11.528	239.149	26
27	1.597	.6260	.0293	.0468	34.141	21.372	11.952	255.425	27
28	1.625	.6152	.0280	.0455	35.738	21.987	12.373	272.036	28
29	1.654	.6046	.0268	.0443	37.363	22.592	12.791	288.967	29
30	1.683	.5942	.0256	.0431	39.017	23.186	13.206	306.200	30
36	1.867	.5355	.0202	.0377	49.566	26.543	15.640	415.130	36
40	2.002	.4996	.0175	.0350	57.234	28.594	17.207	492.017	40
48	2.300	.4349	.0135	.0310	74.263	32.294	20.209	652.612	48
50	2.381	.4200	.0127	.0302	78.903	33.141	20.932	693.708	50
52	2.465	.4057	.0119	.0294	83.706	33.960	21.644	735.039	52
60	2.832	.3531	.00955	.0271	104.676	36.964	24.389	901.503	60
70	3.368	.2969	.00739	.0249	135.331	40.178	27.586	1 108.34	70
72	3.487	.2868	.00704	.0245	142.127	40.757	28.195	1 149.12	72
80	4.006	.2496	.00582	.0233	171.795	42.880	30.533	1 309.25	80
84	4.294	.2329	.00531	.0228	188.246	43.836	31.644	1 387.16	84
90	4.765	.2098	.00465	.0221	215.166	45.152	33.241	1 500.88	90
96	5.288	.1891	.00408	.0216	245.039	46.337	34.756	1 610.48	96
100	5.668	.1764	.00375	.0212	266.753	47.062	35.721	1 681.09	100
104	6.075	.1646	.00345	.0209	290.028	47.737	36.652	1 749.68	104
120	8.019	.1247	.00249	.0200	401.099	50.017	40.047	2 003.03	120
240	64.308	.0156	.00028	.0178	3 617.6	56.254	53.352	3 001.27	240
360	515.702	.00194	.00003	.0175	29 411.5	57.032	56.443	3 219.08	360
480	4 135.500	.00024		.0175	236 259.0	57.129	57.027	3 257.88	480

	Single Payment		Uniform Payment Series				Arithmetic Gradient		
n	Compound Amount Factor Find F Given P F/P	Present Worth Factor Find P Given F P/F	Sinking Fund Factor Find A Given F A/F	Capital Recovery Factor Find A Given P A/P	Compound Amount Factor Find F Given A F/A	Present Worth Factor Find P Given A P/A	Gradient Uniform Series Find A Given G A/G	Gradient Present Worth Find P Given G P/G	n
1	1.020	.9804	1.0000	1.0200	1.000	0.980	0	0	1
2	1.040	.9612	.4951	.5151	2.020	1.942	0.495	0.961	2
3	1.061	.9423	.3268	.3468	3.060	2.884	0.987	2.846	3
4	1.082	.9238	.2426	.2626	4.122	3.808	1.475	5.617	4
5	1.104	.9057	.1922	.2122	5.204	4.713	1.960	9.240	5
6	1.126	.8880	.1585	.1785	6.308	5.601	2.442	13.679	6
7	1.149	.8706	.1345	.1545	7.434	6.472	2.921	18.903	7
8	1.172	.8535	.1165	.1365	8.583	7.325	3.396	24.877	8
9	1.195	.8368	.1025	.1225	9.755	8.162	3.868	31.571	9
10	1.219	.8203	.0913	.1113	10.950	8.983	4.337	38.954	10
11	1.243	.8043	.0822	.1022	12.169	9.787	4.802	46.996	11
12	1.268	.7885	.0746	.0946	13.412	10.575	5.264	55.669	12
13	1.294	.7730	.0681	.0881	14.680	11.348	5.723	64.946	13
14	1.319	.7579	.0626	.0826	15.974	12.106	6.178	74.798	14
15	1.346	.7430	.0578	.0778	17.293	12.849	6.631	85.200	15
16	1.373	.7284	.0537	.0737	18.639	13.578	7.080	96.127	16
17	1.400	.7142	.0500	.0700	20.012	14.292	7.526	107.553	17
18	1.428	.7002	.0467	.0667	21.412	14.992	7.968	119.456	18
19	1.457	.6864	.0438	.0638	22.840	15.678	8.407	131.812	19
20	1.486	.6730	.0412	.0612	24.297	16.351	8.843	144.598	20
21	1.516	.6598	.0388	.0588	25.783	17.011	9.276	157.793	21
22	1.546	.6468	.0366	.0566	27.299	17.658	9.705	171.377	22
23	1.577	.6342	.0347	.0547	28.845	18.292	10.132	185.328	23
24	1.608	.6217	.0329	.0529	30.422	18.914	10.555	199.628	24
25	1.641	.6095	.0312	.0512	32.030	19.523	10.974	214.256	25
26	1.673	.5976	.0297	.0497	33.671	20.121	11.391	229.196	26
27	1.707	.5859	.0283	.0483	35.344	20.707	11.804	244.428	27
28	1.741	.5744	.0270	.0470	37.051	21.281	12.214	259.936	28
29	1.776	.5631	.0258	.0458	38.792	21.844	12.621	275.703	29
30	1.811	.5521	.0247	.0447	40.568	22.396	13.025	291.713	30
36	2.040	.4902	.0192	.0392	51.994	25.489	15.381	392.036	36
40	2.208	.4529	.0166	.0366	60.402	27.355	16.888	461.989	40
48	2.587	.3865	.0126	.0326	79.353	30.673	19.755	605.961	48
50	2.692	.3715	.0118	.0318	84.579	31.424	20.442	642.355	50
52	2.800	.3571	.0111	.0311	90.016	32.145	21.116	678.779	52
60	3.281	.3048	.00877	.0288	114.051	34.761	23.696	823.692	60
70	4.000	.2500	.00667	.0267	149.977	37.499	26.663	999.829	70
72	4.161	.2403	.00633	.0263	158.056	37.984	27.223	1 034.050	72
80	4.875	.2051	.00516	.0252	193.771	39.744	29.357	1 166.781	80
84	5.277	.1895	.00468	.0247	213.865	40.525	30.361	1 230.413	84
90	5.943	.1683	.00405	.0240	247.155	41.587	31.793	1 322.164	90
96	6.693	.1494	.00351	.0235	284.645	42.529	33.137	1 409.291	96
100	7.245	.1380	.00320	.0232	312.230	43.098	33.986	1 464.747	100
104	7.842	.1275	.00292	.0229	342.090	43.624	34.799	1 518.082	104
120	10.765	.0929	.00205	.0220	488.255	45.355	37.711	1 710.411	120
240	115.887	.00863	.00017	.0202	5 744.4	49.569	47.911	2 374.878	240
360	1 247.500	.00080	.00002	.0200	62 326.8	49.960	49.711	2 483.567	360
480	13 429.800	.00007		.0200	671 442.0	49.996	49.964	2 498.027	480

	Single Payment		Uniform Payment Series				Arithmetic Gradient		
	Compound Amount Factor	Present Worth Factor	Sinking Fund Factor	Capital Recovery Factor	Compound Amount Factor	Present Worth Factor	Gradient Uniform Series	Gradient Present Worth	
n	Find F Given P F/P	Find P Given F P/F	Find A Given F A/F	Find A Given P A/P	Find F Given A F/A	Find P Given A P/A	Find A Given G A/G	Find P Given G P/G	n
1	1.025	.9756	1.0000	1.0250	1.000	0.976	0	0	1
2	1.051	.9518	.4938	.5188	2.025	1.927	0.494	0.952	2
3	1.077	.9286	.3251	.3501	3.076	2.856	0.984	2.809	3
4	1.104	.9060	.2408	.2658	4.153	3.762	1.469	5.527	4
5	1.131	.8839	.1902	.2152	5.256	4.646	1.951	9.062	5
6	1.160	.8623	.1566	.1816	6.388	5.508	2.428	13.374	6
7	1.189	.8413	.1325	.1575	7.547	6.349	2.901	18.421	7
8	1.218	.8207	.1145	.1395	8.736	7.170	3.370	24.166	8
9	1.249	.8007	.1005	.1255	9.955	7.971	3.835	30.572	9
10	1.280	.7812	.0893	.1143	11.203	8.752	4.296	37.603	10
11	1.312	.7621	.0801	.1051	12.483	9.514	4.753	45.224	11
12	1.345	.7436	.0725	.0975	13.796	10.258	5.206	53.403	12
13	1.379	.7254	.0660	.0910	15.140	10.983	5.655	62.108	13
14	1.413	.7077	.0605	.0855	16.519	11.691	6.100	71.309	14
15	1.448	.6905	.0558	.0808	17.932	12.381	6.540	80.975	15
16	1.485	.6736	.0516	.0766	19.380	13.055	6.977	91.080	16
17	1.522	.6572	.0479	.0729	20.865	13.712	7.409	101.595	17
18	1.560	.6412	.0447	.0697	22.386	14.353	7.838	112.495	18
19	1.599	.6255	.0418	.0668	23.946	14.979	8.262	123.754	19
20	1.639	.6103	.0391	.0641	25.545	15.589	8.682	135.349	20
21	1.680	.5954	.0368	.0618	27.183	16.185	9.099	147.257	21
22	1.722	.5809	.0346	.0596	28.863	16.765	9.511	159.455	22
23	1.765	.5667	.0327	.0577	30.584	17.332	9.919	171.922	23
24	1.809	.5529	.0309	.0559	32.349	17.885	10.324	184.638	24
25	1.854	.5394	.0293	.0543	34.158	18.424	10.724	197.584	25
26	1.900	.5262	.0278	.0528	36.012	18.951	11.120	210.740	26
27	1.948	.5134	.0264	.0514	37.912	19.464	11.513	224.088	27
28	1.996	.5009	.0251	.0501	39.860	19.965	11.901	237.612	28
29	2.046	.4887	.0239	.0489	41.856	20.454	12.286	251.294	29
30	2.098	.4767	.0228	.0478	43.903	20.930	12.667	265.120	30
31	2.150	.4651	.0217	.0467	46.000	21.395	13.044	279.073	31
32	2.204	.4538	.0208	.0458	48.150	21.849	13.417	293.140	32
33	2.259	.4427	.0199	.0449	50.354	22.292	13.786	307.306	33
34	2.315	.4319	.0190	.0440	52.613	22.724	14.151	321.559	34
35	2.373	.4214	.0182	.0432	54.928	23.145	14.512	335.886	35
40	2.685	.3724	.0148	.0398	67.402	25.103	16.262	408.221	40
45	3.038	.3292	.0123	.0373	81.516	26.833	17.918	480.806	45
50	3.437	.2909	.0103	.0353	97.484	28.362	19.484	552.607	50
55	3.889	.2572	.00865	.0337	115.551	29.714	20.961	622.827	55
60	4.400	.2273	.00735	.0324	135.991	30.909	22.352	690.865	60
65	4.978	.2009	.00628	.0313	159.118	31.965	23.660	756.280	65
70	5.632	.1776	.00540	.0304	185.284	32.898	24.888	818.763	70
75	6.372	.1569	.00465	.0297	214.888	33.723	26.039	878.114	75
80	7.210	.1387	.00403	.0290	248.382	34.452	27.117	934.217	80
85	8.157	.1226	.00349	.0285	286.278	35.096	28.123	987.026	85
90	9.229	.1084	.00304	.0280	329.154	35.666	29.063	1 036.54	90
95	10.442	.0958	.00265	.0276	377.663	36.169	29.938	1 082.83	95
100	11.814	.0846	.00231	.0273	432.548	36.614	30.752	1 125.97	100

	Single Payment		Uniform Payment Series				Arithmetic Gradient		
	Compound Amount Factor	Present Worth Factor	Sinking Fund Factor	Capital Recovery Factor	Compound Amount Factor	Present Worth Factor	Gradient Uniform Series	Gradient Present Worth	
n	Find F Given P F/P	Find P Given F P/F	Find A Given F A/F	Find A Given P A/P	Find F Given A F/A	Find P Given A P/A	Find A Given G A/G	Find P Given G P/G	n
1	1.030	.9709	1.0000	1.0300	1.000	0.971	0	0	1
2	1.061	.9426	.4926	.5226	2.030	1.913	0.493	0.943	2
3	1.093	.9151	.3235	.3535	3.091	2.829	0.980	2.773	3
4	1.126	.8885	.2390	.2690	4.184	3.717	1.463	5.438	4
5	1.159	.8626	.1884	.2184	5.309	4.580	1.941	8.889	5
6	1.194	.8375	.1546	.1846	6.468	5.417	2.414	13.076	6
7	1.230	.8131	.1305	.1605	7.662	6.230	2.882	17.955	7
8	1.267	.7894	.1125	.1425	8.892	7.020	3.345	23.481	8
9	1.305	.7664	.0984	.1284	10.159	7.786	3.803	29.612	9
10	1.344	.7441	.0872	.1172	11.464	8.530	4.256	36.309	10
11	1.384	.7224	.0781	.1081	12.808	9.253	4.705	43.533	11
12	1.426	.7014	.0705	.1005	14.192	9.954	5.148	51.248	12
13	1.469	.6810	.0640	.0940	15.618	10.635	5.587	59.419	13
14	1.513	.6611	.0585	.0885	17.086	11.296	6.021	68.014	14
15	1.558	.6419	.0538	.0838	18.599	11.938	6.450	77.000	15
16	1.605	.6232	.0496	.0796	20.157	12.561	6.874	86.348	16
17	1.653	.6050	.0460	.0760	21.762	13.166	7.294	96.028	17
18	1.702	.5874	.0427	.0727	23.414	13.754	7.708	106.014	18
19	1.754	.5703	.0398	.0698	25.117	14.324	8.118	116.279	19
20	1.806	.5537	.0372	.0672	26.870	14.877	8.523	126.799	20
21	1.860	.5375	.0349	.0649	28.676	15.415	8.923	137.549	21
22	1.916	.5219	.0327	.0627	30.537	15.937	9.319	148.509	22
23	1.974	.5067	.0308	.0608	32.453	16.444	9.709	159.656	23
24	2.033	.4919	.0290	.0590	34.426	16.936	10.095	170.971	24
25	2.094	.4776	.0274	.0574	36.459	17.413	10.477	182.433	25
26	2.157	.4637	.0259	.0559	38.553	17.877	10.853	194.026	26
27	2.221	.4502	.0246	.0546	40.710	18.327	11.226	205.731	27
28	2.288	.4371	.0233	.0533	42.931	18.764	11.593	217.532	28
29	2.357	.4243	.0221	.0521	45.219	19.188	11.956	229.413	29
30	2.427	.4120	.0210	.0510	47.575	19.600	12.314	241.361	30
31	2.500	.4000	.0200	.0500	50.003	20.000	12.668	253.361	31
32	2.575	.3883	.0190	.0490	52.503	20.389	13.017	265.399	32
33	2.652	.3770	.0182	.0482	55.078	20.766	13.362	277.464	33
34	2.732	.3660	.0173	.0473	57.730	21.132	13.702	289.544	34
35	2.814	.3554	.0165	.0465	60.462	21.487	14.037	301.627	35
40	3.262	.3066	.0133	.0433	75.401	23.115	15.650	361.750	40
45	3.782	.2644	.0108	.0408	92.720	24.519	17.156	420.632	45
50	4.384	.2281	.00887	.0389	112.797	25.730	18.558	477.480	50
55	5.082	.1968	.00735	.0373	136.072	26.774	19.860	531.741	55
60	5.892	.1697	.00613	.0361	163.053	27.676	21.067	583.052	60
65	6.830	.1464	.00515	.0351	194.333	28.453	22.184	631.201	65
70	7.918	.1263	.00434	.0343	230.594	29.123	23.215	676.087	70
75	9.179	.1089	.00367	.0337	272.631	29.702	24.163	717.698	75
80	10.641	.0940	.00311	.0331	321.363	30.201	25.035	756.086	80
85	12.336	.0811	.00265	.0326	377.857	30.631	25.835	791.353	85
90	14.300	.0699	.00226	.0323	443.349	31.002	26.567	823.630	90
95	16.578	.0603	.00193	.0319	519.272	31.323	27.235	853.074	95
100	19.219	.0520	.00165	.0316	607.287	31.599	27.844	879.854	100

	Single Payment		Uniform Payment Series				Arithmetic Gradient		
	Compound Amount Factor	Present Worth Factor	Sinking Fund Factor	Capital Recovery Factor	Compound Amount Factor	Present Worth Factor	Gradient Uniform Series	Gradient Present Worth	
n	Find *F* Given *P* *F/P*	Find *P* Given *F* *P/F*	Find *A* Given *F* *A/F*	Find *A* Given *P* *A/P*	Find *F* Given *A* *F/A*	Find *P* Given *A* *P/A*	Find *A* Given *G* *A/G*	Find *P* Given *G* *P/G*	*n*
1	1.035	.9662	1.0000	1.0350	1.000	0.966	0	0	1
2	1.071	.9335	.4914	.5264	2.035	1.900	0.491	0.933	2
3	1.109	.9019	.3219	.3569	3.106	2.802	0.977	2.737	3
4	1.148	.8714	.2373	.2723	4.215	3.673	1.457	5.352	4
5	1.188	.8420	.1865	.2215	5.362	4.515	1.931	8.719	5
6	1.229	.8135	.1527	.1877	6.550	5.329	2.400	12.787	6
7	1.272	.7860	.1285	.1635	7.779	6.115	2.862	17.503	7
8	1.317	.7594	.1105	.1455	9.052	6.874	3.320	22.819	8
9	1.363	.7337	.0964	.1314	10.368	7.608	3.771	28.688	9
10	1.411	.7089	.0852	.1202	11.731	8.317	4.217	35.069	10
11	1.460	.6849	.0761	.1111	13.142	9.002	4.657	41.918	11
12	1.511	.6618	.0685	.1035	14.602	9.663	5.091	49.198	12
13	1.564	.6394	.0621	.0971	16.113	10.303	5.520	56.871	13
14	1.619	.6178	.0566	.0916	17.677	10.921	5.943	64.902	14
15	1.675	.5969	.0518	.0868	19.296	11.517	6.361	73.258	15
16	1.734	.5767	.0477	.0827	20.971	12.094	6.773	81.909	16
17	1.795	.5572	.0440	.0790	22.705	12.651	7.179	90.824	17
18	1.857	.5384	.0408	.0758	24.500	13.190	7.580	99.976	18
19	1.922	.5202	.0379	.0729	26.357	13.710	7.975	109.339	19
20	1.990	.5026	.0354	.0704	28.280	14.212	8.365	118.888	20
21	2.059	.4856	.0330	.0680	30.269	14.698	8.749	128.599	21
22	2.132	.4692	.0309	.0659	32.329	15.167	9.128	138.451	22
23	2.206	.4533	.0290	.0640	34.460	15.620	9.502	148.423	23
24	2.283	.4380	.0273	.0623	36.666	16.058	9.870	158.496	24
25	2.363	.4231	.0257	.0607	38.950	16.482	10.233	168.652	25
26	2.446	.4088	.0242	.0592	41.313	16.890	10.590	178.873	26
27	2.532	.3950	.0229	.0579	43.759	17.285	10.942	189.143	27
28	2.620	.3817	.0216	.0566	46.291	17.667	11.289	199.448	28
29	2.712	.3687	.0204	.0554	48.911	18.036	11.631	209.773	29
30	2.807	.3563	.0194	.0544	51.623	18.392	11.967	220.105	30
31	2.905	.3442	.0184	.0534	54.429	18.736	12.299	230.432	31
32	3.007	.3326	.0174	.0524	57.334	19.069	12.625	240.742	32
33	3.112	.3213	.0166	.0516	60.341	19.390	12.946	251.025	33
34	3.221	.3105	.0158	.0508	63.453	19.701	13.262	261.271	34
35	3.334	.3000	.0150	.0500	66.674	20.001	13.573	271.470	35
40	3.959	.2526	.0118	.0468	84.550	21.355	15.055	321.490	40
45	4.702	.2127	.00945	.0445	105.781	22.495	16.417	369.307	45
50	5.585	.1791	.00763	.0426	130.998	23.456	17.666	414.369	50
55	6.633	.1508	.00621	.0412	160.946	24.264	18.808	456.352	55
60	7.878	.1269	.00509	.0401	196.516	24.945	19.848	495.104	60
65	9.357	.1069	.00419	.0392	238.762	25.518	20.793	530.598	65
70	11.113	.0900	.00346	.0385	288.937	26.000	21.650	562.895	70
75	13.199	.0758	.00287	.0379	348.529	26.407	22.423	592.121	75
80	15.676	.0638	.00238	.0374	419.305	26.749	23.120	618.438	80
85	18.618	.0537	.00199	.0370	503.365	27.037	23.747	642.036	85
90	22.112	.0452	.00166	.0367	603.202	27.279	24.308	663.118	90
95	26.262	.0381	.00139	.0364	721.778	27.483	24.811	681.890	95
100	31.191	.0321	.00116	.0362	862.608	27.655	25.259	698.554	100

4% Compound Interest Factors 4%

	Single Payment		Uniform Payment Series				Arithmetic Gradient		
	Compound Amount Factor	Present Worth Factor	Sinking Fund Factor	Capital Recovery Factor	Compound Amount Factor	Present Worth Factor	Gradient Uniform Series	Gradient Present Worth	
	Find F Given P	Find P Given F	Find A Given F	Find A Given P	Find F Given A	Find P Given A	Find A Given G	Find P Given G	
n	F/P	P/F	A/F	A/P	F/A	P/A	A/G	P/G	n
1	1.040	.9615	1.0000	1.0400	1.000	0.962	0	0	1
2	1.082	.9246	.4902	.5302	2.040	1.886	0.490	0.925	2
3	1.125	.8890	.3203	.3603	3.122	2.775	0.974	2.702	3
4	1.170	.8548	.2355	.2755	4.246	3.630	1.451	5.267	4
5	1.217	.8219	.1846	.2246	5.416	4.452	1.922	8.555	5
6	1.265	.7903	.1508	.1908	6.633	5.242	2.386	12.506	6
7	1.316	.7599	.1266	.1666	7.898	6.002	2.843	17.066	7
8	1.369	.7307	.1085	.1485	9.214	6.733	3.294	22.180	8
9	1.423	.7026	.0945	.1345	10.583	7.435	3.739	27.801	9
10	1.480	.6756	.0833	.1233	12.006	8.111	4.177	33.881	10
11	1.539	.6496	.0741	.1141	13.486	8.760	4.609	40.377	11
12	1.601	.6246	.0666	.1066	15.026	9.385	5.034	47.248	12
13	1.665	.6006	.0601	.1001	16.627	9.986	5.453	54.454	13
14	1.732	.5775	.0547	.0947	18.292	10.563	5.866	61.962	14
15	1.801	.5553	.0499	.0899	20.024	11.118	6.272	69.735	15
16	1.873	.5339	.0458	.0858	21.825	11.652	6.672	77.744	16
17	1.948	.5134	.0422	.0822	23.697	12.166	7.066	85.958	17
18	2.026	.4936	.0390	.0790	25.645	12.659	7.453	94.350	18
19	2.107	.4746	.0361	.0761	27.671	13.134	7.834	102.893	19
20	2.191	.4564	.0336	.0736	29.778	13.590	8.209	111.564	20
21	2.279	.4388	.0313	.0713	31.969	14.029	8.578	120.341	21
22	2.370	.4220	.0292	.0692	34.248	14.451	8.941	129.202	22
23	2.465	.4057	.0273	.0673	36.618	14.857	9.297	138.128	23
24	2.563	.3901	.0256	.0656	39.083	15.247	9.648	147.101	24
25	2.666	.3751	.0240	.0640	41.646	15.622	9.993	156.104	25
26	2.772	.3607	.0226	.0626	44.312	15.983	10.331	165.121	26
27	2.883	.3468	.0212	.0612	47.084	16.330	10.664	174.138	27
28	2.999	.3335	.0200	.0600	49.968	16.663	10.991	183.142	28
29	3.119	.3207	.0189	.0589	52.966	16.984	11.312	192.120	29
30	3.243	.3083	.0178	.0578	56.085	17.292	11.627	201.062	30
31	3.373	.2965	.0169	.0569	59.328	17.588	11.937	209.955	31
32	3.508	.2851	.0159	.0559	62.701	17.874	12.241	218.792	32
33	3.648	.2741	.0151	.0551	66.209	18.148	12.540	227.563	33
34	3.794	.2636	.0143	.0543	69.858	18.411	12.832	236.260	34
35	3.946	.2534	.0136	.0536	73.652	18.665	13.120	244.876	35
40	4.801	.2083	.0105	.0505	95.025	19.793	14.476	286.530	40
45	5.841	.1712	.00826	.0483	121.029	20.720	15.705	325.402	45
50	7.107	.1407	.00655	.0466	152.667	21.482	16.812	361.163	50
55	8.646	.1157	.00523	.0452	191.159	22.109	17.807	393.689	55
60	10.520	.0951	.00420	.0442	237.990	22.623	18.697	422.996	60
65	12.799	.0781	.00339	.0434	294.968	23.047	19.491	449.201	65
70	15.572	.0642	.00275	.0427	364.290	23.395	20.196	472.479	70
75	18.945	.0528	.00223	.0422	448.630	23.680	20.821	493.041	75
80	23.050	.0434	.00181	.0418	551.244	23.915	21.372	511.116	80
85	28.044	.0357	.00148	.0415	676.089	24.109	21.857	526.938	85
90	34.119	.0293	.00121	.0412	827.981	24.267	22.283	540.737	90
95	41.511	.0241	.00099	.0410	1 012.8	24.398	22.655	552.730	95
100	50.505	.0198	.00081	.0408	1 237.6	24.505	22.980	563.125	100

	Single Payment		Uniform Payment Series				Arithmetic Gradient		
	Compound Amount Factor	Present Worth Factor	Sinking Fund Factor	Capital Recovery Factor	Compound Amount Factor	Present Worth Factor	Gradient Uniform Series	Gradient Present Worth	
	Find *F* Given *P*	Find *P* Given *F*	Find *A* Given *F*	Find *A* Given *P*	Find *F* Given *A*	Find *P* Given *A*	Find *A* Given *G*	Find *P* Given *G*	
n	*F/P*	*P/F*	*A/F*	*A/P*	*F/A*	*P/A*	*A/G*	*P/G*	*n*
1	1.045	.9569	1.0000	1.0450	1.000	0.957	0	0	1
2	1.092	.9157	.4890	.5340	2.045	1.873	0.489	0.916	2
3	1.141	.8763	.3188	.3638	3.137	2.749	0.971	2.668	3
4	1.193	.8386	.2337	.2787	4.278	3.588	1.445	5.184	4
5	1.246	.8025	.1828	.2278	5.471	4.390	1.912	8.394	5
6	1.302	.7679	.1489	.1939	6.717	5.158	2.372	12.233	6
7	1.361	.7348	.1247	.1697	8.019	5.893	2.824	16.642	7
8	1.422	.7032	.1066	.1516	9.380	6.596	3.269	21.564	8
9	1.486	.6729	.0926	.1376	10.802	7.269	3.707	26.948	9
10	1.553	.6439	.0814	.1264	12.288	7.913	4.138	32.743	10
11	1.623	.6162	.0722	.1172	13.841	8.529	4.562	38.905	11
12	1.696	.5897	.0647	.1097	15.464	9.119	4.978	45.391	12
13	1.772	.5643	.0583	.1033	17.160	9.683	5.387	52.163	13
14	1.852	.5400	.0528	.0978	18.932	10.223	5.789	59.182	14
15	1.935	.5167	.0481	.0931	20.784	10.740	6.184	66.416	15
16	2.022	.4945	.0440	.0890	22.719	11.234	6.572	73.833	16
17	2.113	.4732	.0404	.0854	24.742	11.707	6.953	81.404	17
18	2.208	.4528	.0372	.0822	26.855	12.160	7.327	89.102	18
19	2.308	.4333	.0344	.0794	29.064	12.593	7.695	96.901	19
20	2.412	.4146	.0319	.0769	31.371	13.008	8.055	104.779	20
21	2.520	.3968	.0296	.0746	33.783	13.405	8.409	112.715	21
22	2.634	.3797	.0275	.0725	36.303	13.784	8.755	120.689	22
23	2.752	.3634	.0257	.0707	38.937	14.148	9.096	128.682	23
24	2.876	.3477	.0240	.0690	41.689	14.495	9.429	136.680	24
25	3.005	.3327	.0224	.0674	44.565	14.828	9.756	144.665	25
26	3.141	.3184	.0210	.0660	47.571	15.147	10.077	152.625	26
27	3.282	.3047	.0197	.0647	50.711	15.451	10.391	160.547	27
28	3.430	.2916	.0185	.0635	53.993	15.743	10.698	168.420	28
29	3.584	.2790	.0174	.0624	57.423	16.022	10.999	176.232	29
30	3.745	.2670	.0164	.0614	61.007	16.289	11.295	183.975	30
31	3.914	.2555	.0154	.0604	64.752	16.544	11.583	191.640	31
32	4.090	.2445	.0146	.0596	68.666	16.789	11.866	199.220	32
33	4.274	.2340	.0137	.0587	72.756	17.023	12.143	206.707	33
34	4.466	.2239	.0130	.0580	77.030	17.247	12.414	214.095	34
35	4.667	.2143	.0123	.0573	81.497	17.461	12.679	221.380	35
40	5.816	.1719	.00934	.0543	107.030	18.402	13.917	256.098	40
45	7.248	.1380	.00720	.0522	138.850	19.156	15.020	287.732	45
50	9.033	.1107	.00560	.0506	178.503	19.762	15.998	316.145	50
55	11.256	.0888	.00439	.0494	227.918	20.248	16.860	341.375	55
60	14.027	.0713	.00345	.0485	289.497	20.638	17.617	363.571	60
65	17.481	.0572	.00273	.0477	366.237	20.951	18.278	382.946	65
70	21.784	.0459	.00217	.0472	461.869	21.202	18.854	399.750	70
75	27.147	.0368	.00172	.0467	581.043	21.404	19.354	414.242	75
80	33.830	.0296	.00137	.0464	729.556	21.565	19.785	426.680	80
85	42.158	.0237	.00109	.0461	914.630	21.695	20.157	437.309	85
90	52.537	.0190	.00087	.0459	1 145.3	21.799	20.476	446.359	90
95	65.471	.0153	.00070	.0457	1 432.7	21.883	20.749	454.039	95
100	81.588	.0123	.00056	.0456	1 790.9	21.950	20.981	460.537	100

	Single Payment		Uniform Payment Series				Arithmetic Gradient		
	Compound Amount Factor	Present Worth Factor	Sinking Fund Factor	Capital Recovery Factor	Compound Amount Factor	Present Worth Factor	Gradient Uniform Series	Gradient Present Worth	
n	Find *F* Given *P* *F/P*	Find *P* Given *F* *P/F*	Find *A* Given *F* *A/F*	Find *A* Given *P* *A/P*	Find *F* Given *A* *F/A*	Find *P* Given *A* *P/A*	Find *A* Given *G* *A/G*	Find *P* Given *G* *P/G*	*n*
1	1.050	.9524	1.0000	1.0500	1.000	0.952	0	0	1
2	1.102	.9070	.4878	.5378	2.050	1.859	0.488	0.907	2
3	1.158	.8638	.3172	.3672	3.152	2.723	0.967	2.635	3
4	1.216	.8227	.2320	.2820	4.310	3.546	1.439	5.103	4
5	1.276	.7835	.1810	.2310	5.526	4.329	1.902	8.237	5
6	1.340	.7462	.1470	.1970	6.802	5.076	2.358	11.968	6
7	1.407	.7107	.1228	.1728	8.142	5.786	2.805	16.232	7
8	1.477	.6768	.1047	.1547	9.549	6.463	3.244	20.970	8
9	1.551	.6446	.0907	.1407	11.027	7.108	3.676	26.127	9
10	1.629	.6139	.0795	.1295	12.578	7.722	4.099	31.652	10
11	1.710	.5847	.0704	.1204	14.207	8.306	4.514	37.499	11
12	1.796	.5568	.0628	.1128	15.917	8.863	4.922	43.624	12
13	1.886	.5303	.0565	.1065	17.713	9.394	5.321	49.988	13
14	1.980	.5051	.0510	.1010	19.599	9.899	5.713	56.553	14
15	2.079	.4810	.0463	.0963	21.579	10.380	6.097	63.288	15
16	2.183	.4581	.0423	.0923	23.657	10.838	6.474	70.159	16
17	2.292	.4363	.0387	.0887	25.840	11.274	6.842	77.140	17
18	2.407	.4155	.0355	.0855	28.132	11.690	7.203	84.204	18
19	2.527	.3957	.0327	.0827	30.539	12.085	7.557	91.327	19
20	2.653	.3769	.0302	.0802	33.066	12.462	7.903	98.488	20
21	2.786	.3589	.0280	.0780	35.719	12.821	8.242	105.667	21
22	2.925	.3419	.0260	.0760	38.505	13.163	8.573	112.846	22
23	3.072	.3256	.0241	.0741	41.430	13.489	8.897	120.008	23
24	3.225	.3101	.0225	.0725	44.502	13.799	9.214	127.140	24
25	3.386	.2953	.0210	.0710	47.727	14.094	9.524	134.227	25
26	3.556	.2812	.0196	.0696	51.113	14.375	9.827	141.258	26
27	3.733	.2678	.0183	.0683	54.669	14.643	10.122	148.222	27
28	3.920	.2551	.0171	.0671	58.402	14.898	10.411	155.110	28
29	4.116	.2429	.0160	.0660	62.323	15.141	10.694	161.912	29
30	4.322	.2314	.0151	.0651	66.439	15.372	10.969	168.622	30
31	4.538	.2204	.0141	.0641	70.761	15.593	11.238	175.233	31
32	4.765	.2099	.0133	.0633	75.299	15.803	11.501	181.739	32
33	5.003	.1999	.0125	.0625	80.063	16.003	11.757	188.135	33
34	5.253	.1904	.0118	.0618	85.067	16.193	12.006	194.416	34
35	5.516	.1813	.0111	.0611	90.320	16.374	12.250	200.580	35
40	7.040	.1420	.00828	.0583	120.799	17.159	13.377	229.545	40
45	8.985	.1113	.00626	.0563	159.699	17.774	14.364	255.314	45
50	11.467	.0872	.00478	.0548	209.347	18.256	15.223	277.914	50
55	14.636	.0683	.00367	.0537	272.711	18.633	15.966	297.510	55
60	18.679	.0535	.00283	.0528	353.582	18.929	16.606	314.343	60
65	23.840	.0419	.00219	.0522	456.795	19.161	17.154	328.691	65
70	30.426	.0329	.00170	.0517	588.525	19.343	17.621	340.841	70
75	38.832	.0258	.00132	.0513	756.649	19.485	18.018	351.072	75
80	49.561	.0202	.00103	.0510	971.222	19.596	18.353	359.646	80
85	63.254	.0158	.00080	.0508	1 245.1	19.684	18.635	366.800	85
90	80.730	.0124	.00063	.0506	1 594.6	19.752	18.871	372.749	90
95	103.034	.00971	.00049	.0505	2 040.7	19.806	19.069	377.677	95
100	131.500	.00760	.00038	.0504	2 610.0	19.848	19.234	381.749	100

	Single Payment		Uniform Payment Series				Arithmetic Gradient		
	Compound Amount Factor	Present Worth Factor	Sinking Fund Factor	Capital Recovery Factor	Compound Amount Factor	Present Worth Factor	Gradient Uniform Series	Gradient Present Worth	
n	Find F Given P F/P	Find P Given F P/F	Find A Given F A/F	Find A Given P A/P	Find F Given A F/A	Find P Given A P/A	Find A Given G A/G	Find P Given G P/G	n
1	1.060	.9434	1.0000	1.0600	1.000	0.943	0	0	1
2	1.124	.8900	.4854	.5454	2.060	1.833	0.485	0.890	2
3	1.191	.8396	.3141	.3741	3.184	2.673	0.961	2.569	3
4	1.262	.7921	.2286	.2886	4.375	3.465	1.427	4.945	4
5	1.338	.7473	.1774	.2374	5.637	4.212	1.884	7.934	5
6	1.419	.7050	.1434	.2034	6.975	4.917	2.330	11.459	6
7	1.504	.6651	.1191	.1791	8.394	5.582	2.768	15.450	7
8	1.594	.6274	.1010	.1610	9.897	6.210	3.195	19.841	8
9	1.689	.5919	.0870	.1470	11.491	6.802	3.613	24.577	9
10	1.791	.5584	.0759	.1359	13.181	7.360	4.022	29.602	10
11	1.898	.5268	.0668	.1268	14.972	7.887	4.421	34.870	11
12	2.012	.4970	.0593	.1193	16.870	8.384	4.811	40.337	12
13	2.133	.4688	.0530	.1130	18.882	8.853	5.192	45.963	13
14	2.261	.4423	.0476	.1076	21.015	9.295	5.564	51.713	14
15	2.397	.4173	.0430	.1030	23.276	9.712	5.926	57.554	15
16	2.540	.3936	.0390	.0990	25.672	10.106	6.279	63.459	16
17	2.693	.3714	.0354	.0954	28.213	10.477	6.624	69.401	17
18	2.854	.3503	.0324	.0924	30.906	10.828	6.960	75.357	18
19	3.026	.3305	.0296	.0896	33.760	11.158	7.287	81.306	19
20	3.207	.3118	.0272	.0872	36.786	11.470	7.605	87.230	20
21	3.400	.2942	.0250	.0850	39.993	11.764	7.915	93.113	21
22	3.604	.2775	.0230	.0830	43.392	12.042	8.217	98.941	22
23	3.820	.2618	.0213	.0813	46.996	12.303	8.510	104.700	23
24	4.049	.2470	.0197	.0797	50.815	12.550	8.795	110.381	24
25	4.292	.2330	.0182	.0782	54.864	12.783	9.072	115.973	25
26	4.549	.2198	.0169	.0769	59.156	13.003	9.341	121.468	26
27	4.822	.2074	.0157	.0757	63.706	13.211	9.603	126.860	27
28	5.112	.1956	.0146	.0746	68.528	13.406	9.857	132.142	28
29	5.418	.1846	.0136	.0736	73.640	13.591	10.103	137.309	29
30	5.743	.1741	.0126	.0726	79.058	13.765	10.342	142.359	30
31	6.088	.1643	.0118	.0718	84.801	13.929	10.574	147.286	31
32	6.453	.1550	.0110	.0710	90.890	14.084	10.799	152.090	32
33	6.841	.1462	.0103	.0703	97.343	14.230	11.017	156.768	33
34	7.251	.1379	.00960	.0696	104.184	14.368	11.228	161.319	34
35	7.686	.1301	.00897	.0690	111.435	14.498	11.432	165.743	35
40	10.286	.0972	.00646	.0665	154.762	15.046	12.359	185.957	40
45	13.765	.0727	.00470	.0647	212.743	15.456	13.141	203.109	45
50	18.420	.0543	.00344	.0634	290.335	15.762	13.796	217.457	50
55	24.650	.0406	.00254	.0625	394.171	15.991	14.341	229.322	55
60	32.988	.0303	.00188	.0619	533.126	16.161	14.791	239.043	60
65	44.145	.0227	.00139	.0614	719.080	16.289	15.160	246.945	65
70	59.076	.0169	.00103	.0610	967.928	16.385	15.461	253.327	70
75	79.057	.0126	.00077	.0608	1 300.9	16.456	15.706	258.453	75
80	105.796	.00945	.00057	.0606	1 746.6	16.509	15.903	262.549	80
85	141.578	.00706	.00043	.0604	2 343.0	16.549	16.062	265.810	85
90	189.464	.00528	.00032	.0603	3 141.1	16.579	16.189	268.395	90
95	253.545	.00394	.00024	.0602	4 209.1	16.601	16.290	270.437	95
100	339.300	.00295	.00018	.0602	5 638.3	16.618	16.371	272.047	100

	Single Payment		Uniform Payment Series				Arithmetic Gradient		
	Compound Amount Factor	Present Worth Factor	Sinking Fund Factor	Capital Recovery Factor	Compound Amount Factor	Present Worth Factor	Gradient Uniform Series	Gradient Present Worth	
	Find *F* Given *P*	Find *P* Given *F*	Find *A* Given *F*	Find *A* Given *P*	Find *F* Given *A*	Find *P* Given *A*	Find *A* Given *G*	Find *P* Given *G*	
n	*F/P*	*P/F*	*A/F*	*A/P*	*F/A*	*P/A*	*A/G*	*P/G*	*n*
1	1.070	.9346	1.0000	1.0700	1.000	0.935	0	0	1
2	1.145	.8734	.4831	.5531	2.070	1.808	0.483	0.873	2
3	1.225	.8163	.3111	.3811	3.215	2.624	0.955	2.506	3
4	1.311	.7629	.2252	.2952	4.440	3.387	1.416	4.795	4
5	1.403	.7130	.1739	.2439	5.751	4.100	1.865	7.647	5
6	1.501	.6663	.1398	.2098	7.153	4.767	2.303	10.978	6
7	1.606	.6227	.1156	.1856	8.654	5.389	2.730	14.715	7
8	1.718	.5820	.0975	.1675	10.260	5.971	3.147	18.789	8
9	1.838	.5439	.0835	.1535	11.978	6.515	3.552	23.140	9
10	1.967	.5083	.0724	.1424	13.816	7.024	3.946	27.716	10
11	2.105	.4751	.0634	.1334	15.784	7.499	4.330	32.467	11
12	2.252	.4440	.0559	.1259	17.888	7.943	4.703	37.351	12
13	2.410	.4150	.0497	.1197	20.141	8.358	5.065	42.330	13
14	2.579	.3878	.0443	.1143	22.551	8.745	5.417	47.372	14
15	2.759	.3624	.0398	.1098	25.129	9.108	5.758	52.446	15
16	2.952	.3387	.0359	.1059	27.888	9.447	6.090	57.527	16
17	3.159	.3166	.0324	.1024	30.840	9.763	6.411	62.592	17
18	3.380	.2959	.0294	.0994	33.999	10.059	6.722	67.622	18
19	3.617	.2765	.0268	.0968	37.379	10.336	7.024	72.599	19
20	3.870	.2584	.0244	.0944	40.996	10.594	7.316	77.509	20
21	4.141	.2415	.0223	.0923	44.865	10.836	7.599	82.339	21
22	4.430	.2257	.0204	.0904	49.006	11.061	7.872	87.079	22
23	4.741	.2109	.0187	.0887	53.436	11.272	8.137	91.720	23
24	5.072	.1971	.0172	.0872	58.177	11.469	8.392	96.255	24
25	5.427	.1842	.0158	.0858	63.249	11.654	8.639	100.677	25
26	5.807	.1722	.0146	.0846	68.677	11.826	8.877	104.981	26
27	6.214	.1609	.0134	.0834	74.484	11.987	9.107	109.166	27
28	6.649	.1504	.0124	.0824	80.698	12.137	9.329	113.227	28
29	7.114	.1406	.0114	.0814	87.347	12.278	9.543	117.162	29
30	7.612	.1314	.0106	.0806	94.461	12.409	9.749	120.972	30
31	8.145	.1228	.00980	.0798	102.073	12.532	9.947	124.655	31
32	8.715	.1147	.00907	.0791	110.218	12.647	10.138	128.212	32
33	9.325	.1072	.00841	.0784	118.934	12.754	10.322	131.644	33
34	9.978	.1002	.00780	.0778	128.259	12.854	10.499	134.951	34
35	10.677	.0937	.00723	.0772	138.237	12.948	10.669	138.135	35
40	14.974	.0668	.00501	.0750	199.636	13.332	11.423	152.293	40
45	21.002	.0476	.00350	.0735	285.750	13.606	12.036	163.756	45
50	29.457	.0339	.00246	.0725	406.530	13.801	12.529	172.905	50
55	41.315	.0242	.00174	.0717	575.930	13.940	12.921	180.124	55
60	57.947	.0173	.00123	.0712	813.523	14.039	13.232	185.768	60
65	81.273	.0123	.00087	.0709	1 146.8	14.110	13.476	190.145	65
70	113.990	.00877	.00062	.0706	1 614.1	14.160	13.666	193.519	70
75	159.877	.00625	.00044	.0704	2 269.7	14.196	13.814	196.104	75
80	224.235	.00446	.00031	.0703	3 189.1	14.222	13.927	198.075	80
85	314.502	.00318	.00022	.0702	4 478.6	14.240	14.015	199.572	85
90	441.105	.00227	.00016	.0702	6 287.2	14.253	14.081	200.704	90
95	618.673	.00162	.00011	.0701	8 823.9	14.263	14.132	201.558	95
100	867.720	.00115	.00008	.0701	12 381.7	14.269	14.170	202.200	100

	Single Payment		Uniform Payment Series				Arithmetic Gradient		
n	Compound Amount Factor Find F Given P F/P	Present Worth Factor Find P Given F P/F	Sinking Fund Factor Find A Given F A/F	Capital Recovery Factor Find A Given P A/P	Compound Amount Factor Find F Given A F/A	Present Worth Factor Find P Given A P/A	Gradient Uniform Series Find A Given G A/G	Gradient Present Worth Find P Given G P/G	n
1	1.080	.9259	1.0000	1.0800	1.000	0.926	0	0	1
2	1.166	.8573	.4808	.5608	2.080	1.783	0.481	0.857	2
3	1.260	.7938	.3080	.3880	3.246	2.577	0.949	2.445	3
4	1.360	.7350	.2219	.3019	4.506	3.312	1.404	4.650	4
5	1.469	.6806	.1705	.2505	5.867	3.993	1.846	7.372	5
6	1.587	.6302	.1363	.2163	7.336	4.623	2.276	10.523	6
7	1.714	.5835	.1121	.1921	8.923	5.206	2.694	14.024	7
8	1.851	.5403	.0940	.1740	10.637	5.747	3.099	17.806	8
9	1.999	.5002	.0801	.1601	12.488	6.247	3.491	21.808	9
10	2.159	.4632	.0690	.1490	14.487	6.710	3.871	25.977	10
11	2.332	.4289	.0601	.1401	16.645	7.139	4.240	30.266	11
12	2.518	.3971	.0527	.1327	18.977	7.536	4.596	34.634	12
13	2.720	.3677	.0465	.1265	21.495	7.904	4.940	39.046	13
14	2.937	.3405	.0413	.1213	24.215	8.244	5.273	43.472	14
15	3.172	.3152	.0368	.1168	27.152	8.559	5.594	47.886	15
16	3.426	.2919	.0330	.1130	30.324	8.851	5.905	52.264	16
17	3.700	.2703	.0296	.1096	33.750	9.122	6.204	56.588	17
18	3.996	.2502	.0267	.1067	37.450	9.372	6.492	60.843	18
19	4.316	.2317	.0241	.1041	41.446	9.604	6.770	65.013	19
20	4.661	.2145	.0219	.1019	45.762	9.818	7.037	69.090	20
21	5.034	.1987	.0198	.0998	50.423	10.017	7.294	73.063	21
22	5.437	.1839	.0180	.0980	55.457	10.201	7.541	76.926	22
23	5.871	.1703	.0164	.0964	60.893	10.371	7.779	80.673	23
24	6.341	.1577	.0150	.0950	66.765	10.529	8.007	84.300	24
25	6.848	.1460	.0137	.0937	73.106	10.675	8.225	87.804	25
26	7.396	.1352	.0125	.0925	79.954	10.810	8.435	91.184	26
27	7.988	.1252	.0114	.0914	87.351	10.935	8.636	94.439	27
28	8.627	.1159	.0105	.0905	95.339	11.051	8.829	97.569	28
29	9.317	.1073	.00962	.0896	103.966	11.158	9.013	100.574	29
30	10.063	.0994	.00883	.0888	113.283	11.258	9.190	103.456	30
31	10.868	.0920	.00811	.0881	123.346	11.350	9.358	106.216	31
32	11.737	.0852	.00745	.0875	134.214	11.435	9.520	108.858	32
33	12.676	.0789	.00685	.0869	145.951	11.514	9.674	111.382	33
34	13.690	.0730	.00630	.0863	158.627	11.587	9.821	113.792	34
35	14.785	.0676	.00580	.0858	172.317	11.655	9.961	116.092	35
40	21.725	.0460	.00386	.0839	259.057	11.925	10.570	126.042	40
45	31.920	.0313	.00259	.0826	386.506	12.108	11.045	133.733	45
50	46.902	.0213	.00174	.0817	573.771	12.233	11.411	139.593	50
55	68.914	.0145	.00118	.0812	848.925	12.319	11.690	144.006	55
60	101.257	.00988	.00080	.0808	1 253.2	12.377	11.902	147.300	60
65	148.780	.00672	.00054	.0805	1 847.3	12.416	12.060	149.739	65
70	218.607	.00457	.00037	.0804	2 720.1	12.443	12.178	151.533	70
75	321.205	.00311	.00025	.0802	4 002.6	12.461	12.266	152.845	75
80	471.956	.00212	.00017	.0802	5 887.0	12.474	12.330	153.800	80
85	693.458	.00144	.00012	.0801	8 655.7	12.482	12.377	154.492	85
90	1 018.9	.00098	.00008	.0801	12 724.0	12.488	12.412	154.993	90
95	1 497.1	.00067	.00005	.0801	18 701.6	12.492	12.437	155.352	95
100	2 199.8	.00045	.00004	.0800	27 484.6	12.494	12.455	155.611	100

	Single Payment		Uniform Payment Series				Arithmetic Gradient		
n	Compound Amount Factor Find F Given P F/P	Present Worth Factor Find P Given F P/F	Sinking Fund Factor Find A Given F A/F	Capital Recovery Factor Find A Given P A/P	Compound Amount Factor Find F Given A F/A	Present Worth Factor Find P Given A P/A	Gradient Uniform Series Find A Given G A/G	Gradient Present Worth Find P Given G P/G	n
1	1.090	.9174	1.0000	1.0900	1.000	0.917	0	0	1
2	1.188	.8417	.4785	.5685	2.090	1.759	0.478	0.842	2
3	1.295	.7722	.3051	.3951	3.278	2.531	0.943	2.386	3
4	1.412	.7084	.2187	.3087	4.573	3.240	1.393	4.511	4
5	1.539	.6499	.1671	.2571	5.985	3.890	1.828	7.111	5
6	1.677	.5963	.1329	.2229	7.523	4.486	2.250	10.092	6
7	1.828	.5470	.1087	.1987	9.200	5.033	2.657	13.375	7
8	1.993	.5019	.0907	.1807	11.028	5.535	3.051	16.888	8
9	2.172	.4604	.0768	.1668	13.021	5.995	3.431	20.571	9
10	2.367	.4224	.0658	.1558	15.193	6.418	3.798	24.373	10
11	2.580	.3875	.0569	.1469	17.560	6.805	4.151	28.248	11
12	2.813	.3555	.0497	.1397	20.141	7.161	4.491	32.159	12
13	3.066	.3262	.0436	.1336	22.953	7.487	4.818	36.073	13
14	3.342	.2992	.0384	.1284	26.019	7.786	5.133	39.963	14
15	3.642	.2745	.0341	.1241	29.361	8.061	5.435	43.807	15
16	3.970	.2519	.0303	.1203	33.003	8.313	5.724	47.585	16
17	4.328	.2311	.0270	.1170	36.974	8.544	6.002	51.282	17
18	4.717	.2120	.0242	.1142	41.301	8.756	6.269	54.886	18
19	5.142	.1945	.0217	.1117	46.019	8.950	6.524	58.387	19
20	5.604	.1784	.0195	.1095	51.160	9.129	6.767	61.777	20
21	6.109	.1637	.0176	.1076	56.765	9.292	7.001	65.051	21
22	6.659	.1502	.0159	.1059	62.873	9.442	7.223	68.205	22
23	7.258	.1378	.0144	.1044	69.532	9.580	7.436	71.236	23
24	7.911	.1264	.0130	.1030	76.790	9.707	7.638	74.143	24
25	8.623	.1160	.0118	.1018	84.701	9.823	7.832	76.927	25
26	9.399	.1064	.0107	.1007	93.324	9.929	8.016	79.586	26
27	10.245	.0976	.00973	.0997	102.723	10.027	8.191	82.124	27
28	11.167	.0895	.00885	.0989	112.968	10.116	8.357	84.542	28
29	12.172	.0822	.00806	.0981	124.136	10.198	8.515	86.842	29
30	13.268	.0754	.00734	.0973	136.308	10.274	8.666	89.028	30
31	14.462	.0691	.00669	.0967	149.575	10.343	8.808	91.102	31
32	15.763	.0634	.00610	.0961	164.037	10.406	8.944	93.069	32
33	17.182	.0582	.00556	.0956	179.801	10.464	9.072	94.931	33
34	18.728	.0534	.00508	.0951	196.983	10.518	9.193	96.693	34
35	20.414	.0490	.00464	.0946	215.711	10.567	9.308	98.359	35
40	31.409	.0318	.00296	.0930	337.883	10.757	9.796	105.376	40
45	48.327	.0207	.00190	.0919	525.860	10.881	10.160	110.556	45
50	74.358	.0134	.00123	.0912	815.085	10.962	10.430	114.325	50
55	114.409	.00874	.00079	.0908	1 260.1	11.014	10.626	117.036	55
60	176.032	.00568	.00051	.0905	1 944.8	11.048	10.768	118.968	60
65	270.847	.00369	.00033	.0903	2 998.3	11.070	10.870	120.334	65
70	416.731	.00240	.00022	.0902	4 619.2	11.084	10.943	121.294	70
75	641.193	.00156	.00014	.0901	7 113.3	11.094	10.994	121.965	75
80	986.555	.00101	.00009	.0901	10 950.6	11.100	11.030	122.431	80
85	1 517.9	.00066	.00006	.0901	16 854.9	11.104	11.055	122.753	85
90	2 335.5	.00043	.00004	.0900	25 939.3	11.106	11.073	122.976	90
95	3 593.5	.00028	.00003	.0900	39 916.8	11.108	11.085	123.129	95
100	5 529.1	.00018	.00002	.0900	61 422.9	11.109	11.093	123.233	100

	Single Payment		Uniform Payment Series				Arithmetic Gradient		
n	Compound Amount Factor Find *F* Given *P* *F/P*	Present Worth Factor Find *P* Given *F* *P/F*	Sinking Fund Factor Find *A* Given *F* *A/F*	Capital Recovery Factor Find *A* Given *P* *A/P*	Compound Amount Factor Find *F* Given *A* *F/A*	Present Worth Factor Find *P* Given *A* *P/A*	Gradient Uniform Series Find *A* Given *G* *A/G*	Gradient Present Worth Find *P* Given *G* *P/G*	*n*
1	1.100	.9091	1.0000	1.1000	1.000	0.909	0	0	1
2	1.210	.8264	.4762	.5762	2.100	1.736	0.476	0.826	2
3	1.331	.7513	.3021	.4021	3.310	2.487	0.937	2.329	3
4	1.464	.6830	.2155	.3155	4.641	3.170	1.381	4.378	4
5	1.611	.6209	.1638	.2638	6.105	3.791	1.810	6.862	5
6	1.772	.5645	.1296	.2296	7.716	4.355	2.224	9.684	6
7	1.949	.5132	.1054	.2054	9.487	4.868	2.622	12.763	7
8	2.144	.4665	.0874	.1874	11.436	5.335	3.004	16.029	8
9	2.358	.4241	.0736	.1736	13.579	5.759	3.372	19.421	9
10	2.594	.3855	.0627	.1627	15.937	6.145	3.725	22.891	10
11	2.853	.3505	.0540	.1540	18.531	6.495	4.064	26.396	11
12	3.138	.3186	.0468	.1468	21.384	6.814	4.388	29.901	12
13	3.452	.2897	.0408	.1408	24.523	7.103	4.699	33.377	13
14	3.797	.2633	.0357	.1357	27.975	7.367	4.996	36.801	14
15	4.177	.2394	.0315	.1315	31.772	7.606	5.279	40.152	15
16	4.595	.2176	.0278	.1278	35.950	7.824	5.549	43.416	16
17	5.054	.1978	.0247	.1247	40.545	8.022	5.807	46.582	17
18	5.560	.1799	.0219	.1219	45.599	8.201	6.053	49.640	18
19	6.116	.1635	.0195	.1195	51.159	8.365	6.286	52.583	19
20	6.728	.1486	.0175	.1175	57.275	8.514	6.508	55.407	20
21	7.400	.1351	.0156	.1156	64.003	8.649	6.719	58.110	21
22	8.140	.1228	.0140	.1140	71.403	8.772	6.919	60.689	22
23	8.954	.1117	.0126	.1126	79.543	8.883	7.108	63.146	23
24	9.850	.1015	.0113	.1113	88.497	8.985	7.288	65.481	24
25	10.835	.0923	.0102	.1102	98.347	9.077	7.458	67.696	25
26	11.918	.0839	.00916	.1092	109.182	9.161	7.619	69.794	26
27	13.110	.0763	.00826	.1083	121.100	9.237	7.770	71.777	27
28	14.421	.0693	.00745	.1075	134.210	9.307	7.914	73.650	28
29	15.863	.0630	.00673	.1067	148.631	9.370	8.049	75.415	29
30	17.449	.0573	.00608	.1061	164.494	9.427	8.176	77.077	30
31	19.194	.0521	.00550	.1055	181.944	9.479	8.296	78.640	31
32	21.114	.0474	.00497	.1050	201.138	9.526	8.409	80.108	32
33	23.225	.0431	.00450	.1045	222.252	9.569	8.515	81.486	33
34	25.548	.0391	.00407	.1041	245.477	9.609	8.615	82.777	34
35	28.102	.0356	.00369	.1037	271.025	9.644	8.709	83.987	35
40	45.259	.0221	.00226	.1023	442.593	9.779	9.096	88.953	40
45	72.891	.0137	.00139	.1014	718.905	9.863	9.374	92.454	45
50	117.391	.00852	.00086	.1009	1 163.9	9.915	9.570	94.889	50
55	189.059	.00529	.00053	.1005	1 880.6	9.947	9.708	96.562	55
60	304.482	.00328	.00033	.1003	3034.8	9.967	9.802	97.701	60
65	490.371	.00204	.00020	.1002	4893.7	9.980	9.867	98.471	65
70	789.748	.00127	.00013	.1001	7887.5	9.987	9.911	98.987	70
75	1 271.9	.00079	.00008	.1001	12 709.0	9.992	9.941	99.332	75
80	2 048.4	.00049	.00005	.1000	20 474.0	9.995	9.961	99.561	80
85	3 299.0	.00030	.00003	.1000	32 979.7	9.997	9.974	99.712	85
90	5 313.0	.00019	.00002	.1000	53 120.3	9.998	9.983	99.812	90
95	8 556.7	.00012	.00001	.1000	85 556.9	9.999	9.989	99.877	95
100	13 780.6	.00007	.00001	.1000	137 796.3	9.999	9.993	99.920	100

n	Single Payment: Compound Amount Factor, Find F Given P, F/P	Single Payment: Present Worth Factor, Find P Given F, P/F	Uniform Payment Series: Sinking Fund Factor, Find A Given F, A/F	Uniform Payment Series: Capital Recovery Factor, Find A Given P, A/P	Uniform Payment Series: Compound Amount Factor, Find F Given A, F/A	Uniform Payment Series: Present Worth Factor, Find P Given A, P/A	Arithmetic Gradient: Gradient Uniform Series, Find A Given G, A/G	Arithmetic Gradient: Gradient Present Worth, Find P Given G, P/G	n
1	1.120	.8929	1.0000	1.1200	1.000	0.893	0	0	1
2	1.254	.7972	.4717	.5917	2.120	1.690	0.472	0.797	2
3	1.405	.7118	.2963	.4163	3.374	2.402	0.925	2.221	3
4	1.574	.6355	.2092	.3292	4.779	3.037	1.359	4.127	4
5	1.762	.5674	.1574	.2774	6.353	3.605	1.775	6.397	5
6	1.974	.5066	.1232	.2432	8.115	4.111	2.172	8.930	6
7	2.211	.4523	.0991	.2191	10.089	4.564	2.551	11.644	7
8	2.476	.4039	.0813	.2013	12.300	4.968	2.913	14.471	8
9	2.773	.3606	.0677	.1877	14.776	5.328	3.257	17.356	9
10	3.106	.3220	.0570	.1770	17.549	5.650	3.585	20.254	10
11	3.479	.2875	.0484	.1684	20.655	5.938	3.895	23.129	11
12	3.896	.2567	.0414	.1614	24.133	6.194	4.190	25.952	12
13	4.363	.2292	.0357	.1557	28.029	6.424	4.468	28.702	13
14	4.887	.2046	.0309	.1509	32.393	6.628	4.732	31.362	14
15	5.474	.1827	.0268	.1468	37.280	6.811	4.980	33.920	15
16	6.130	.1631	.0234	.1434	42.753	6.974	5.215	36.367	16
17	6.866	.1456	.0205	.1405	48.884	7.120	5.435	38.697	17
18	7.690	.1300	.0179	.1379	55.750	7.250	5.643	40.908	18
19	8.613	.1161	.0158	.1358	63.440	7.366	5.838	42.998	19
20	9.646	.1037	.0139	.1339	72.052	7.469	6.020	44.968	20
21	10.804	.0926	.0122	.1322	81.699	7.562	6.191	46.819	21
22	12.100	.0826	.0108	.1308	92.503	7.645	6.351	48.554	22
23	13.552	.0738	.00956	.1296	104.603	7.718	6.501	50.178	23
24	15.179	.0659	.00846	.1285	118.155	7.784	6.641	51.693	24
25	17.000	.0588	.00750	.1275	133.334	7.843	6.771	53.105	25
26	19.040	.0525	.00665	.1267	150.334	7.896	6.892	54.418	26
27	21.325	.0469	.00590	.1259	169.374	7.943	7.005	55.637	27
28	23.884	.0419	.00524	.1252	190.699	7.984	7.110	56.767	28
29	26.750	.0374	.00466	.1247	214.583	8.022	7.207	57.814	29
30	29.960	.0334	.00414	.1241	241.333	8.055	7.297	58.782	30
31	33.555	.0298	.00369	.1237	271.293	8.085	7.381	59.676	31
32	37.582	.0266	.00328	.1233	304.848	8.112	7.459	60.501	32
33	42.092	.0238	.00292	.1229	342.429	8.135	7.530	61.261	33
34	47.143	.0212	.00260	.1226	384.521	8.157	7.596	61.961	34
35	52.800	.0189	.00232	.1223	431.663	8.176	7.658	62.605	35
40	93.051	.0107	.00130	.1213	767.091	8.244	7.899	65.116	40
45	163.988	.00610	.00074	.1207	1 358.2	8.283	8.057	66.734	45
50	289.002	.00346	.00042	.1204	2 400.0	8.304	8.160	67.762	50
55	509.321	.00196	.00024	.1202	4 236.0	8.317	8.225	68.408	55
60	897.597	.00111	.00013	.1201	7 471.6	8.324	8.266	68.810	60
65	1 581.9	.00063	.00008	.1201	13 173.9	8.328	8.292	69.058	65
70	2 787.8	.00036	.00004	.1200	23 223.3	8.330	8.308	69.210	70
75	4 913.1	.00020	.00002	.1200	40 933.8	8.332	8.318	69.303	75
80	8 658.5	.00012	.00001	.1200	72 145.7	8.332	8.324	69.359	80
85	15 259.2	.00007	.00001	.1200	127 151.7	8.333	8.328	69.393	85
90	26 891.9	.00004		.1200	224 091.1	8.333	8.330	69.414	90
95	47 392.8	.00002		.1200	394 931.4	8.333	8.331	69.426	95
100	83 522.3	.00001		.1200	696 010.5	8.333	8.332	69.434	100

	Single Payment		Uniform Payment Series				Arithmetic Gradient		
	Compound Amount Factor	Present Worth Factor	Sinking Fund Factor	Capital Recovery Factor	Compound Amount Factor	Present Worth Factor	Gradient Uniform Series	Gradient Present Worth	
	Find *F* Given *P*	Find *P* Given *F*	Find *A* Given *F*	Find *A* Given *P*	Find *F* Given *A*	Find *P* Given *A*	Find *A* Given *G*	Find *P* Given *G*	
n	*F/P*	*P/F*	*A/F*	*A/P*	*F/A*	*P/A*	*A/G*	*P/G*	*n*
1	1.150	.8696	1.0000	1.1500	1.000	0.870	0	0	1
2	1.322	.7561	.4651	.6151	2.150	1.626	0.465	0.756	2
3	1.521	.6575	.2880	.4380	3.472	2.283	0.907	2.071	3
4	1.749	.5718	.2003	.3503	4.993	2.855	1.326	3.786	4
5	2.011	.4972	.1483	.2983	6.742	3.352	1.723	5.775	5
6	2.313	.4323	.1142	.2642	8.754	3.784	2.097	7.937	6
7	2.660	.3759	.0904	.2404	11.067	4.160	2.450	10.192	7
8	3.059	.3269	.0729	.2229	13.727	4.487	2.781	12.481	8
9	3.518	.2843	.0596	.2096	16.786	4.772	3.092	14.755	9
10	4.046	.2472	.0493	.1993	20.304	5.019	3.383	16.979	10
11	4.652	.2149	.0411	.1911	24.349	5.234	3.655	19.129	11
12	5.350	.1869	.0345	.1845	29.002	5.421	3.908	21.185	12
13	6.153	.1625	.0291	.1791	34.352	5.583	4.144	23.135	13
14	7.076	.1413	.0247	.1747	40.505	5.724	4.362	24.972	14
15	8.137	.1229	.0210	.1710	47.580	5.847	4.565	26.693	15
16	9.358	.1069	.0179	.1679	55.717	5.954	4.752	28.296	16
17	10.761	.0929	.0154	.1654	65.075	6.047	4.925	29.783	17
18	12.375	.0808	.0132	.1632	75.836	6.128	5.084	31.156	18
19	14.232	.0703	.0113	.1613	88.212	6.198	5.231	32.421	19
20	16.367	.0611	.00976	.1598	102.444	6.259	5.365	33.582	20
21	18.822	.0531	.00842	.1584	118.810	6.312	5.488	34.645	21
22	21.645	.0462	.00727	.1573	137.632	6.359	5.601	35.615	22
23	24.891	.0402	.00628	.1563	159.276	6.399	5.704	36.499	23
24	28.625	.0349	.00543	.1554	184.168	6.434	5.798	37.302	24
25	32.919	.0304	.00470	.1547	212.793	6.464	5.883	38.031	25
26	37.857	.0264	.00407	.1541	245.712	6.491	5.961	38.692	26
27	43.535	.0230	.00353	.1535	283.569	6.514	6.032	39.289	27
28	50.066	.0200	.00306	.1531	327.104	6.534	6.096	39.828	28
29	57.575	.0174	.00265	.1527	377.170	6.551	6.154	40.315	29
30	66.212	.0151	.00230	.1523	434.745	6.566	6.207	40.753	30
31	76.144	.0131	.00200	.1520	500.957	6.579	6.254	41.147	31
32	87.565	.0114	.00173	.1517	577.100	6.591	6.297	41.501	32
33	100.700	.00993	.00150	.1515	664.666	6.600	6.336	41.818	33
34	115.805	.00864	.00131	.1513	765.365	6.609	6.371	42.103	34
35	133.176	.00751	.00113	.1511	881.170	6.617	6.402	42.359	35
40	267.864	.00373	.00056	.1506	1 779.1	6.642	6.517	43.283	40
45	538.769	.00186	.00028	.1503	3 585.1	6.654	6.583	43.805	45
50	1 083.7	.00092	.00014	.1501	7 217.7	6.661	6.620	44.096	50
55	2 179.6	.00046	.00007	.1501	14 524.1	6.664	6.641	44.256	55
60	4 384.0	.00023	.00003	.1500	29 220.0	6.665	6.653	44.343	60
65	8 817.8	.00011	.00002	.1500	58 778.6	6.666	6.659	44.390	65
70	17 735.7	.00006	.00001	.1500	118 231.5	6.666	6.663	44.416	70
75	35 672.9	.00003		.1500	237 812.5	6.666	6.665	44.429	75
80	71 750.9	.00001		.1500	478 332.6	6.667	6.666	44.436	80
85	144 316.7	.00001		.1500	962 104.4	6.667	6.666	44.440	85

18% Compound Interest Factors 18%

	Single Payment		Uniform Payment Series				Arithmetic Gradient		
	Compound Amount Factor	Present Worth Factor	Sinking Fund Factor	Capital Recovery Factor	Compound Amount Factor	Present Worth Factor	Gradient Uniform Series	Gradient Present Worth	
n	Find *F* Given *P* *F/P*	Find *P* Given *F* *P/F*	Find *A* Given *F* *A/F*	Find *A* Given *P* *A/P*	Find *F* Given *A* *F/A*	Find *P* Given *A* *P/A*	Find *A* Given *G* *A/G*	Find *P* Given *G* *P/G*	*n*
1	1.180	.8475	1.0000	1.1800	1.000	0.847	0	0	1
2	1.392	.7182	.4587	.6387	2.180	1.566	0.459	0.718	2
3	1.643	.6086	.2799	.4599	3.572	2.174	0.890	1.935	3
4	1.939	.5158	.1917	.3717	5.215	2.690	1.295	3.483	4
5	2.288	.4371	.1398	.3198	7.154	3.127	1.673	5.231	5
6	2.700	.3704	.1059	.2859	9.442	3.498	2.025	7.083	6
7	3.185	.3139	.0824	.2624	12.142	3.812	2.353	8.967	7
8	3.759	.2660	.0652	.2452	15.327	4.078	2.656	10.829	8
9	4.435	.2255	.0524	.2324	19.086	4.303	2.936	12.633	9
10	5.234	.1911	.0425	.2225	23.521	4.494	3.194	14.352	10
11	6.176	.1619	.0348	.2148	28.755	4.656	3.430	15.972	11
12	7.288	.1372	.0286	.2086	34.931	4.793	3.647	17.481	12
13	8.599	.1163	.0237	.2037	42.219	4.910	3.845	18.877	13
14	10.147	.0985	.0197	.1997	50.818	5.008	4.025	20.158	14
15	11.974	.0835	.0164	.1964	60.965	5.092	4.189	21.327	15
16	14.129	.0708	.0137	.1937	72.939	5.162	4.337	22.389	16
17	16.672	.0600	.0115	.1915	87.068	5.222	4.471	23.348	17
18	19.673	.0508	.00964	.1896	103.740	5.273	4.592	24.212	18
19	23.214	.0431	.00810	.1881	123.413	5.316	4.700	24.988	19
20	27.393	.0365	.00682	.1868	146.628	5.353	4.798	25.681	20
21	32.324	.0309	.00575	.1857	174.021	5.384	4.885	26.300	21
22	38.142	.0262	.00485	.1848	206.345	5.410	4.963	26.851	22
23	45.008	.0222	.00409	.1841	244.487	5.432	5.033	27.339	23
24	53.109	.0188	.00345	.1835	289.494	5.451	5.095	27.772	24
25	62.669	.0160	.00292	.1829	342.603	5.467	5.150	28.155	25
26	73.949	.0135	.00247	.1825	405.272	5.480	5.199	28.494	26
27	87.260	.0115	.00209	.1821	479.221	5.492	5.243	28.791	27
28	102.966	.00971	.00177	.1818	566.480	5.502	5.281	29.054	28
29	121.500	.00823	.00149	.1815	669.447	5.510	5.315	29.284	29
30	143.370	.00697	.00126	.1813	790.947	5.517	5.345	29.486	30
31	169.177	.00591	.00107	.1811	934.317	5.523	5.371	29.664	31
32	199.629	.00501	.00091	.1809	1 103.5	5.528	5.394	29.819	32
33	235.562	.00425	.00077	.1808	1 303.1	5.532	5.415	29.955	33
34	277.963	.00360	.00065	.1806	1 538.7	5.536	5.433	30.074	34
35	327.997	.00305	.00055	.1806	1 816.6	5.539	5.449	30.177	35
40	750.377	.00133	.00024	.1802	4 163.2	5.548	5.502	30.527	40
45	1 716.7	.00058	.00010	.1801	9 531.6	5.552	5.529	30.701	45
50	3 927.3	.00025	.00005	.1800	21 813.0	5.554	5.543	30.786	50
55	8 984.8	.00011	.00002	.1800	49 910.1	5.555	5.549	30.827	55
60	20 555.1	.00005	.00001	.1800	114 189.4	5.555	5.553	30.846	60
65	47 025.1	.00002		.1800	261 244.7	5.555	5.554	30.856	65
70	107 581.9	.00001		.1800	597 671.1	5.556	5.555	30.860	70

	Single Payment		Uniform Payment Series				Arithmetic Gradient		
	Compound Amount Factor	Present Worth Factor	Sinking Fund Factor	Capital Recovery Factor	Compound Amount Factor	Present Worth Factor	Gradient Uniform Series	Gradient Present Worth	
	Find *F* Given *P*	Find *P* Given *F*	Find *A* Given *F*	Find *A* Given *P*	Find *F* Given *A*	Find *P* Given *A*	Find *A* Given *G*	Find *P* Given *G*	
n	*F/P*	*P/F*	*A/F*	*A/P*	*F/A*	*P/A*	*A/G*	*P/G*	*n*
1	1.200	.8333	1.0000	1.2000	1.000	0.833	0	0	1
2	1.440	.6944	.4545	.6545	2.200	1.528	0.455	0.694	2
3	1.728	.5787	.2747	.4747	3.640	2.106	0.879	1.852	3
4	2.074	.4823	.1863	.3863	5.368	2.589	1.274	3.299	4
5	2.488	.4019	.1344	.3344	7.442	2.991	1.641	4.906	5
6	2.986	.3349	.1007	.3007	9.930	3.326	1.979	6.581	6
7	3.583	.2791	.0774	.2774	12.916	3.605	2.290	8.255	7
8	4.300	.2326	.0606	.2606	16.499	3.837	2.576	9.883	8
9	5.160	.1938	.0481	.2481	20.799	4.031	2.836	11.434	9
10	6.192	.1615	.0385	.2385	25.959	4.192	3.074	12.887	10
11	7.430	.1346	.0311	.2311	32.150	4.327	3.289	14.233	11
12	8.916	.1122	.0253	.2253	39.581	4.439	3.484	15.467	12
13	10.699	.0935	.0206	.2206	48.497	4.533	3.660	16.588	13
14	12.839	.0779	.0169	.2169	59.196	4.611	3.817	17.601	14
15	15.407	.0649	.0139	.2139	72.035	4.675	3.959	18.509	15
16	18.488	.0541	.0114	.2114	87.442	4.730	4.085	19.321	16
17	22.186	.0451	.00944	.2094	105.931	4.775	4.198	20.042	17
18	26.623	.0376	.00781	.2078	128.117	4.812	4.298	20.680	18
19	31.948	.0313	.00646	.2065	154.740	4.843	4.386	21.244	19
20	38.338	.0261	.00536	.2054	186.688	4.870	4.464	21.739	20
21	46.005	.0217	.00444	.2044	225.026	4.891	4.533	22.174	21
22	55.206	.0181	.00369	.2037	271.031	4.909	4.594	22.555	22
23	66.247	.0151	.00307	.2031	326.237	4.925	4.647	22.887	23
24	79.497	.0126	.00255	.2025	392.484	4.937	4.694	23.176	24
25	95.396	.0105	.00212	.2021	471.981	4.948	4.735	23.428	25
26	114.475	.00874	.00176	.2018	567.377	4.956	4.771	23.646	26
27	137.371	.00728	.00147	.2015	681.853	4.964	4.802	23.835	27
28	164.845	.00607	.00122	.2012	819.223	4.970	4.829	23.999	28
29	197.814	.00506	.00102	.2010	984.068	4.975	4.853	24.141	29
30	237.376	.00421	.00085	.2008	1 181.9	4.979	4.873	24.263	30
31	284.852	.00351	.00070	.2007	1 419.3	4.982	4.891	24.368	31
32	341.822	.00293	.00059	.2006	1 704.1	4.985	4.906	24.459	32
33	410.186	.00244	.00049	.2005	2 045.9	4.988	4.919	24.537	33
34	492.224	.00203	.00041	.2004	2 456.1	4.990	4.931	24.604	34
35	590.668	.00169	.00034	.2003	2 948.3	4.992	4.941	24.661	35
40	1 469.8	.00068	.00014	.2001	7 343.9	4.997	4.973	24.847	40
45	3 657.3	.00027	.00005	.2001	18 281.3	4.999	4.988	24.932	45
50	9 100.4	.00011	.00002	.2000	45 497.2	4.999	4.995	24.970	50
55	22 644.8	.00004	.00001	.2000	113 219.0	5.000	4.998	24.987	55
60	56 347.5	.00002		.2000	281 732.6	5.000	4.999	24.994	60

	Single Payment		Uniform Payment Series				Arithmetic Gradient		
n	Compound Amount Factor Find F Given P F/P	Present Worth Factor Find P Given F P/F	Sinking Fund Factor Find A Given F A/F	Capital Recovery Factor Find A Given P A/P	Compound Amount Factor Find F Given A F/A	Present Worth Factor Find P Given A P/A	Gradient Uniform Series Find A Given G A/G	Gradient Present Worth Find P Given G P/G	n
1	1.250	.8000	1.0000	1.2500	1.000	0.800	0	0	1
2	1.563	.6400	.4444	.6944	2.250	1.440	0.444	0.640	2
3	1.953	.5120	.2623	.5123	3.813	1.952	0.852	1.664	3
4	2.441	.4096	.1734	.4234	5.766	2.362	1.225	2.893	4
5	3.052	.3277	.1218	.3718	8.207	2.689	1.563	4.204	5
6	3.815	.2621	.0888	.3388	11.259	2.951	1.868	5.514	6
7	4.768	.2097	.0663	.3163	15.073	3.161	2.142	6.773	7
8	5.960	.1678	.0504	.3004	19.842	3.329	2.387	7.947	8
9	7.451	.1342	.0388	.2888	25.802	3.463	2.605	9.021	9
10	9.313	.1074	.0301	.2801	33.253	3.571	2.797	9.987	10
11	11.642	.0859	.0235	.2735	42.566	3.656	2.966	10.846	11
12	14.552	.0687	.0184	.2684	54.208	3.725	3.115	11.602	12
13	18.190	.0550	.0145	.2645	68.760	3.780	3.244	12.262	13
14	22.737	.0440	.0115	.2615	86.949	3.824	3.356	12.833	14
15	28.422	.0352	.00912	.2591	109.687	3.859	3.453	13.326	15
16	35.527	.0281	.00724	.2572	138.109	3.887	3.537	13.748	16
17	44.409	.0225	.00576	.2558	173.636	3.910	3.608	14.108	17
18	55.511	.0180	.00459	.2546	218.045	3.928	3.670	14.415	18
19	69.389	.0144	.00366	.2537	273.556	3.942	3.722	14.674	19
20	86.736	.0115	.00292	.2529	342.945	3.954	3.767	14.893	20
21	108.420	.00922	.00233	.2523	429.681	3.963	3.805	15.078	21
22	135.525	.00738	.00186	.2519	538.101	3.970	3.836	15.233	22
23	169.407	.00590	.00148	.2515	673.626	3.976	3.863	15.362	23
24	211.758	.00472	.00119	.2512	843.033	3.981	3.886	15.471	24
25	264.698	.00378	.00095	.2509	1 054.8	3.985	3.905	15.562	25
26	330.872	.00302	.00076	.2508	1 319.5	3.988	3.921	15.637	26
27	413.590	.00242	.00061	.2506	1 650.4	3.990	3.935	15.700	27
28	516.988	.00193	.00048	.2505	2 064.0	3.992	3.946	15.752	28
29	646.235	.00155	.00039	.2504	2 580.9	3.994	3.955	15.796	29
30	807.794	.00124	.00031	.2503	3 227.2	3.995	3.963	15.832	30
31	1 009.7	.00099	.00025	.2502	4 035.0	3.996	3.969	15.861	31
32	1 262.2	.00079	.00020	.2502	5 044.7	3.997	3.975	15.886	32
33	1 577.7	.00063	.00016	.2502	6 306.9	3.997	3.979	15.906	33
34	1 972.2	.00051	.00013	.2501	7 884.6	3.998	3.983	15.923	34
35	2 465.2	.00041	.00010	.2501	9 856.8	3.998	3.986	15.937	35
40	7 523.2	.00013	.00003	.2500	30 088.7	3.999	3.995	15.977	40
45	22 958.9	.00004	.00001	.2500	91 831.5	4.000	3.998	15.991	45
50	70 064.9	.00001		.2500	280 255.7	4.000	3.999	15.997	50
55	213 821.2			.2500	855 280.7	4.000	4.000	15.999	55

	Single Payment		Uniform Payment Series				Arithmetic Gradient		
	Compound Amount Factor	Present Worth Factor	Sinking Fund Factor	Capital Recovery Factor	Compound Amount Factor	Present Worth Factor	Gradient Uniform Series	Gradient Present Worth	
n	Find *F* Given *P* *F/P*	Find *P* Given *F* *P/F*	Find *A* Given *F* *A/F*	Find *A* Given *P* *A/P*	Find *F* Given *A* *F/A*	Find *P* Given *A* *P/A*	Find *A* Given *G* *A/G*	Find *P* Given *G* *P/G*	*n*
1	1.300	.7692	1.0000	1.3000	1.000	0.769	0	0	1
2	1.690	.5917	.4348	.7348	2.300	1.361	0.435	0.592	2
3	2.197	.4552	.2506	.5506	3.990	1.816	0.827	1.502	3
4	2.856	.3501	.1616	.4616	6.187	2.166	1.178	2.552	4
5	3.713	.2693	.1106	.4106	9.043	2.436	1.490	3.630	5
6	4.827	.2072	.0784	.3784	12.756	2.643	1.765	4.666	6
7	6.275	.1594	.0569	.3569	17.583	2.802	2.006	5.622	7
8	8.157	.1226	.0419	.3419	23.858	2.925	2.216	6.480	8
9	10.604	.0943	.0312	.3312	32.015	3.019	2.396	7.234	9
10	13.786	.0725	.0235	.3235	42.619	3.092	2.551	7.887	10
11	17.922	.0558	.0177	.3177	56.405	3.147	2.683	8.445	11
12	23.298	.0429	.0135	.3135	74.327	3.190	2.795	8.917	12
13	30.287	.0330	.0102	.3102	97.625	3.223	2.889	9.314	13
14	39.374	.0254	.00782	.3078	127.912	3.249	2.969	9.644	14
15	51.186	.0195	.00598	.3060	167.286	3.268	3.034	9.917	15
16	66.542	.0150	.00458	.3046	218.472	3.283	3.089	10.143	16
17	86.504	.0116	.00351	.3035	285.014	3.295	3.135	10.328	17
18	112.455	.00889	.00269	.3027	371.518	3.304	3.172	10.479	18
19	146.192	.00684	.00207	.3021	483.973	3.311	3.202	10.602	19
20	190.049	.00526	.00159	.3016	630.165	3.316	3.228	10.702	20
21	247.064	.00405	.00122	.3012	820.214	3.320	3.248	10.783	21
22	321.184	.00311	.00094	.3009	1 067.3	3.323	3.265	10.848	22
23	417.539	.00239	.00072	.3007	1 388.5	3.325	3.278	10.901	23
24	542.800	.00184	.00055	.3006	1 806.0	3.327	3.289	10.943	24
25	705.640	.00142	.00043	.3004	2 348.8	3.329	3.298	10.977	25
26	917.332	.00109	.00033	.3003	3 054.4	3.330	3.305	11.005	26
27	1 192.5	.00084	.00025	.3003	3 971.8	3.331	3.311	11.026	27
28	1 550.3	.00065	.00019	.3002	5 164.3	3.331	3.315	11.044	28
29	2 015.4	.00050	.00015	.3001	6 714.6	3.332	3.319	11.058	29
30	2 620.0	.00038	.00011	.3001	8 730.0	3.332	3.322	11.069	30
31	3 406.0	.00029	.00009	.3001	11 350.0	3.332	3.324	11.078	31
32	4 427.8	.00023	.00007	.3001	14 756.0	3.333	3.326	11.085	32
33	5 756.1	.00017	.00005	.3001	19 183.7	3.333	3.328	11.090	33
34	7 483.0	.00013	.00004	.3000	24 939.9	3.333	3.329	11.094	34
35	9 727.8	.00010	.00003	.3000	32 422.8	3.333	3.330	11.098	35
40	36 118.8	.00003	.00001	.3000	120 392.6	3.333	3.332	11.107	40
45	134 106.5	.00001		.3000	447 018.3	3.333	3.333	11.110	45

	Single Payment		Uniform Payment Series				Arithmetic Gradient		
	Compound Amount Factor	Present Worth Factor	Sinking Fund Factor	Capital Recovery Factor	Compound Amount Factor	Present Worth Factor	Gradient Uniform Series	Gradient Present Worth	
n	Find *F* Given *P* *F/P*	Find *P* Given *F* *P/F*	Find *A* Given *F* *A/F*	Find *A* Given *P* *A/P*	Find *F* Given *A* *F/A*	Find *P* Given *A* *P/A*	Find *A* Given *G* *A/G*	Find *P* Given *G* *P/G*	*n*
1	1.350	.7407	1.0000	1.3500	1.000	0.741	0	0	1
2	1.822	.5487	.4255	.7755	2.350	1.289	0.426	0.549	2
3	2.460	.4064	.2397	.5897	4.173	1.696	0.803	1.362	3
4	3.322	.3011	.1508	.5008	6.633	1.997	1.134	2.265	4
5	4.484	.2230	.1005	.4505	9.954	2.220	1.422	3.157	5
6	6.053	.1652	.0693	.4193	14.438	2.385	1.670	3.983	6
7	8.172	.1224	.0488	.3988	20.492	2.508	1.881	4.717	7
8	11.032	.0906	.0349	.3849	28.664	2.598	2.060	5.352	8
9	14.894	.0671	.0252	.3752	39.696	2.665	2.209	5.889	9
10	20.107	.0497	.0183	.3683	54.590	2.715	2.334	6.336	10
11	27.144	.0368	.0134	.3634	74.697	2.752	2.436	6.705	11
12	36.644	.0273	.00982	.3598	101.841	2.779	2.520	7.005	12
13	49.470	.0202	.00722	.3572	138.485	2.799	2.589	7.247	13
14	66.784	.0150	.00532	.3553	187.954	2.814	2.644	7.442	14
15	90.158	.0111	.00393	.3539	254.739	2.825	2.689	7.597	15
16	121.714	.00822	.00290	.3529	344.897	2.834	2.725	7.721	16
17	164.314	.00609	.00214	.3521	466.611	2.840	2.753	7.818	17
18	221.824	.00451	.00158	.3516	630.925	2.844	2.776	7.895	18
19	299.462	.00334	.00117	.3512	852.748	2.848	2.793	7.955	19
20	404.274	.00247	.00087	.3509	1 152.2	2.850	2.808	8.002	20
21	545.769	.00183	.00064	.3506	1 556.5	2.852	2.819	8.038	21
22	736.789	.00136	.00048	.3505	2 102.3	2.853	2.827	8.067	22
23	994.665	.00101	.00035	.3504	2 839.0	2.854	2.834	8.089	23
24	1 342.8	.00074	.00026	.3503	3 833.7	2.855	2.839	8.106	24
25	1 812.8	.00055	.00019	.3502	5 176.5	2.856	2.843	8.119	25
26	2 447.2	.00041	.00014	.3501	6 989.3	2.856	2.847	8.130	26
27	3 303.8	.00030	.00011	.3501	9 436.5	2.856	2.849	8.137	27
28	4 460.1	.00022	.00008	.3501	12 740.3	2.857	2.851	8.143	28
29	6 021.1	.00017	.00006	.3501	17 200.4	2.857	2.852	8.148	29
30	8 128.5	.00012	.00004	.3500	23 221.6	2.857	2.853	8.152	30
31	10 973.5	.00009	.00003	.3500	31 350.1	2.857	2.854	8.154	31
32	14 814.3	.00007	.00002	.3500	42 323.7	2.857	2.855	8.157	32
33	19 999.3	.00005	.00002	.3500	57 137.9	2.857	2.855	8.158	33
34	26 999.0	.00004	.00001	.3500	77 137.2	2.857	2.856	8.159	34
35	36 448.7	.00003	.00001	.3500	104 136.3	2.857	2.856	8.160	35

	Single Payment		Uniform Payment Series				Arithmetic Gradient		
	Compound Amount Factor	Present Worth Factor	Sinking Fund Factor	Capital Recovery Factor	Compound Amount Factor	Present Worth Factor	Gradient Uniform Series	Gradient Present Worth	
n	Find *F* Given *P* *F/P*	Find *P* Given *F* *P/F*	Find *A* Given *F* *A/F*	Find *A* Given *P* *A/P*	Find *F* Given *A* *F/A*	Find *P* Given *A* *P/A*	Find *A* Given *G* *A/G*	Find *P* Given *G* *P/G*	*n*
1	1.400	.7143	1.0000	1.4000	1.000	0.714	0	0	1
2	1.960	.5102	.4167	.8167	2.400	1.224	0.417	0.510	2
3	2.744	.3644	.2294	.6294	4.360	1.589	0.780	1.239	3
4	3.842	.2603	.1408	.5408	7.104	1.849	1.092	2.020	4
5	5.378	.1859	.0914	.4914	10.946	2.035	1.358	2.764	5
6	7.530	.1328	.0613	.4613	16.324	2.168	1.581	3.428	6
7	10.541	.0949	.0419	.4419	23.853	2.263	1.766	3.997	7
8	14.758	.0678	.0291	.4291	34.395	2.331	1.919	4.471	8
9	20.661	.0484	.0203	.4203	49.153	2.379	2.042	4.858	9
10	28.925	.0346	.0143	.4143	69.814	2.414	2.142	5.170	10
11	40.496	.0247	.0101	.4101	98.739	2.438	2.221	5.417	11
12	56.694	.0176	.00718	.4072	139.235	2.456	2.285	5.611	12
13	79.371	.0126	.00510	.4051	195.929	2.469	2.334	5.762	13
14	111.120	.00900	.00363	.4036	275.300	2.478	2.373	5.879	14
15	155.568	.00643	.00259	.4026	386.420	2.484	2.403	5.969	15
16	217.795	.00459	.00185	.4018	541.988	2.489	2.426	6.038	16
17	304.913	.00328	.00132	.4013	759.783	2.492	2.444	6.090	17
18	426.879	.00234	.00094	.4009	1 064.7	2.494	2.458	6.130	18
19	597.630	.00167	.00067	.4007	1 419.6	2.496	2.468	6.160	19
20	836.682	.00120	.00048	.4005	2 089.2	2.497	2.476	6.183	20
21	1 171.4	.00085	.00034	.4003	2 925.9	2.498	2.482	6.200	21
22	1 639.9	.00061	.00024	.4002	4 097.2	2.498	2.487	6.213	22
23	2 295.9	.00044	.00017	.4002	5 737.1	2.499	2.490	6.222	23
24	3 214.2	.00031	.00012	.4001	8 033.0	2.499	2.493	6.229	24
25	4 499.9	.00022	.00009	.4001	11 247.2	2.499	2.494	6.235	25
26	6 299.8	.00016	.00006	.4001	15 747.1	2.500	2.496	6.239	26
27	8 819.8	.00011	.00005	.4000	22 046.9	2.500	2.497	6.242	27
28	12 347.7	.00008	.00003	.4000	30 866.7	2.500	2.498	6.244	28
29	17 286.7	.00006	.00002	.4000	43 214.3	2.500	2.498	6.245	29
30	24 201.4	.00004	.00002	.4000	60 501.0	2.500	2.499	6.247	30
31	33 882.0	.00003	.00001	.4000	84 702.5	2.500	2.499	6.248	31
32	47 434.8	.00002	.00001	.4000	118 584.4	2.500	2.499	6.248	32
33	66 408.7	.00002	.00001	.4000	166 019.2	2.500	2.500	6.249	33
34	92 972.1	.00001		.4000	232 427.9	2.500	2.500	6.249	34
35	130 161.0	.00001		.4000	325 400.0	2.500	2.500	6.249	35

	Single Payment		Uniform Payment Series				Arithmetic Gradient		
n	Compound Amount Factor Find F Given P F/P	Present Worth Factor Find P Given F P/F	Sinking Fund Factor Find A Given F A/F	Capital Recovery Factor Find A Given P A/P	Compound Amount Factor Find F Given A F/A	Present Worth Factor Find P Given A P/A	Gradient Uniform Series Find A Given G A/G	Gradient Present Worth Find P Given G P/G	n
1	1.450	.6897	1.0000	1.4500	1.000	0.690	0	0	1
2	2.103	.4756	.4082	.8582	2.450	1.165	0.408	0.476	2
3	3.049	.3280	.2197	.6697	4.553	1.493	0.758	1.132	3
4	4.421	.2262	.1316	.5816	7.601	1.720	1.053	1.810	4
5	6.410	.1560	.0832	.5332	12.022	1.876	1.298	2.434	5
6	9.294	.1076	.0543	.5043	18.431	1.983	1.499	2.972	6
7	13.476	.0742	.0361	.4861	27.725	2.057	1.661	3.418	7
8	19.541	.0512	.0243	.4743	41.202	2.109	1.791	3.776	8
9	28.334	.0353	.0165	.4665	60.743	2.144	1.893	4.058	9
10	41.085	.0243	.0112	.4612	89.077	2.168	1.973	4.277	10
11	59.573	.0168	.00768	.4577	130.162	2.185	2.034	4.445	11
12	86.381	.0116	.00527	.4553	189.735	2.196	2.082	4.572	12
13	125.252	.00798	.00362	.4536	276.115	2.204	2.118	4.668	13
14	181.615	.00551	.00249	.4525	401.367	2.210	2.145	4.740	14
15	263.342	.00380	.00172	.4517	582.982	2.214	2.165	4.793	15
16	381.846	.00262	.00118	.4512	846.325	2.216	2.180	4.832	16
17	553.677	.00181	.00081	.4508	1 228.2	2.218	2.191	4.861	17
18	802.831	.00125	.00056	.4506	1 781.8	2.219	2.200	4.882	18
19	1 164.1	.00086	.00039	.4504	2 584.7	2.220	2.206	4.898	19
20	1 688.0	.00059	.00027	.4503	3 748.8	2.221	2.210	4.909	20
21	2 447.5	.00041	.00018	.4502	5 436.7	2.221	2.214	4.917	21
22	3 548.9	.00028	.00013	.4501	7 884.3	2.222	2.216	4.923	22
23	5 145.9	.00019	.00009	.4501	11 433.2	2.222	2.218	4.927	23
24	7 461.6	.00013	.00006	.4501	16 579.1	2.222	2.219	4.930	24
25	10 819.3	.00009	.00004	.4500	24 040.7	2.222	2.220	4.933	25
26	15 688.0	.00006	.00003	.4500	34 860.1	2.222	2.221	4.934	26
27	22 747.7	.00004	.00002	.4500	50 548.1	2.222	2.221	4.935	27
28	32 984.1	.00003	.00001	.4500	73 295.8	2.222	2.221	4.936	28
29	47 826.9	.00002	.00001	.4500	106 279.9	2.222	2.222	4.937	29
30	69 349.1	.00001	.00001	.4500	154 106.8	2.222	2.222	4.937	30
31	100 556.1	.00001		.4500	223 455.9	2.222	2.222	4.938	31
32	145 806.4	.00001		.4500	324 012.0	2.222	2.222	4.938	32
33	211 419.3			.4500	469 818.5	2.222	2.222	4.938	33
34	306 558.0			.4500	681 237.8	2.222	2.222	4.938	34
35	444 509.2			.4500	987 795.9	2.222	2.222	4.938	35

	Single Payment		Uniform Payment Series				Arithmetic Gradient		
	Compound Amount Factor	Present Worth Factor	Sinking Fund Factor	Capital Recovery Factor	Compound Amount Factor	Present Worth Factor	Gradient Uniform Series	Gradient Present Worth	
n	Find F Given P F/P	Find P Given F P/F	Find A Given F A/F	Find A Given P A/P	Find F Given A F/A	Find P Given A P/A	Find A Given G A/G	Find P Given G P/G	n
1	1.500	.6667	1.0000	1.5000	1.000	0.667	0	0	1
2	2.250	.4444	.4000	.9000	2.500	1.111	0.400	0.444	2
3	3.375	.2963	.2105	.7105	4.750	1.407	0.737	1.037	3
4	5.063	.1975	.1231	.6231	8.125	1.605	1.015	1.630	4
5	7.594	.1317	.0758	.5758	13.188	1.737	1.242	2.156	5
6	11.391	.0878	.0481	.5481	20.781	1.824	1.423	2.595	6
7	17.086	.0585	.0311	.5311	32.172	1.883	1.565	2.947	7
8	25.629	.0390	.0203	.5203	49.258	1.922	1.675	3.220	8
9	38.443	.0260	.0134	.5134	74.887	1.948	1.760	3.428	9
10	57.665	.0173	.00882	.5088	113.330	1.965	1.824	3.584	10
11	86.498	.0116	.00585	.5058	170.995	1.977	1.871	3.699	11
12	129.746	.00771	.00388	.5039	257.493	1.985	1.907	3.784	12
13	194.620	.00514	.00258	.5026	387.239	1.990	1.933	3.846	13
14	291.929	.00343	.00172	.5017	581.859	1.993	1.952	3.890	14
15	437.894	.00228	.00114	.5011	873.788	1.995	1.966	3.922	15
16	656.841	.00152	.00076	.5008	1 311.7	1.997	1.976	3.945	16
17	985.261	.00101	.00051	.5005	1 968.5	1.998	1.983	3.961	17
18	1 477.9	.00068	.00034	.5003	2 953.8	1.999	1.988	3.973	18
19	2 216.8	.00045	.00023	.5002	4 431.7	1.999	1.991	3.981	19
20	3 325.3	.00030	.00015	.5002	6 648.5	1.999	1.994	3.987	20
21	4 987.9	.00020	.00010	.5001	9 973.8	2.000	1.996	3.991	21
22	7 481.8	.00013	.00007	.5001	14 961.7	2.000	1.997	3.994	22
23	11 222.7	.00009	.00004	.5000	22 443.5	2.000	1.998	3.996	23
24	16 834.1	.00006	.00003	.5000	33 666.2	2.000	1.999	3.997	24
25	25 251.2	.00004	.00002	.5000	50 500.3	2.000	1.999	3.998	25
26	37 876.8	.00003	.00001	.5000	75 751.5	2.000	1.999	3.999	26
27	56 815.1	.00002	.00001	.5000	113 628.3	2.000	2.000	3.999	27
28	85 222.7	.00001	.00001	.5000	170 443.4	2.000	2.000	3.999	28
29	127 834.0	.00001		.5000	255 666.1	2.000	2.000	4.000	29
30	191 751.1	.00001		.5000	383 500.1	2.000	2.000	4.000	30
31	287 626.6			.5000	575 251.2	2.000	2.000	4.000	31
32	431 439.9			.5000	862 877.8	2.000	2.000	4.000	32

CHAPTER NINE

Rate of Return

IN THIS CHAPTER YOU WILL LEARN ABOUT

- The meaning of *rate of return (ROR)*
- How to calculate a project's ROR
- Application of the ROR criterion
- Difficulties in evaluating ROR
- The concept of "pure" investment
- Hybrid cashflows
- Internal and external rates of return
- The cashflow sign-change rule
- The incremental method of analysis
- Analysis period for the ROR method

So far in Part II we have discussed four analysis methods, namely payback period, PW, FW, and AW, in Chapters 5, 6, 7, and 8 respectively. In this chapter, another method, called rate of return (ROR), is presented. An obvious question is why there are so many methods. This question is answered in Chapter 11, where we compare the merits and limitations of all the methods and discuss when one is preferred over the other.

In industry, ROR is the most widely used method. It is easier to comprehend, but its analysis is relatively complex, and the pertaining calculations are lengthy. In PW, FW, or AW analysis, the interest rate i is known; in ROR analysis, i is the unknown.

We evaluate a project's ROR to ensure that it is higher than the cost of capital, as expressed by the *minimum acceptable rate of return* (MARR). The MARR is a cut-off value, determined on the basis of financial market conditions and company's business "health," below which investment is undesirable. In general[1], the MARR is significantly higher than what financial institutions charge for lending capital. The term *significantly*

[1]For marketing and other technical reasons, a project may sometimes be funded even if its ROR is less than the MARR.

is befitting, since investing in engineering projects is risky. If the MARR is only marginally higher, doing engineering business does not make much of a business sense.

9.1 MEANING

Of the three commonly used measures—PW, AW, and ROR—of investment worthiness, rate of return is the most comprehensive and easily understood. This is primarily because ROR is expressed as a percentage, which is an *absolute* measure, being independent of the amount of investment. Besides, we are accustomed to the percentage measure in our daily lives in so many different ways. The PW and AW measures of a project, on the other hand, being in monetary units, are *relative*; they must be judged in conjunction with the amount of investment.

Economists explain the meaning of ROR in several ways. According to one meaning, ROR is the interest earned on the remaining balance of an investment. By another, ROR is *the value of i that renders PW zero*. In other words, ROR is that interest rate for which cash inflows (benefits) equal cash outflows (costs). Of course, the time value of money must be taken into account in this equivalency.

9.2 EVALUATION

The ROR is basically the interest rate for which a project's costs are fully recovered through its benefits. As explained earlier, the costs and benefits of a project can be considered as weights on the two pans of a balance. By this analogy, ROR is that interest rate which renders the balance in balance. We illustrate this meaning of ROR through Example 9.1.

EXAMPLE 9.1

An investment of \$5,000 is paid back after 5 years as a lump sum of \$7,345. What is the ROR assuming annual compounding?

Solution

By equating (or balancing) the single cash outflow P (\$5,000) with the single cash inflow F (\$7,345), from $F = P(F/P, i, n)$ for $n = 5$,

$$7{,}345 = 5{,}000(F/P, i, 5)$$

Transferring the unknown, which is the factor containing i, to the left side of the equal sign,

$$(F/P, i, 5) = 7{,}345/5{,}000 = 1.469$$

Thumbing through the interest tables, we see that $(F/P, i, 5)$ is 1.469 when i is 8%. Thus, $i = 8\%$ per year. For this interest rate, the \$7,345 payback (benefit) fully compensates for the investment (cost); hence the ROR $= i = 8\%$ per year.

Example 9.1 was set up intentionally to make the evaluation of ROR easier. The value of i was simply read off the tables. ROR evaluation is rarely so simple. In fact, it is usually more involved, even for simple problems, as illustrated in Example 9.2.

EXAMPLE 9.2

Evaluate ROR for the project whose cashflows are diagramed in Fig. 9.1.

Solution

From the diagram, an investment (or cost) of \$500 now brings in benefits of \$200 at period 4 and \$500 at period 6. We can easily express the cost–benefit equivalency[2] by transferring the two F-values to period 0, as

$$500 = 200(P/F, i, 4) + 500\,(P/F, i, 6)$$

Our task is to evaluate i in this expression. There are two approaches[3] to this. In one, we substitute the equations for functional notations and solve for i. The substitution (Equation 3.7b) yields

$$500 = 200(1 + i)^{-4} + 500\,(1 + i)^{-6}$$

Multiplying throughout by $(1 + i)^6$ and dividing by 500, and rearranging the terms, we get

$$(1 + i)^6 - 0.4\,(1 + i)^2 - 1 = 0$$

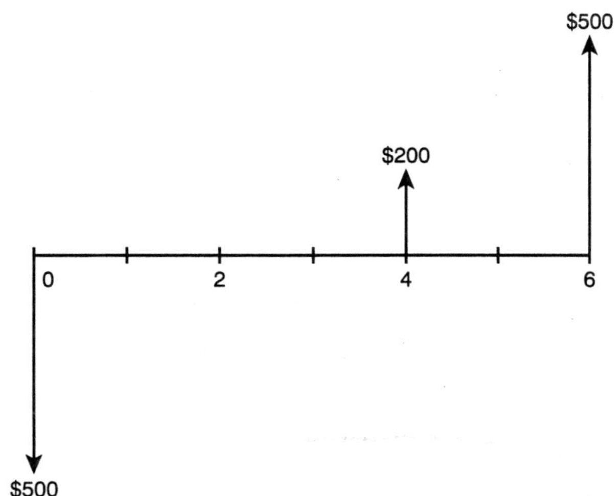

Figure 9.1 Evaluation of ROR

[2]Note that equivalency is valid only if all the costs and benefits pertain to the same time period, for example to period 0 in this case.

[3]Readers will notice by the end of this section that the two approaches are basically the same.

This relationship can be solved for i in several ways, including iteration[4] methods. However, we skip this solution approach and use instead the table-based functional notations emphasized throughout the text. This is the second approach of evaluating i in the original expression: $500 = 200(P/F, i, 4) + 500(P/F, i, 6)$.

Rearranging the terms, we get

$$500(P/F, i, 6) + 200(P/F, i, 4) - 500 = 0$$

This expression[5] can be solved for the unknown i through iteration and interpolation. The iteration effort can be reduced by shrinking the large numbers 200 and 500 by dividing the expression throughout by 500. This yields

$$(P/F, i, 6) + 0.4(P/F, i, 4) - 1 = 0$$

One could have divided by 200 instead to get

$$2.5(P/F, i, 6) + (P/F, i, 4) - 2.5 = 0$$

The extent and type of shrinking to achieve smaller multipliers is your own preference, since the evaluated i will not be affected by such manipulations.

Let us consider the first expression for further analysis, namely

$$(P/F, i, 6) + 0.4(P/F, i, 4) - 1 = 0$$

---

[4] In some simple cases we may be able to evaluate i directly. As an illustration, if there was only one benefit in this example, say of \$750 at period 6, then we would have

$$500 = 750\,(1 + i)^{-6}$$
$$(1 + i)^6 = 750/500 = 1.5$$

Using natural logarithms, this gives

$$\begin{aligned}
6\,\ell\text{n}(1 + i) &= \ell\text{n}\,1.5 \\
6\,\ell\text{n}(1 + i) &= 0.405465 \\
\ell\text{n}(1 + i) &= 0.06758 \\
1 + i &= e^{0.06758} \\
1 + i &\approx 1.07 \\
i &= 1.07 - 1 \\
&= 0.07 \\
&= 7\%
\end{aligned}$$

However, problems where such a direct evaluation of i is possible are rare, and hence, in ROR analysis the equation-based solution approach is limited in scope.

[5] Many analysts prefer the format

$$500(P/F, i, 6) + 200(P/F, i, 4) = 500$$

instead of the one used here. Either format is acceptable, since the result will be the same. So use the format you feel comfortable with.

Note that the left side of this expression is related to the PW of the cashflows. We therefore call it the PW function. There is no simple way to solve the PW function for i, except trial and error, in which different values of i are tried to see which one satisfies it. For each i tried, the functional factors' values are read off the interest tables and the PW function evaluated. The trials and errors[6] in this case are as follows.

i(%)	$(P/F, i, 6) + 0.4(P/F, i, 4) - 1$
4	$0.7903 + 0.4 \times 0.8548 - 1 = 0.132$
5	$0.7462 + 0.4 \times 0.8227 - 1 = 0.075$
6	$0.7050 + 0.4 \times 0.7921 - 1 = 0.022$
7	$0.6663 + 0.4 \times 0.7621 - 1 = -0.028$
8	$0.6302 + 0.4 \times 0.7350 - 1 = -0.076$

We notice that the function changes sign when i changes from 6% to 7%. Thus, i that will render the function zero (the right side of the expression) must lie in between 6% and 7%. We therefore interpolate[7] between these two interest rates[8]. From the interpolation[9],

$$\begin{aligned} \text{ROR} = i &= 6 + 0.022/(0.022 + 0.028) \\ &= 6 + 0.44 \\ &= 6.44\% \end{aligned}$$

As seen in Example 9.2, evaluation of ROR even for a simple problem can be lengthy. This is the major drawback of the ROR method. Programmable calculators, spreadsheets, and canned software make the task of ROR evaluation easier. Their advent is one reason why the ROR method has recently gained popularity in industry.

9.2.1 Where To Begin

In evaluating the ROR in Example 9.2 we began the trial with i as 4%. Why did we begin with 4%? Why not 15% or 35%? Also why did we increment i by 1%? Answers to these questions are important for efficient analysis. Unfortunately, there are no simple answers except that the analyst should use intelligent guesses about where to start and how much to increment or decrement i.

[6]Error is the difference between the value of the expression, for the tried i, on the left and that on the right which is zero. For example, for $i = 5\%$, the error is $0.075 - 0 = 0.075$

[7]Interpolation assumes a linear relationship between the PW function and i, which is not true as illustrated in Fig. 9.2. Thus some sacrifice in the accuracy of results is made. For highest accuracy, keep the interest boundaries within which interpolation is done as close to each other as the interest tables allow.

[8]See Chapter 3 to recall the interpolation procedure.

[9]As an alternative to algebraic interpolation, one can draw a graph between the values of the PW function and i, and read i (=ROR) off the graph where the function is zero. Fig. 9.3 shows the graph for this case; such a graph is called a *PW profile*. Note that additional points had to be calculated to get a smooth curve. A programmable calculator is very useful in evaluating ROR this way. With graphic capability such a calculator can also display the PW profile.

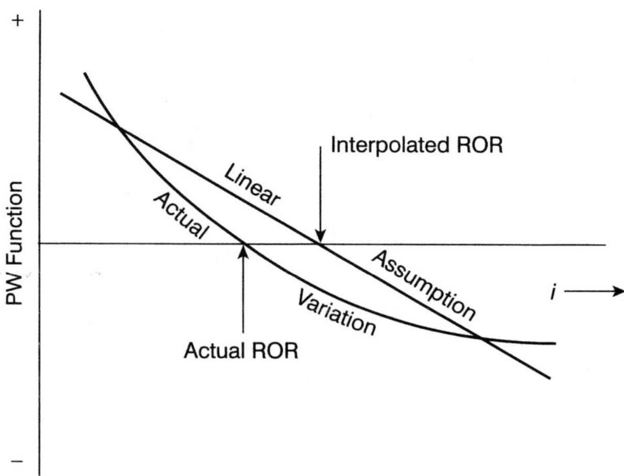

Figure 9.2 Approximation of Nonlinear Variation

Start with a *reasonable* value of i. If the value of the PW function is getting closer to zero[10], increment i by as little as the interest tables allow. Should you increment or decrement? Here the rule is, Do whichever will lead toward zero. Notice this in the trial results of Example 9.2, where the PW function's value of 0.132 for $i = 4\%$ decreased to 0.075 as i was incremented to 5%. This decrease confirmed that incrementing was helping (leading toward zero) in the search. Thus, keep watching the function's value as the search for i continues, to ensure that you are heading in the right direction, and also to decide whether to slow down the incrementing of i, say from 1% to 0.5%.

The value of i with which the trial is begun can be approximated by analyzing the diagram mentally. Let us do this for Example 9.2 with reference to Fig. 9.1. Neglecting the time value of money for this approximation, a payback of $700 ($200 plus $500) for an investment of $500 means that the total interest is $700 − $500 = $200. Assuming period 5 as the approximate time of payback, since it lies between periods 4 and 6, the $200 total interest is equivalent to 200/5 = $40 interest per year. For a $500 investment, this $40 annual interest works[11] out as a rate of 8% per year. So one could have begun in this case with $i = 8\%$. The resulting negative value of the PW function from trying $i = 8\%$ would have suggested decrementing i for the next trial.

One could have approximated differently[12], getting another value of i to begin the trial. But it would have been in the same range, not 35–50%. To guess the value of i with which to begin the trial,

---

[10]Right-hand side of the evaluated expression

[11]This 8% is higher than the actual 6.44% because, in the approximation, we assumed all $700 to be located at period 5, and also assumed simple interest to make the mental arithmetic easy.

[12]For example, one can apply the Rule of 70 (see Section 3.9) according to which a sum doubles every $70/i$ periods, where i is in percentage per period. For example, if $i = 9\%$ per year, then a sum will double in $70/9 \approx 7.8$ years. So if it takes n periods for a sum to double, then i can be approximated by $70/n$.

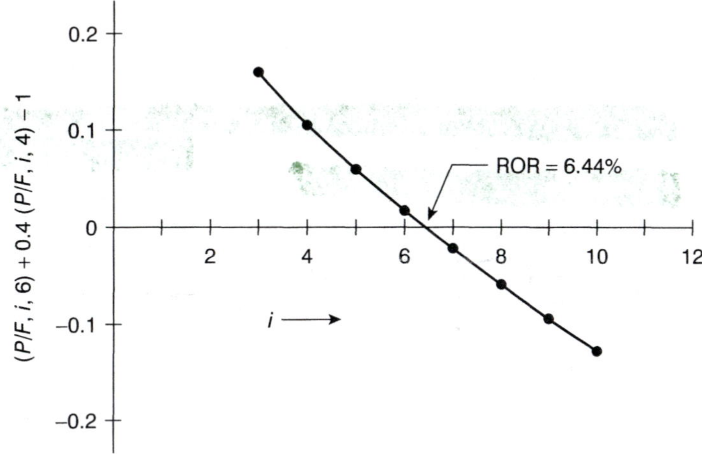

Figure 9.3 Graphical Evaluation of ROR

a. Estimate the approximate total interest and the number of periods over which it is earned, and
b. From the above, work out a rate for the investment.

Here are some tips on determining the beginning value of *i,* on varying *i* during interpolation, and on calculating ROR.

1. Approximate the cashflows and their timings in a way that facilitates estimation of the beginning value.
2. If it is difficult to approximate the beginning *i,* then use a *reasonable* value. Values such as 5% or 10% are more reasonable as an annual interest rate than 50% or 60%.
3. In the beginning of the trials, you may increment or decrement *i* by more than 1%, say 2% or 3%, to reduce the number of iterations.
4. You do not have to only increment or decrement. The search may be either way—forward with higher *i* or backward with lower *i.*
5. Use a programmable calculator to save time and effort in evaluating the ROR. One with graphic capability can plot the PW profile and let you read the ROR off the display.

9.2.2 ROR and *i*

The interest rate *i* is a generic term applicable to all situations. ROR, on the other hand, is specific to a project. As a project-specific parameter, it is used to assess the project's investment worthiness. It answers the question, At what interest rate do the benefits from the project pay for its costs? For a project to be attractive its ROR should be greater than the interest rate *i* prevailing in the financial marketplace, also called the *cost of capital.* If the mousetrap you intend to design, manufacture, and market using your own capital is expected to yield 8% ROR while the local bank pays its savers 9% interest, shouldn't you hand over your capital to the bank and go fishing?

9.2.3 MARR as *i*

Rather than the value of its ROR, we are often interested simply in knowing whether a project is worthy of investment. To judge this, companies set a cut-off limit on the ROR below which projects are not funded. This limiting "bottom-line" ROR is the *minimum acceptable rate of return*, MARR, which should obviously be greater than the cost of borrowed capital. It is also used for rationing capital for competing projects.

If the MARR is given and the task is to determine whether or not the project should be funded, then simply determine the project PW with the MARR as interest rate i. If this PW is positive (benefits outweigh the costs at the MARR), the project[13] is investment worthy; otherwise not. Example 9.3 illustrates the point.

EXAMPLE 9.3

A milling machine's first cost is $25,000. Its likely benefits are $5,000 in the first year, $4,750 in the second year, and so on, decreasing annually by $250. The machine's useful life is 10 years, and its salvage[14] value is $1,000. Based on the ROR criterion should the machine be approved for purchase if the MARR is 15%?

Solution

The MARR is given, and our task is assessing the machine's investment worthiness on the basis of the ROR. Note that the value of the ROR is not being asked for. We can reach the decision simply by judging the sign of PW corresponding to i = MARR = 15%. If the PW is positive, then the machine's ROR will be greater than 15%, and hence it should be purchased.

The given cash inflows form a decreasing arithmetic series, as shown in Fig. 9.4(a). The $1,000 vector at period 10 represents salvage value at the end of useful life.

For determining PW we need to evaluate $PW_{benefits}$ and PW_{costs} from the diagram. The value of PW_{costs} is $25,000, since the first cost is the only cost. Evaluation of $PW_{benefits}$ requires us to transfer all the benefits to period 0. To exploit the approximate pattern in the benefits, we tailor it to the standard pattern for which the arithmetic-series equation and function apply (see Chapter 4).

The tailoring is illustrated in Fig. 9.4(b). It involves adding $5,000 upward vectors (uniform annual benefit) throughout, that is, at each period except period 0. From the added $5,000 vector we subtract at each period the arithmetic-series downward vector of Fig. 9.4(b). The subtraction results in the cash inflows of Fig. 9.4(a). Therefore, the given cashflows of Fig. 9.4(a) are equivalent to the tailored cashflows of Fig. 9.4(b). The $1,000 salvage vector remains unchanged in the two figures at period 10. So does the $25,000 cost at period 0. We can easily determine $PW_{benefits}$ from Fig. 9.4(b) as

$$PW_{benefits} = [5{,}000(P/A, 15\%, 10) - 250(P/G, 15\%, 10)] + 1{,}000(P/F, 15\%, 10)$$

[13]A positive PW corresponding to i = MARR indicates that the project ROR is greater than the MARR; hence the project is investment worthy.

[14]Salvage value is usually positive and hence a cash inflow (benefit). It happens when the market value of the resource at the end of its useful life is more than the cost of its removal and disposal. At times, this may not be true and the salvage value may be negative, in which case treat it as a cash outflow (cost).

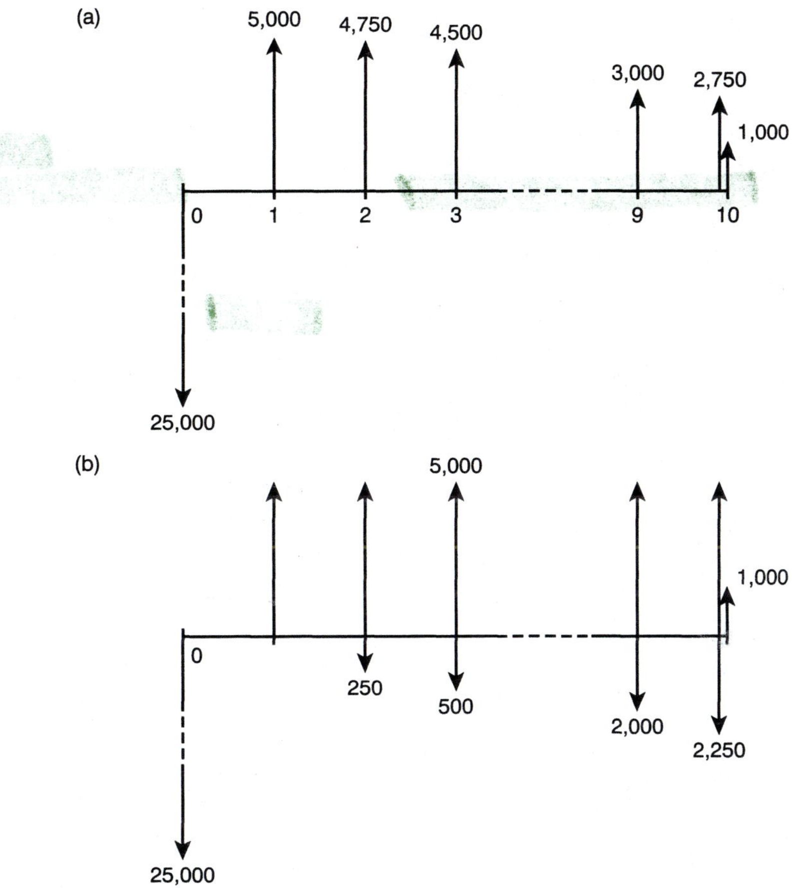

Figure 9.4 Diagram for Example 9.3

The last term in this expression accounts for the salvage value.

Thus, the machine's present worth is

$$
\begin{aligned}
\text{PW} &= PW_{\text{benefits}} - PW_{\text{costs}} \\
&= 5{,}000(P/A, 15\%, 10) - 250(P/G, 15\%, 10) + 1{,}000(P/F, 15\%, 10) \\
&\quad - 25{,}000 \\
&= 5{,}000 \times 5.019 - 250 \times 16.979 + 1{,}000 \times 0.2472 - 25{,}000 \\
&= 25{,}095 - 4{,}245 + 247 - 25{,}000 \\
&= -\$3{,}903
\end{aligned}
$$

Since the PW corresponding to 15% MARR is negative, the machine's ROR is less than the MARR, and hence it should not be approved[15] for purchase.

---

[15]The alternative approach, of evaluating the ROR exactly and comparing it with the given MARR, is lengthier than the one illustrated in this example.

9.3 FORMULAS

As explained in Section 9.2, the ROR can be evaluated from first principles. In this section, we extend this approach to formula-based analysis.

Based on its meaning explained earlier in Section 9.1, the ROR is the interest rate that satisfies $PW_{benefits} = PW_{costs}$. In other words, $PW_{benefits} - PW_{costs} = 0$ is the condition for evaluating the ROR. Since this difference is the PW of the project, the condition can be expressed as

$$PW = 0$$

Since PW is directly related[16] to AW, we can say that the ROR will equal i also when the AW is zero. Thus, the condition is also satisfied by

$$AW = 0$$

and, for the same reason, by

$$FW = 0$$

In most cases, the condition PW = 0 is used to evaluate the ROR or apply the ROR criterion. Where calculations for AW are simpler than those for PW, the condition AW = 0 is used. For the same reason, use FW = 0 where appropriate.

EXAMPLE 9.4

A milling machine's first cost is $30,000. The likely benefits from the machine are $5,000 in the first year, $4,750 in the second year, and so on, decreasing annually by $250. If the machine has a useful life of 10 years and a salvage value of $1,000, what is its ROR?

Solution

We are asked to determine the value of the ROR. To do this, express the PW of the given cashflows and equate it to zero. That means evaluating $PW_{benefits}$ and PW_{costs} from the machine's cashflow diagram. The diagram is in Fig. 9.4(a) except for the first cost, which is $30,000.

From the tailored cash inflows in Fig. 9.4(b), as explained in the previous example,

$$PW_{benefits} = [5{,}000(P/A, i, 10) - 250(P/G, i, 10)] + 1{,}000(P/F, i, 10)$$

Thus,

$$\begin{aligned} PW &= PW_{benefits} - PW_{costs} \\ &= [5{,}000(P/A, i, 10) - 250(P/G, i, 10)] + 1{,}000(P/F, i, 10) - 30{,}000 \end{aligned}$$

[16]In this text we treat PW, AW, or FW as *net*, i.e. as the excess of benefits over costs.

By imposing the condition PW = 0, i is evaluated from

$$5{,}000(P/A, i, 10) - 250(P/G, i, 10) + 1{,}000(P/F, i, 10) - 30{,}000 = 0$$

Dividing throughout by 5,000, we get

$$(P/A, i, 10) - 0.05(P/G, i, 10) + 0.2(P/F, i, 10) - 6 = 0$$

This PW function can be solved[17] for i the way explained in Example 9.2, yielding ROR of approximately 6%.

9.4 ONE PROJECT—TWO RORs!

As mentioned earlier, ROR analysis is both lengthy and complex. It is lengthy because of the repetitive nature of trial and error calculations and interpolation, as seen in Examples 9.2–9.4. Fortunately, programmable calculators and PCs are helpful in coping with the repetitive calculations.

In this section, we discuss the complexity aspect of ROR analysis. The first complexity arises from the fact that there may at times be two or more rates of return for the same project. If so, which ROR is admissible, or are both or all of them inadmissible?

Two or more different RORs for the same project is mind-boggling, at least initially. Before we discuss how to resolve this more-than-one-answer-to-the-same-problem situation, it is helpful to know when such a situation can arise.

9.4.1 Pure Cashflows Yield One ROR

So far we have intentionally discussed only those projects that yielded only one ROR. In this subsection we discuss the characteristics of cashflows of such projects. Cashflows that yield more than one ROR are discussed in the next section.

Cashflows encountered in engineering economics can be divided into two groups: *pure* and *hybrid*. For projects whose cashflows are *pure*, there is only one ROR. For those with hybrid cashflows, there *may exist* more than one ROR. Note the word *may*, since hybrid cashflows can sometimes yield only one ROR.

---

[17]The trial and error results are

i(%)	$(P/A, i, 10) - 0.05(P/G, i, 10) + 0.2(P/F, i, 10) - 6$
4	$8.111 - 0.05 \times 33.881 + 0.2 \times 0.6756 - 6 = 0.552$
5	$7.722 - 0.05 \times 31.652 + 0.2 \times 0.6139 - 6 = 0.262$
6	$7.360 - 0.05 \times 29.602 + 0.2 \times 0.5584 - 6 = -0.008$

Since the sign changes between 5% and 6%, interpolate within this range to get

$$\begin{aligned} \text{ROR} = i &= 5 + 0.262/(0.008 + 0.262) \\ &= 5 + 0.97 \\ &= 5.97\% \end{aligned}$$

In *pure* cashflows, one or more costs are followed by several benefits, or one or more benefits are followed by several costs. The former[18] represents *pure investment*, while the latter represents *pure borrowing*.

The following cashflows are an example[19] of *pure investment*

Year	Cashflow
0	−$5,000
1	1,500
2	1,500
3	1,300
4	900
5	3,000

The following cashflows are an example[20] of *pure borrowing*.

Year	Cashflow
0	$10,000
1	−2,500
2	−2,200
3	−2,000
4	−4,000

The basic characteristic of *pure* cashflows is that there is *only one* sign change in the data. In the *pure* investment above, this occurs[21] at year 1 when the cashflow changes from −$5,000 to $1,500 (a change from negative sign to positive). In the *pure* borrowing above,

---

[18]In engineering economics, we come across pure investments more often than pure borrowing. Metaphorically, an engineering project can be considered a borrower and the company a lender (or investor). We usually analyze a project from the company's viewpoint, and hence as an investment problem.

[19]Another example is:

Year	Cashflow
0	−$100
1	−400
2	0
3	300
4	350

[20]Another example is:

Year	Cashflow
0	$500
1	100
2	−200
3	−250
4	−350

[21]To determine whether the cashflow has changed sign, compare its sign with that of the *previous* one. In this case, at period 1 the sign is positive, which was negative at the previous period (at 0).

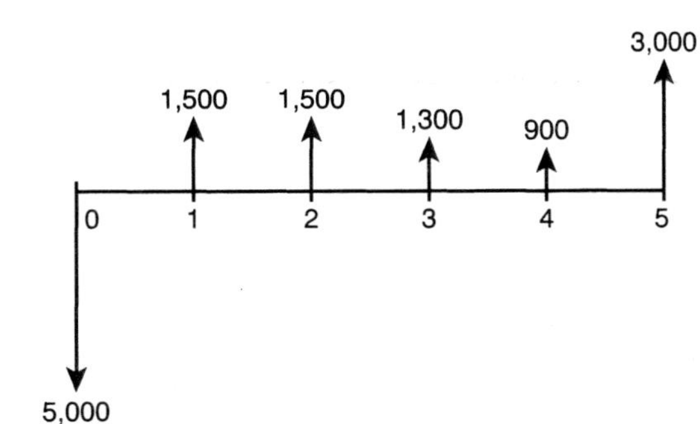

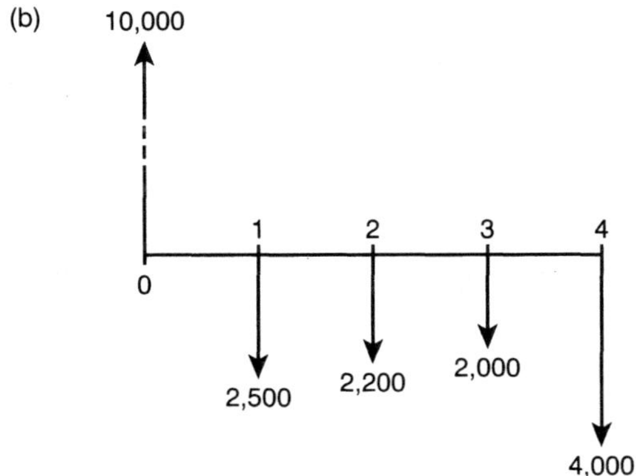

Figure 9.5 (a) Pure Investment, (b) Pure Borrowing

this occurs when the cashflow changes from \$10,000 to −\$2,500 (a change from positive sign to negative). In pure cashflows, once the sign has changed it remains that way.

The sign change is visible in the cashflow diagram too. For the two pure cashflows just discussed, it is obvious in their diagrams in Fig. 9.5. Note that once the vector has changed direction, as at period 1 with respect to period 0, it continues to stay that way.

In these illustrations, the sign changed immediately following the first cashflow. This is not essential for the cashflows to be pure. There may be more than one cashflow, as long as they are all of the same sign, before the sign changes. Thus, there may be several costs in *pure* investment problems, as shown in Fig. 9.6(a), as long as they are incurred before the benefits begin to accrue. The only characteristic essential for the cashflows to be *pure* to yield one ROR is that once they change sign they should continue to be that way. In Fig. 9.6(a) the sign changes at period 3 (from downward vector at period 2 to upward vector at 3), beyond which all vectors remain upward.

A similar situation can occur in the case of pure borrowing, that is, there may be several benefits preceding one or more costs as long as all the benefits are realized prior to the costs.

Another point to note is that for being pure the cashflows do not need to exist at each period, as in Fig. 9.6(b). For this diagram the cashflow table is

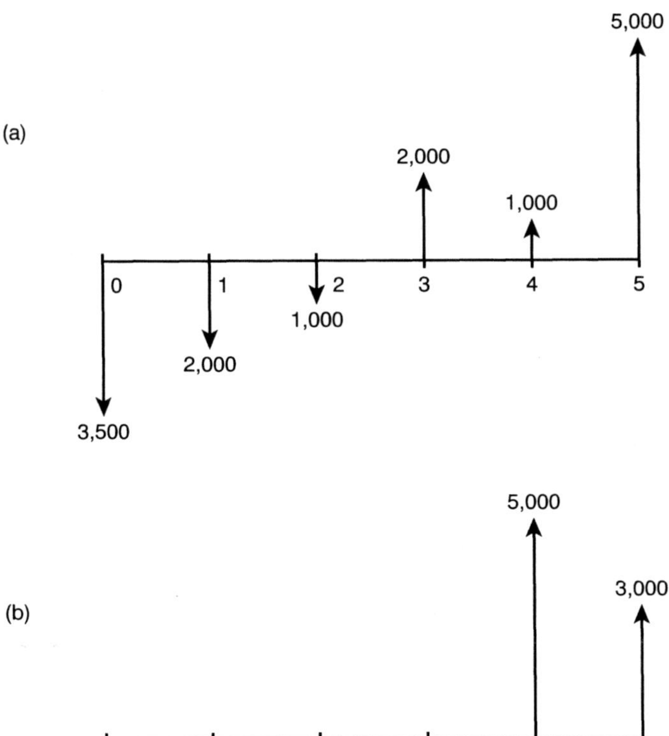

Figure 9.6 Multiple Cashflows Preceding the Sign Change

Year	Cashflow
0	−\$3, 500
1	0
2	−1,500
3	0
4	5,000
5	3,000

The existence of zero(s) in a cashflow table, as here, can be confusing in judging the number of sign changes. The zeros are simply ignored[22], since they represent the ab-

[22]The zero can be considered to carry the same sign as the previous nonzero cashflow. Thus, the zero at period 1 is negative.

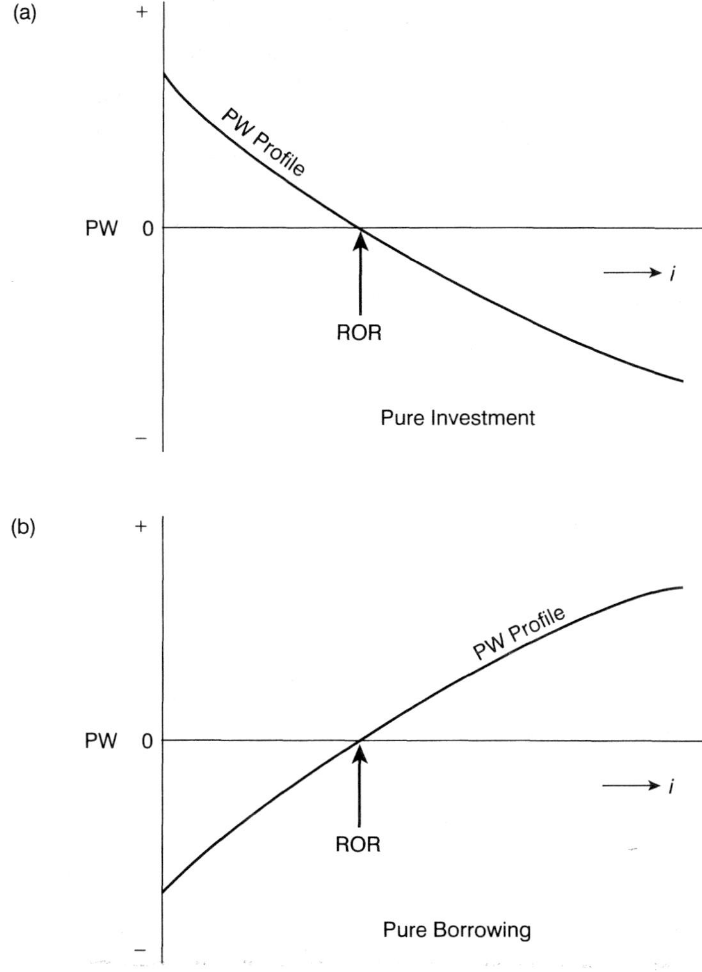

Figure 9.7 PW Profiles of Pure Cashflows

sence of vectors in the diagram. In the preceding table, therefore, there is only one sign change, at period 4, and thus only one ROR for the project the cashflows pertain to.

Why do *pure* cashflows yield only one ROR? This can be explained through the PW profile, a plot between PW values versus interest rates. For *pure* investments, PW profiles are as in Fig. 9.7(a), and for *pure* borrowing as in Fig. 9.7(b). As obvious in these figures, the profile[23] curve for *pure* cashflows crosses the x-axis (i-axis) only once. The crossover point represents the ROR, since PW = 0 there.

[23]In a PW profile, the value of PW corresponding to $i = 0$ (where the profile meets the y-axis) is simply cash inflows minus outflows, with no regard to the time value of money. For example, for the cashflows in Fig. 9.6(b), this will be (5,000 + 3,000) − (3,500 + 1,500) = 3,000. This characteristic of the PW profile can be used to check whether the profile seems correct. Also note that at $i = 0$ all functional factors become unity, since $i = 0$ represents interest-free transactions.

9.5 HYBRID CASHFLOWS

Several engineering projects generate cashflows that are not *pure*, with costs and benefits so occurring that there is more than one sign change. We call such cashflows *hybrid*; the following is an example.

Year	Cashflow
0	−$3,000
1	2,000
2	500
3	−1,000
4	1,500
5	2,500

Note in this table, as well as in the corresponding diagram in Fig. 9.8, that there are three sign changes: one at period 1, another at period 3, and the third at period 4. For this cashflow pattern, the number of RORs may be as many as three, that is, none, one, two, or three. In general, *the number of RORs "may be" as many as the number of sign changes*. This is called the *cashflow-sign-change rule*.

Hybrid cashflows typically arise in contractual projects where contractors are paid on "as-you-go" basis. Examples are construction of interstate highways and manufacture of fighter planes or space shuttles. In such projects, the contractor is paid (benefit) a portion of the cost in the beginning, usually on signing the contract. Other payments follow as the work progresses.

Prior to any ROR analysis, it is always a good practice to check[24] the cashflow signs to ascertain whether the cashflows are of the *pure* type. If so, then make a decision by determining[25] the ROR or checking the sign of the PW corresponding to the given MARR. If the cashflows are of the hybrid type, then the project may have more than

[24]In fact, even before checking the cashflows for their purity, first determine whether they will yield an admissible (positive) ROR at all. This is done by totaling the benefits *as they are*, i.e., ignoring the time value of money. Similarly total the costs as they are. If the total of the benefits is less than the total of costs, obviously no positive ROR can exist, since the costs outweigh the benefits, as in the following cashflows.

Year	Cashflows
0	−$500
1	−50
2	100
3	150
4	175

Here the benefits' total ($425) is less than costs' total ($550). In the special case where their totals are equal, the ROR is obviously zero.

To summarize, check the algebraic sum of the given cashflows. If this sum is negative (−$125 in the above example), no admissible ROR exists; if it is zero, the ROR is zero. Only if the algebraic sum is positive, is one or more ROR likely.

[25]As explained earlier, you can avoid the evaluation of ROR if the MARR is given; simply determine the MARR-based PW and make the decision based on its sign.

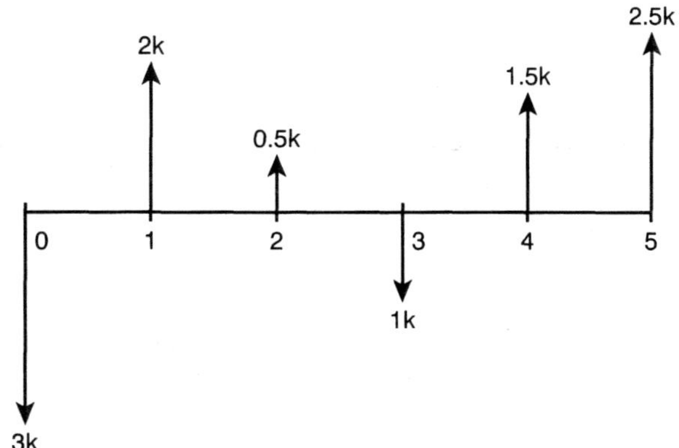

Figure 9.8 Hybrid Cashflows

one ROR. The determination of the correct ROR in the case of hybrid cashflows is presented in Subsection 9.5.2.

9.5.1 Internal ROR

A company rarely undertakes only one project at a time; several projects run concurrently. In general, projects are allowed to interact financially by pooling together capital funds allocated to various projects. In the rare case where the project is financially shielded, its cashflows are isolated from those of the others. In the pooled projects, any temporary surplus cash in one project is used for temporary needs of the others. Thus, a project with surplus cash conceptually becomes a short-term lender to another project.

Consider project A with hybrid cashflows as follows:

Year	Cashflow
0	−$3,000
1	10,000
2	−2,000
3	−3,000

There is a $10,000 cash inflow at period 1. A portion of this money can be loaned to another project as long as project A will get back $2,000 at period 2 and $3,000 at period 3 to pay for its own costs.

The interproject shuffling of funds gives rise to what is called *external* ROR. The *external ROR* of a project is the rate of return from lending to another project within the company or to financial institutions outside. It simply is external to the project. If the surplus cash of a project is loaned at an external ROR, then the resulting ROR of the project is called the *internal* ROR (IROR), to differentiate it from the ROR discussed so far in this chapter. Wherever external ROR is meant, the adjective *external* is explicitly used. But this is not so with internal ROR; we often use the term *ROR* to mean IROR.

As discussed earlier, companies fix the MARR to reflect the financial climate of the marketplace as well as their own financial health. The MARR is the "bottom line" rate of return below which projects are considered investment unworthy. The external ROR may at times be the MARR to ensure that the surplus funds are wisely shuffled among the projects. The external ROR may differ from the MARR; for example, if the financial marketplace is cash-hungry the company may lend the surplus to outside borrowers rather than to the other projects within.

9.5.2 ROR Evaluation[26]

As mentioned earlier, hybrid cashflows in which more than one sign change takes place can give rise to more than one ROR. The number of likely RORs is obvious in a PW profile, for example two in Fig. 9.9, provided the profile is complete, that is, covers a wide range of interest rates.

When more than one ROR is likely for a project, two choices exist:

1. Abandon the ROR analysis, since it is getting complex, and use one of the other five methods, such as PW or AW, to make the decision.
2. Carry out a *comprehensive ROR analysis*.

A comprehensive ROR analysis requires the analyst to conduct four tests on the data. These tests throw more light on the characteristics of the cashflows and lead to the determination of the correct ROR. By *correct ROR* we mean the one that assures profitability. The four comprehensive tests are

Test 1	Algebraic Sum of the Cashflows
Test 2	Cashflow Sign Change
Test 3	Accumulated Cashflow Sign Change
Test 4	Net Investment

According to the first test, if the algebraic sum of the cashflows is positive, then only one or more RORs are likely. Note that this test does not predict the likelihood of *only one* positive ROR. A negative algebraic sum usually[27] means no admissible (positive) ROR. If the algebraic sum is zero, the ROR will be zero.

The second test, based on the cashflow-sign-change rule, has already been discussed in Sections 9.4 and 9.5. According to this test there can be as many RORs as the number of sign changes in the cashflows.

To conduct the third test, accumulated cashflows corresponding to each time period are posted under a new column in the cashflow table. They are then subjected to two subtests:

[26]The discussions in this subsection are of an advanced nature. This subsection can be skipped if projects with hybrid cashflows are unlikely to be encountered.

[27]Merritt, A. J., and Allen Sykes. *The Finance and Analysis of Capital Projects*. Longman. 1973. p. 135

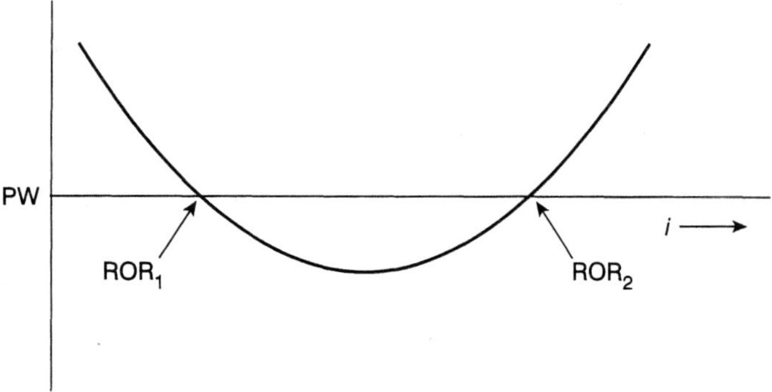

Figure 9.9 PW Profile Showing Two RORs

Subtest 3.1 *Is the accumulated cashflow at the last time period positive?*

Subtest 3.2 *Is there only one sign change in the accumulated cashflows?*

If the answers to these two questions are both yes, then the hybrid cashflows yield only one ROR. These two conditions are thus more discriminating than the first test which can predict the likelihood of *one or more RORs*. Note that the third test combines the first two tests but on the *accumulated cashflows*, rather than on the *original (given) cashflows* to which Tests 1 and 2 are applied.

The fourth (net investment) test is applied after the ROR has been evaluated. It confirms the uniqueness of the evaluated ROR. According to this test, the evaluated ROR is the correct rate of return if the given cashflows pass the following two subtests.

Subtest 4.1 Is the net investment zero at the last time period?

Subtest 4.2 Is there net investment throughout the project life?

These two subtests ensure that the project is always an investment[28], and at the end of its life there is no fund left, that is, the benefits have exactly paid for all the costs.

As you may have realized by now, a comprehensive ROR analysis is quite complex compared to the other five methods. Example 9.5 illustrates such an analysis by considering hybrid cashflows.

EXAMPLE 9.5

Determine the internal rate of return for the project whose cashflows are given below. If required, use an external investment interest rate of 5%.

[28]Being a net investment means that at each period the current and preceding cashflows yield a negative balance. In other words, the project never has a surplus cash.

Year	Cashflow
0	−$3,000
1	2,300
2	900
3	−200
4	322

Solution

Since the internal ROR is to be evaluated, a comprehensive analysis is called for. We evaluate the IROR and also confirm its uniqueness through the fourth test, discussed earlier. For a comprehensive analysis we conduct all the four tests, the first three in the beginning on the given cashflows and the fourth one at the end for the evaluated ROR.

Test 1: Algebraic Sum of the Cashflows

Here, the algebraic sum of the given cashflows is $-3{,}000 + 2{,}300 + 900 - 200 + 322 = 322$. Since it is positive, the benefits outweigh the costs; hence, one or more positive RORs are likely. Note that this test ignores the time value of money.

Test 2: Cashflow Sign Change

Examine the cashflows to determine whether they are pure or hybrid. The cashflows are hybrid, since there is more than one sign change. Had the cashflows been pure, we would have stopped here and determined the sole ROR for the project[29]. The second test predicts the maximum number of RORs possible for the project. In this case, the cashflows undergo three sign changes, which means that the project can have up to three RORs. The second test is more discriminating than the first since it can predict the likely number of RORs.

Test 3: Accumulated Cashflow Sign Change

The third test involves accumulated cashflows. To generate and examine them, we widen the original table by creating an additional column for the accumulated cashflows, as below.

Year	Cashflow	Accumulated Cashflows
0	−$3,000	−$3,000
1	2,300	−700
2	900	200
3	−200	0
4	322	322

[29]The first two tests are conducted in all ROR analyses. In fact, the first test should be conducted prior to all economic analyses, by any of the methods discussed in the text, since it reveals any inadmissibility of the given data. A negative algebraic sum of the given cashflows means that the project is not investment worthy at all, since the costs outweigh the benefits.

The entries in the last column have been computed as

Period	Accumulated Cashflow
0	$= -3{,}000$
1	$2{,}300 + (-3{,}000) = -700$
2	$900 + (-700) = 200$
3	$-200 + 200 = 0$
4	$322 + 0 = 322$

Once the accumulated cashflows have been posted we apply on them the two subtests.

Subtest 3.1: We note that the accumulated cashflow for the last period[30], being 322, is positive.

Subtest 3.2: We also note that there is only one sign change in the accumulated cashflows.

On the basis of these two sufficient conditions we can say that only one (positive) ROR is likely for the project.

Test 4: Net Investment

The fourth test is conducted later after the ROR has been evaluated.

We now proceed to evaluate the ROR. Following the procedure discussed earlier and illustrated in Example 9.2, the PW function in this case is

$$2{,}300(P/F, i, 1) + 900(P/F, i, 2) + 322(P/F, i, 4) - 200(P/F, i, 3) - 3{,}000 = 0$$

Dividing[31] throughout by 200 to dwarf the multipliers, we get

$$11.5(P/F, i, 1) + 4.5(P/F, i, 2) + 1.61(P/F, i, 4) - (P/F, i, 3) - 15 = 0$$

We solve this by trial and error, as explained earlier. The solution begins with a guessed beginning value of i. We can mentally work out such an i by relating the total interest with investment.

The total interest from the project, ignoring the time value of money, is \$322 (obtained in Test 1 or 3) over four years. This is approximately \$80 per year. For an initial investment of \$3,000, this yields an approximate interest rate of $80/3{,}000 \approx 3\%$. We therefore begin the trial with $i = 3\%$.

---

[30]Note that it must equal the algebraic sum of Test 1. Thus, Subtest 3.1 and Test 1 are the same. Test 1 is conducted earlier for the benefit of pure-cashflow problems, which don't need to undergo Tests 3 and 4.

[31]After such a division, the left side of this function is no longer the present worth.

With $i = 3\%$, the left-side of the function is

$$\begin{aligned} &11.5(P/F, 3\%, 1) + 4.5(P/F, 3\%, 2) + 1.61(P/F, 3\%, 4) - (P/F, 3\%, 3) - 15 \\ &= 11.5 \times 0.9709 + 4.5 \times 0.9426 + 1.61 \times 0.8885 - 0.9151 - 15 \\ &= 11.1654 + 4.2417 + 1.4305 - 15.9151 \\ &= 0.9225 \end{aligned}$$

Since it is positive, we need to increase i to reduce it to zero (to equal the right side of the PW function). So we try $i = 4\%$, which yields 0.7051. The decrease from 0.9225 to 0.7051 for a 1% increase in i seems slow, so we increase i for the next trial by 2%. With i as 6%, the left-side becomes 0.2898. Next, we try 8%, and so on. The results of the trial and error are

i	Calculations	Function's Left Side
3%	$11.5 \times 0.9709 + 4.5 \times 0.9426 + 1.61 \times 0.8885 - 15.9151$	$= 0.9225$
4%	$11.5 \times 0.9615 + 4.5 \times 0.9246 + 1.61 \times 0.8548 - 15.8890$	$= 0.7051$
6%	$11.5 \times 0.9434 + 4.5 \times 0.8900 + 1.61 \times 0.7921 - 15.8396$	$= 0.2898$
8%	$11.5 \times 0.9259 + 4.5 \times 0.8573 + 1.61 \times 0.7350 - 15.7938$	$= -0.1048$

As soon as the function's value changes sign, we stop the calculations, and interpolate the value of i that will render the left side of the equality zero. In the present case, this happens between 6% and 8%. The interpolation[32] yields the value of i, which is the ROR, as

$$\begin{aligned} \text{ROR} = i &= 6 + 2\,(0.2898)/(0.2898 + 0.1048) \\ &= 6 + 1.47 \\ &= 7.47\% \end{aligned}$$

Finally, we conduct the fourth test to confirm the correctness of the evaluated 7.47% ROR. We do this by examining whether or not the net cashflows are investments. This accounts for the time value of money at an interest rate equal to the computed ROR—7.47% in this case. The calculations and their results are

Year	Cashflow	Calculations	Net Cashflow
0	−\$3,000		= −\$3,000
1	2,300	−3,000 (1 + 0.0747) + 2,300 =	−924
2	900	−924 (1 + 0.0747) + 900 =	−93
3	−200	−93 (1 + 0.0747) − 200 =	−300
4	322	−300 (1 + 0.0747) + 322 =	0

[32]The accuracy of the results could have been improved by trying $i = 7\%$, and interpolating between 7% and 8%, as discussed earlier in this chapter.

For period 1, net cashflow is the cashflow for year 1 ($2,300) plus the time-valued period 0 cashflow $[-3{,}000(1 + 0.0747)]$. Other net cashflows have been obtained the same way. For example, for period 3, it is the cashflow for period 3 (−$200) plus the time-valued net cashflow up to period 2 $[-93(1 + 0.0747)]$.

After posting the net cashflows (last column), we examine them to see whether the two conditions of the fourth test are met. First, the net investment at project end should be zero, which is met in this case, since the net cashflow at period 4 is zero. Second, the net cashflow at each time period should be negative, so that the project continues to be a net investment. This is also true in the last column data, since all the cashflows there are negative.

Thus, at 7.47%, the project has continued to be an investment throughout its life, never generating any surplus that could have been invested externally. With the two conditions of Test 4 satisfied, the computed 7.47% is the correct ROR for the project. In other words, this value of ROR ensures profitability as long as it is greater[33] than or equal to the MARR. Thus, 7.47% is the *internal rate of return* (*IROR*).

Example 9.5 was intentionally framed to keep the comprehensive analysis simple. Real-world ROR analysis can be more difficult. The following comments pertain to some of the difficulties that may arise in such analyses. The comments should help you appreciate the extent of complexities and in coping with them.

1. The net investment test (Test 4) fails, that is, one or more net cashflows are positive. This means that there is surplus cash in the project at certain periods. Due to the pooling of investment capital, this surplus is taken out of the project and invested elsewhere at the given external ROR (or MARR).

   The ROR analysis in such a case should invest externally *only enough*[34] to render the net cashflows negative throughout the project, still achieving zero net cashflow at project end. This manipulation allows for external investment of surpluses without "starving" the project of funds, leading to the *true* ROR (or internal ROR). For instance, if in Example 9.5 the net cashflow for year 2 is $150 rather than −$93, the $150 surplus would be invested externally at 5%. The net investment for year 3 would then be $\$150(1 + 0.05) - 200 = -\$43$.

   External investment of the project surplus adds further realism to the decision. If allowed to "sit idle" in the project, the surplus would have been assumed to earn interest at the internal rate of return. This is not true in the real world, where funds are pooled together.
2. The aim of manipulating the hybrid cashflows is not to achieve one sign change, but to satisfy Test 4, namely zero net cashflow at project end and net investment throughout the project life.

---

[33]In the absence of any given value we can assume that the MARR equals the external rate of return of 5%, since had there been a surplus in the project it would have earned interest at 5%.

[34]Invest externally as little as necessary to render the project a net investment throughout its life.

3. In some problems, Test 2, based on the accumulated-cashflow-sign-change rule may predict fewer RORs than Test 1 does based on the cashflow-sign-change rule. This is due to the sharper discriminating power of Test 2.
4. Don't stop at Test 3 just because one positive ROR could be computed. Only when Test 4 has been gone through and its two conditions met can we be sure of the computed ROR as a measure of profitability.
5. A project may yield a ROR that is lower than the external ROR (or MARR), signifying its investment unworthiness.
6. If in Test 2 the accumulated cashflow at the last period is zero, along with more than one sign change in the accumulated cashflows, it simply means that one of the RORs is zero.
7. If in Test 2 the accumulated cashflow at the last period is negative, along with more than one sign change in the accumulated cashflows, it means that one of the RORs is negative.

9.6 MULTIPLE ALTERNATIVES

Engineering economics projects can be categorized into one of two groups: no-alternative or multialternative. In no-alternative projects, there is only one candidate, and hence no choice. We have extensively discussed ROR analysis of such a project in the previous sections. It involves determining the project ROR and comparing it with the MARR set by the management. If the project ROR is financially attractive, the project is funded. As an alternative to evaluating the project ROR exactly, we may evaluate its PW with MARR as the interest rate and decide on the basis of the PW criterion (Example 9.3).

Let us begin the discussion of multialternative projects by first considering two-alternative[35] projects.

9.6.1 Two Alternatives

In Chapters 6, 7, and 8 we have discussed PW, FW, and AW methods of analysis. There the logical criterion in selecting the better of the two alternatives has been higher PW, FW, or AW. Unfortunately, this logic is unsound in the case of ROR analysis. The alternative with the higher ROR may not be better. The ROR analysis of two (or more) alternatives is based on a technique called *incremental analysis*.

9.6.1.1 Incremental analysis

Under incremental analysis the two alternatives are compared for the differences in their cashflows. The incremental data are obtained by subtracting the data of the lower-investment alternative (lower first cost) from those of the other alternative (higher first

[35]A two-alternative project has two candidates, each being an alternative to the other. Purists may call it a one-alternative project.

cost). The difference in the first cost so obtained is positive, representing an *increment*, and hence the name *incremental analysis*. Using the *incremental data* (or *cashflows*) we determine the *incremental rate of return*, denoted by ΔROR. The computed ΔROR is then compared with the given MARR for making decisions the usual way, that is, if ΔROR ≥ MARR, then select the higher-investment alternative, otherwise select the lower-investment alternative. The concept underlying incremental analysis is explained next.

Suppose machine *A* costs $4,000 and yields certain benefits. Another machine *B* can also do the job, but it costs $6,000 and yields benefits that *look superior*. The question is; Is *B* better than *A* from ROR viewpoint? In other words, we want to know whether the additional cost of $2,000 for *B* is worth the additional benefits expected from it. So our goal is to determine the *incremental rate of return* ΔROR by analyzing the additional cost against the additional benefits. If ΔROR is greater than the MARR, that is, the additional $2,000 cost brings in return in excess of the MARR, then machine *B* is selected. Otherwise, alternative *B* is discarded; and *A* is selected provided $ROR_A \geq$ MARR. Conceptually, the alternative with the higher first cost is like a *challenger*. The incremental technique judges the challenger's investment worthiness not on its own but in comparison to that of the lower-investment alternative (defender), which would have been selected in the absence of the challenger. In the illustration under discussion, *A* is the defender because in the absence of its challenger *B*, we would have selected *A* due to its lower first cost. Example 9.6 applies incremental analysis to a two-alternative project.

EXAMPLE 9.6

Two mutually exclusive alternatives *A* and *B* exist for a project. Which alternative should be selected on the basis of ROR if the MARR = 10%?

n	A	B
0	−$4,500	−$18,000
1	2,000	6,500
2	2,700	9,250
3	2,250	9,500

Solution

Since there are two alternatives and the decision criterion is ROR, the solution must be based on incremental analysis.

Which of the two alternatives costs less? It is *A*. So we would prefer to invest in *A*. But since alternative *B* exists, we would like to analyze it for its additional cost and benefits to see whether investing in *B* makes sense. The analysis is carried out therefore on the incremental data *B*−*A*. We begin by creating another column to post the *B*−*A* data.

n	A	B	$B-A$
0	−$4,500	−$18,000	−$13,500
1	2,000	6,500	4,500
2	2,700	9,250	6,550
3	2,250	9,500	7,250

From the $B-A$ data for period 0, investment in B is $13,500 more than A. Other $B-A$ data are the benefits from B over that from A.
At this point of the solution procedure there are two choices:

1. Determine ΔROR for the $B-A$ incremental cashflows, or
2. Since the MARR is given, check whether PW for the $B-A$ incremental cashflows is positive.

The second choice is usually easier in terms of calculations, and is preferred if the MARR is given, as here. So we opt for this choice. PW for the $B-A$ cashflows corresponding to 10% MARR is given by

$$\begin{aligned} \text{PW}_{B-A} &= -13{,}500 + 4{,}500(P/F, 10\%, 1) + 6{,}550(P/F, 10\%, 2) + 7{,}250(P/F, i, 3) \\ &= -13{,}500 + 4{,}500 \times 0.9091 + 6{,}550 \times 0.8264 + 7{,}250 \times 0.7513 \\ &= -13{,}500 + 4{,}091 + 5{,}413 + 5{,}447 \\ &= 1{,}451 \end{aligned}$$

Since PW_{B-A} corresponding to the MARR is positive, the actual value of ΔROR will be greater than MARR. Hence, B is selected[36].

Were we to evaluate ΔROR in the above example we would have used the trial and error method discussed earlier. This would have involved evaluating i that would render the PW function zero. The result would have yielded[37] ΔROR_{B-A} greater than the MARR, selecting B over A.

---

[36]One must however check that PW_B is positive, as below.

$$\begin{aligned} \text{PW}_B &= -18{,}000 + 6{,}500(P/F, 10\%, 1) + 9{,}250(P/F, 10\%, 2) + 9{,}500(P/F, i, 3) \\ &= -18{,}000 + 6{,}500 \times 0.9091 + 9{,}250 \times 0.8264 + 9{,}500 \times 0.7513 \\ &= -18{,}500 + 5{,}909 + 7{,}644 + 7{,}137 \\ &\approx 2{,}690 \end{aligned}$$

[37]Applying Tests 1 and 2 on the incremental $B-A$ cashflows, we can see that only one ΔROR is likely. The relationship PW = 0 is

$$-13{,}500 + 4{,}500(P/F, i, 1) + 6{,}550(P/F, i, 2) + 7{,}250(P/F, i, 3) = 0$$

Dividing throughout by 500, we have

$$-27 + 9(P/F, i, 1) + 13.1(P/F, i, 2) + 14.5(P/F, i, 3) = 0$$

The beginning i for the trial can be approximated in the way explained in earlier examples. The approximate total incremental interest is $(4{,}500 + 6{,}550 + 7{,}250) - 13{,}500 = \$4{,}800$ over three years. So yearly interest is $1,600, giving an approximate incremental interest rate of $1{,}600/13{,}500 \approx 12\%$. Thus, the trial and error analysis is begun with $i = 12\%$. With a few trials, the exact value of $i(\Delta\text{ROR}_{B-A})$ is computed as 15.6%.

9.6.2 More Than Two Alternatives

ROR analysis of projects with more than two alternatives is basically an extension of that for two alternatives discussed in the preceding subsection. Here too, the selection based on the highest *individual ROR* can lead to erroneous decisions. Hence incremental analysis is carried out, as illustrated in Example 9.7.

EXAMPLE 9.7

If each of the following four mutually exclusive alternatives has 8 years of useful life, which one should be selected based on ROR if the MARR = 8%?

	A	*B*	*C*	*D*
First cost	\$600	\$500	\$965	\$800
Annual benefit	100	120	130	110
Salvage value	375	40	800	747

Solution
Incremental analysis is essential, since ROR is the criterion in a multialternative project. The procedure comprises two phases.

Phase I
The first phase is a screening process. In this phase, we determine each alternative's ROR on its own to check whether it is greater than the given MARR. Alternatives with a ROR less than the MARR are discarded, and hence do not undergo incremental analysis in phase II.

Since the MARR is given, we can opt for the process of checking the sign of alternatives' PW at the MARR. This does away with the need to evaluate the ROR. Alternatives whose PW at the MARR are positive will yield a ROR greater than the MARR, and hence participate in the incremental analysis.

$$\begin{aligned} PW_A &= 100(P/A, 8\%, 8) + 375(P/F, 8\%, 8) - 600 \\ &= 100 \times 5.747 + 375 \times 0.5403 - 600 \\ &= 575 + 203 - 600 \\ &= 178 \end{aligned}$$

$$\begin{aligned} PW_B &= 120(P/A, 8\%, 8) + 40(P/F, 8\%, 8) - 500 \\ &= 120 \times 5.747 + 40 \times 0.5403 - 500 \\ &= 690 + 22 - 500 \\ &= 212 \end{aligned}$$

$$\begin{aligned} PW_C &= 130(P/A, 8\%, 8) + 800(P/F, 8\%, 8) - 965 \\ &= 130 \times 5.747 + 800 \times 0.5403 - 965 \\ &= 747 + 432 - 965 \\ &= 214 \end{aligned}$$

$$\begin{aligned}\text{PW}_D &= 110(P/A, 8\%, 8) + 747(P/F, 8\%, 8) - 800\\ &= 110 \times 5.747 + 747 \times 0.5403 - 800\\ &= 632 + 404 - 800\\ &= 236\end{aligned}$$

Since the present worths of all the four alternatives are positive, each one participates in the incremental analysis.

Phase II

The procedure in this phase is similar to that in Example 9.6. We compare two alternatives at a time. We should prefer to select alternative *B*, since its \$500 first cost is the lowest. The next-higher-first-cost alternative is *A* with its \$600 first cost. So we compare *A* with *B* to determine whether the additional cost of \$100 for *A* brings in enough additional benefits to justify this cost, that is, whether ΔROR_{A-B} is greater than the MARR. The winner between *A* and *B* is compared with the next-higher-first-cost alternative, and the process continues until all the alternatives have been examined incrementally for their worthiness.

The $A-B$ incremental analysis is begun by tabulating the data under a new column, as below. We also include a column for the time period.

Year	*B*	*A*	*A−B*
0	−500	−600	−100
1–8	120	100	−20
8	40	375	335

Again, since the MARR is given, we can use it as *i* and evaluate the incremental PW_{A-B}, based on whose sign either *A* or *B* is selected. As explained in Example 9.6, this approach is usually shorter than evaluating ΔROR exactly[38].

---

[38]The analysis of the incremental $A-B$ cashflows to evaluate ΔROR_{A-B} is more involved. By applying Tests 1 and 2 (Example 9.5) on the cashflows we predict that, with only one sign change, one positive ΔROR_{A-B} is likely. This is determined from the PW function

$$-20(P/A, i, 8) + 335(P/F, i, 8) - 100 = 0$$

A trial and error solution yields $i_{A-B} = 4.8\%(\Delta\text{ROR}_{A-B})$. Since ΔROR_{A-B} is less than the 8% MARR, the additional cost of *A* is not justified. Hence we decide to discard *A* in favor of *B*.

Next, *B* is compared with the next-higher-first-cost alternative, i.e., *D*. The procedure is very similar. The PW function for $D-B$ incremental cashflows is

$$-10(P/A, i, 8) + 707(P/F, i, 8) - 300 = 0$$

From trial and error, ΔROR_{D-B} is found to be 9% . Since this is greater than the MARR of 8%, alternative *D* wins.

Alternative *D* is finally compared with *C* in the same way. The PW function for $C-D$ incremental cashflows is

$$20(P/A, i, 8) + 53(P/F, i, 8) - 165 = 0$$

From trial and error, ΔROR_{C-D} is found to be 5%. Since this is less than the 8% MARR, alternative *C* loses to *D*.

Thus alternative *D* is selected.

Thus, we determine PW_{A-B} corresponding to 8% as

$$\begin{aligned} PW_{A-B} &= 335(P/F, 8\%, 8) - 20(P/A, 8\%, 8) - 100 \\ &= 335 \times 0.5403 - 20 \times 5.747 - 100 \\ &= 181 - 115 - 100 \\ &= -34 \end{aligned}$$

Since PW_{A-B} is negative, ΔROR_{A-B} will be lower than the MARR. So additional investment in A is not worthwhile; B is the winner.

We now compare B with the next-higher-first-cost alternative, which is D. Generate the $D-B$ data the way we did for $A-B$, and determine PW_{C-D} as

$$\begin{aligned} PW_{D-B} &= 707(P/F, 8\%, 8) - 10(P/A, 8\%, 8) - 300 \\ &= 707 \times 0.5403 - 10 \times 5.747 - 300 \\ &= 382 - 57 - 300 \\ &= 25 \end{aligned}$$

Since PW_{D-B} is positive, ΔROR_{D-B} will be greater than the MARR. So additional investment in D is worthwhile; hence D is selected over B.

Next we compare D with C, the next-higher-first-cost alternative. Generating the $C-D$ data, we determine PW_{C-D} as

$$\begin{aligned} PW_{C-D} &= 53(P/F, 8\%, 8) + 20(P/A, 8\%, 8) - 165 \\ &= 53 \times 0.5403 + 20 \times 5.747 - 165 \\ &= 29 + 115 - 165 \\ &= -21 \end{aligned}$$

Since PW_{C-D} is negative, additional investment in C can't be justified. Hence C is discarded.

Therefore, alternative D should be selected.

Note that all the complexities of analyzing hybrid cashflows, some discussed in Section 9.5, can exist in the case of multialternative projects. If at all, they may be more pronounced because of several alternatives.

9.6.3 Analysis Period

Similar[39] to the PW analysis, ROR analysis must take care of differential useful lives of the alternatives. This is done by selecting an analysis period so that all the alternatives terminate at the same time. This may require replacing the alternatives *several times during the analysis period*. The cashflow data of the alternatives replaced are usually

---

[39]See Chapter 6 for an extensive discussion.

assumed to remain unchanged. As discussed in Chapter 6, the analysis period is normally the LCM of the alternatives' useful lives. If the LCM-based analysis period is too long, a shorter period can be used provided the data are carefully prorated. To avoid the analysis complexity, the AW method may be preferred, since unequal lives do not create such a difficulty in this method.

9.7 COMPUTER USE

The discussions presented in this chapter and the accompanying illustrations must have convinced you by now of the complexities of ROR analysis. The repetitive nature of the calculations in the functional-notation method, arising from the trial and error approach, further accentuates the complexity. Programmable calculators and PCs ease the calculation efforts. Software are appropriate tools for solving problems under the ROR criterion. However, the objective of this chapter, indeed of the entire text, is *education*, focusing on the principles of engineering economics, rather than the *training* that can get emphasized in a software-centered learning environment. That is why I chose to take you through the "dirt road."

SUMMARY

The rate of return (ROR) method has been discussed extensively in this chapter. ROR can be defined as the interest rate earned on the remaining balance of an investment. For a project to be investment worthy, its ROR must be greater than or equal to the minimum acceptable rate of return (MARR). The MARR is set by management based on several financial factors, both within and without the company. If the MARR is known as given, a short-cut approach to ROR analysis is through project PW corresponding to the MARR. If such a PW is positive, project ROR will exceed the MARR.

When the cashflows are *pure*, there is only one ROR for a project. For hybrid cashflows, more than one ROR is possible; such cashflows involve sophisticated analysis, demanding four different tests.

A project has an internal rate of return (IROR) if its surplus cash is invested elsewhere at an external rate of return. The incremental technique is essential in ROR analysis of projects with alternatives. For a project with alternatives of unequal lives, an analysis period equal to the LCM of the lives is normally used. Alternatively, the AW method could be used to keep the decision making simpler.

That ROR is easier to comprehend is the attraction of ROR analysis. On the other hand, ROR analysis is both complex and lengthy. The advent of computer technology has simplified ROR analysis to some extent, with the result that this analysis method is the most widely used in industry.

EXERCISES

Discussion Questions

9.1 Explain the difference between interest rate i and ROR.
9.2 Is MARR the same as interest rate i? Explain.
9.3 Why is the evaluation of ROR usually lengthy as well as complex?
9.4 Why is ROR widely used in industry in spite of its analysis complexities?
9.5 Discuss the four tests pertinent to a comprehensive ROR analysis.
9.6 Given a project's cashflows, how can you be sure of only one admissible ROR?
9.7 Explain the concept underlying incremental analysis.
9.8 Should one evaluate the exact value of the ROR in ROR analysis if the MARR is given? If yes, why? If no, why not?
9.9 Explain the concept behind the net investment test.

Multiple-Choice Questions (Circle the *best* answer.)

9.10 The ROR method of analyzing engineering economics problems is
a. complex.
b. popular.
c. both a and b
d. neither a nor b

9.11 For the following cashflows,

Year	Cashflow
0	−\$500
1	0
2	0
3	400
4	−50
5	300

the project may have up to
a. one ROR.
b. two RORs.
c. three RORs.
d. four RORs.

9.12 For the cashflows in Problem 9.11, the function for evaluating the ROR is
a. $500 + 50(P/F, i, 4) - 400(P/F, i, 3) + 300(P/F, i, 5) = 0$
b. $-500 - 50(P/F, i, 2) + 400(P/F, i, 1) + 300(P/F, i, 3) = 0$
c. $500 - 50(P/F, i, 4) + 400(P/F, i, 3) + 300(P/F, i, 5) = 0$
d. none of the above

9.13 Incremental method is *not* used in ROR analysis of projects with
a. no alternative.
b. two alternatives.
c. three alternatives.
d. more than three alternatives.

9.14 In multialternative projects, decision making based on the ROR of individual alternatives is
a. relatively difficult.
b. erroneous.
c. impossible without a computer.
d. widely practiced in industry.

Numerical Problems

9.15 The first cost of a waterjet machine to be used for slicing cheese is $60,000. The machine will generate net annual income of $12,000 during its useful life of 10 years. Determine the rate of return on investment in this machine.

9.16 Mary borrows $80,000 to buy her home. Beginning next month she will pay the lender $850 per month for the next 20 years. What ROR will the lender be enjoying?

9.17 Cutting-Edge R&D Company is considering acquiring a new computing system for $300,000. The estimated economic life of the system is five years with no salvage value. If the benefit for the first year is estimated to be $100,000 and for the subsequent years $150,000 annually, should the system be acquired? Assume MARR = 20%.

9.18 American Indians sold an island in 1630 to a Dutch developer for some glass beads and trinkets worth $24. The island was worth $15 billion in 1999. What rate of return have the developer's beneficiaries enjoyed on the deal?

9.19 For the given cashflows, which of the two alternatives should be selected on the basis of rate of return if MARR = 25%?

Year	A	B
0	−$1,600	−$3,000
1–5	800	1,400

9.20 Which of the following three alternatives should be selected if MARR = 25%? Note that *C* is a "do-nothing" alternative (maintain the status quo).

Year	A	B	C
0	−$1,600	−$3,000	−$0
1–5	800	1,400	0

9.21 Twenty-First Century R&D Company has won a 2-year contract from NASA for a feasibility study to develop a township in space. The company will receive an advance of $30 million dollars on signing the contract and two more payments of $31.5 million and $4 million respectively on the first and second anniversaries of the contract. The costs of the project are estimated to be $35 million now, $20 million during year 1, and another $10 million during year 2. Determine the ROR for the project.

9.22 For the following cashflows what is the internal ROR, if external investment can earn 10%?

n	Cashflow
0	−$1,600
1	1,000
2	1,000
3	−600
4	800
5	550

9.23 Three types of used robots can do a loading and unloading job. Considering these alternatives to be mutually exclusive (the selection of one precludes the others), which one should be selected on the basis of incremental rate of return if the MARR = 15%?

Year	A	B	C
0	−$4,000	−$2,000	−$6,000
1	3,000	1,600	3,000
2	2,000	1,000	4,000
3	1,600	1,000	2,000

9.24 For the following cashflows,

Year	Cashflow
0	−$1,275
1	900
2–9	300
10	−2,700

a. Compute the PW at 10% MARR.
b. Plot the PW profile and determine the two RORs.
c. Compute the internal ROR by considering an external interest rate of 6%, if necessary.

CHAPTER TEN

Benefit–Cost Ratio

IN THIS CHAPTER YOU WILL LEARN ABOUT

- The meaning of *benefit–cost ratio* (BCR)
- Evaluation of a project's BCR
- Application of the BCR criterion
- The concept of disbenefit
- Incremental BCR analysis
- The use of the input–output concept

The benefit–cost ratio (BCR) is the sixth and last method of analysis. It is popular with governments, which make decisions on how to spend taxpayers' money on public projects to maximize the common good. Profit-focused industries seldom use this method.

The BCR is easy to comprehend. It is simply the ratio of benefits to costs. A benefit–cost ratio greater than one means that the benefits outweigh the costs, and hence the project is investment worthy. As in the other criteria discussed in the last four chapters, the BCR accounts for the time value of money.

10.1 MEANING

Consider a project that costs \$100 today and generates a benefit of \$120 next year. We might be tempted to say that the BCR = \$120/\$100 = 1.2. But this is not correct, since this way we have not accounted for the time value of money. We know that a sum of \$120 next year is worth less today due to the interest for the year. Its equivalent value today is only $120(P/F, i, 1)$. Let us say i is 8% per year, for which $(P/F, 8\%, 1) = 0.9259$ from the interest table. The \$120 benefit of the next year is therefore equivalent to $\$120 \times 0.9259 \approx \$111$ of today. Therefore, the correct BCR is $\$111/\$100 \approx 1.11$, since we have accounted for the time value of money.

While evaluating the BCR, both the benefits and costs of the project must refer to the same time period. In the preceding paragraph, the reference time for determining the BCR was the *present,* that is $t = 0$. We could instead have considered $t = 1$ as the

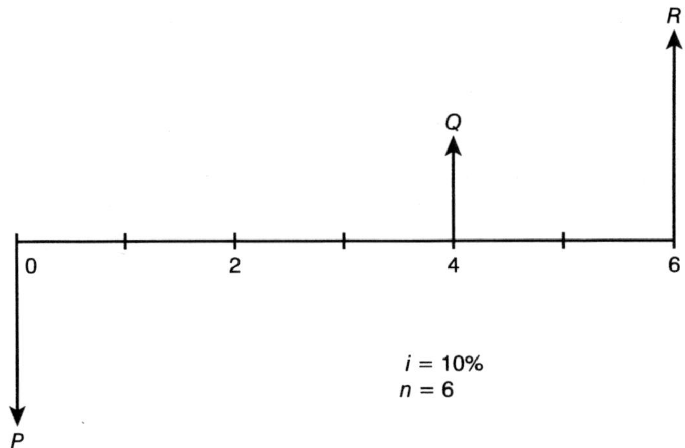

Figure 10.1 Concept of Benefit–Cost Ratio

reference time. In that case, the BCR = \$120/\$100(F/P, 8%, 1) = 120/(100 × 1.080) = 1.11. This is the same as that for $t = 0$. Thus, a project's BCR is unique if the ratio is obtained by referring the costs and benefits to the same time period.

Consider a project whose cashflow diagram is Fig. 10.1. The investment P brings in benefits Q and R. Referring to which time period should the BCR be evaluated—period 0, period 4, or period 6 where there are vectors, or for another period where there is no vector? As explained in the previous paragraphs, the BCR for the project will be the same irrespective of the time period at which it is evaluated. In general, the BCR is evaluated at $t = 0$, referring to the present when investment decision is being made. Doing so for the diagram in Fig. 10.1 for $i = 10\%$,

$$\text{BCR} = \frac{Q(P/F, 10\%, 4) + R(P/F, 10\%, 6)}{P}$$

For projects to be investment worthy, their time-valued benefits must exceed the costs. The BCR of such a project therefore must be greater than one. For no-alternative projects (one candidate only), the higher the BCR more favorable the project.

10.2 EVALUATION

Generally, a project's BCR is determined by evaluating its $\text{PW}_{\text{benefits}}$ and PW_{costs} and dividing the former by the latter. Thus,

$$\text{BCR} = \frac{\text{PW}_{\text{benefits}}}{\text{PW}_{\text{costs}}} \tag{10.1}$$

Where it is easier to evaluate $FW_{benefits}$ and FW_{costs}, or where they are already known, the BCR is determined from

$$\text{BCR} = \frac{FW_{benefits}}{FW_{costs}} \tag{10.2}$$

Likewise where it is easier to evaluate $AW_{benefits}$ and AW_{costs}, or where they are already known, the BCR is determined from

$$\text{BCR} = \frac{AW_{benefits}}{AW_{costs}} \tag{10.3}$$

We have already learned in Chapters 6, 7, and 8 how to evaluate the numerators and denominators of these three equations. Thus, the BCR analysis is a reformatted[1] PW, FW, or AW analysis. Examples 10.1 and 10.2 illustrate the evaluation of the BCR.

EXAMPLE 10.1

Should the project represented by the cashflow diagram in Fig. 10.1 be funded if $P = \$350$, $Q = \$200$, and $R = \$400$? Assume $i = 10\%$ per year.

Solution

The \$350 cost is incurred now, while the two benefits are realized in the future. We need to determine the present value of these benefits, which is

$$\begin{aligned} PW_{benefits} &= Q(P/F, 10\%, 4) + R(P/F, 10\%, 6) \\ &= 200 \times 0.6830 + 400 \times 0.5645 \\ &= 136.60 + 225.80 \\ &= \$362.40 \end{aligned}$$

With the \$350 cost at period 0 being the only cost, we have

$$PW_{costs} = \$350$$

---

[1]Referring to Chapter 6, where the PW criterion was discussed, we had Equation (6.1) as

$$PW = PW_{benefits} - PW_{costs}$$

Since investment-worthy projects must have a PW ≥ 0, we can say

$$PW_{benefits} - PW_{costs} \geq 0$$

Dividing by PW_{costs} this becomes

$$PW_{benefits}/PW_{costs} - 1 \geq 0$$
$$PW_{benefits}/PW_{costs} \geq 1$$

Since we define the BCR as $PW_{benefits}/PW_{costs}$, it can be said that for a project to be investment-worthy, the BCR ≥ 1.

Therefore,

$$\begin{aligned} \text{BCR} &= \frac{\text{PW}_{\text{benefits}}}{\text{PW}_{\text{costs}}} \\ &= 362.40/350 \\ &= 1.04 \end{aligned}$$

Since the BCR is greater than one, the project should be funded.

EXAMPLE 10.2

As part of ongoing industrialization based on external capital, Bangladesh asks the World Bank for a loan to construct a canal to divert the waters of the Ganges. The estimated construction cost is \$5 million, and the annual maintenance cost is \$100,000. The canal will carry water for irrigation, yielding \$400,000 annually as fees from farmers. If $i = 8\%$ per year and the canal is expected to last for 50 years, should the World Bank sanction the loan based on the BCR criterion?

Solution

The one-time construction cost is \$5 million. The other two cashflows are annual. Let us assume that the maintenance cost will be paid out of the fees collected from the farmers. Thus,

$$\begin{aligned} \text{Net annual benefit} &= \text{Fees collected} - \text{Maintenance cost} \\ &= \$400{,}000 - \$100{,}000 \\ &= \$300{,}000 \end{aligned}$$

We can base the BCR evaluation on PW, FW, or AW. Let us follow the PW approach, since the construction cost already given at period zero will not require transferring. We determine the equivalent present value of the 50-year net annual benefits as

$$\begin{aligned} \text{PW}_{\text{benefits}} &= 300{,}000(P/A, 8\%, 50) \\ &= 300{,}000 \times 12.233 \\ &= \$3{,}669{,}900 \end{aligned}$$

From the construction cost, being the only cost, we have

$$\text{PW}_{\text{costs}} = \$5{,}000{,}000$$

Therefore,

$$\text{BCR} = \frac{\text{PW}_{\text{benefits}}}{\text{PW}_{\text{costs}}}$$

$$= 3{,}669{,}900/5{,}000{,}000$$
$$= 0.734$$

Since the BCR is less than one, the loan should be denied. There may, however, be reasons other than economic, such as social, political, or humanitarian[2] that may justify sanctioning the loan.

In Example 10.2 we might have considered the annual maintenance cost as part of the project cost, if so planned. In that case,

$$\begin{aligned} PW_{costs} &= \text{First cost} + \text{Present value of maintenance costs} \\ &= 5{,}000{,}000 + 100{,}000(P/A, 8\%, 50) \\ &= 5{,}000{,}000 + 100{,}000 \times 12.233 \\ &= 5{,}000{,}000 + 1{,}223{,}300 \\ &= \$6{,}223{,}300 \end{aligned}$$

The $400,000 in fees collected annually is realized as benefits. Hence,

$$\begin{aligned} PW_{benefits} &= 400{,}000(P/A, 8\%, 50) \\ &= 400{,}000 \times 12.233 \\ &= \$4{,}893{,}200 \end{aligned}$$

Therefore,

$$\begin{aligned} BCR &= \frac{PW_{benefits}}{PW_{costs}} \\ &= 4{,}893{,}200/6{,}223{,}300 \\ &= 0.786 \end{aligned}$$

Since the BCR is under one, the decision not to sanction the loan remains unchanged. However, consideration of the maintenance cost as part of the project cost changed the BCR from 0.734 (Example 10.2) to 0.786. This illustrates the major weakness of BCR analysis. The BCR is prone to misrepresentation arising from whether the operational costs are recovered from the benefits or treated as part of the first cost.

To minimize such a confusion, we distinguish between the initial (or first) cost and future (running or operational) costs of a project. *In general, only the initial investment is considered project cost.* The operational expenses are paid out of the gross benefits, resulting in net benefits as in Example 10.2. It is for this reason—to avoid confusion—that operational and other future costs are called *disbenefits*. The use of this

---

[2]The Ganges floods the downstream villages each year during the Monsoon season, resulting in thousands of deaths. The canal is expected to reduce the death toll.

terminology distinguishes the first cost from other costs. The disbenefits[3] are subtracted from gross benefits, yielding data for the numerator of the BCR equation. We can thus redefine BCR as

$$\text{BCR} = \frac{\text{Net benefit}}{\text{Initial cost}} \tag{10.4}$$

From Equation (10.4), the BCR is simply the *net benefit per unit initial investment.*

The BCR method is used mostly in analyzing public projects, as illustrated in Examples 10.3 and 10.4.

EXAMPLE 10.3

The twin cities of Laurel and Hattiesburg are planning to build a municipal waste recycling plant at an initial cost of $16 million. During its life of 25 years the plant will require major overhauls every five years at a cost of $300,000 each. Compared to other alternatives of garbage disposal this plant will annually save the 20,000 city households $75 each. Determine the project BCR if the interest rate is 6% per year.

Solution

Refer to the project cashflow diagram in Fig. 10.2. The one-time construction cost is $16 million. Every five years another $300,000 will be spent on overhaul. Thus a total of four overhauls will take place at the 5, 10, 15, and 20th year. The annual benefits are 20,000 × $75 = $1,500,000.

We can base the BCR evaluation either on PW or AW, since the given cashflows are lump as well as annual sums. In the former we convert the 25-year annual benefits to their present value and divide by the total of initial cost and the present value of overhaul costs. In the AW approach we convert the initial and overhaul costs to their equivalent annual cost, and divide the annual benefit by this equivalent cost. Based on what we have learned so far, the AW approach will be lengthier, since the overhaul costs will have to be converted first to their P value and then to the A value. Hence we follow the PW approach, wherefrom

$$\begin{aligned} \text{PW}_{\text{benefits}} &= 1{,}500{,}000(P/A, 6\%, 25) \\ &= 1{,}500{,}000 \times 12.783 \\ &= \$19{,}174{,}500 \end{aligned}$$

---

[3]Where it is difficult to judge whether a cash outflow is cost or disbenefit, we use a modified version of the BCR method. In this, rather than their ratio, the criterion is the difference between benefits and costs, called the *benefit–cost difference* (BCD). This difference is unaffected by whether cash outflows are treated as costs or disbenefits. Note, however, that the BCD is basically PW.

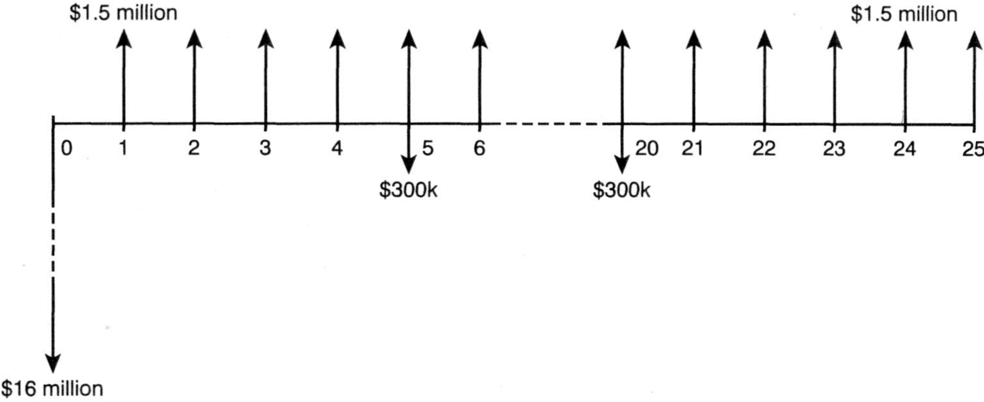

Figure 10.2 Diagram for Example 10.3

The present worth of the costs is

$$\begin{aligned}
\text{PW}_{\text{costs}} &= 16{,}000{,}000 + 300{,}000(P/F, 6\%, 5) + 300{,}000(P/F, 6\%, 10) \\
&\quad + 300{,}000(P/F, 6\%, 15) + 300{,}000(P/F, 6\%, 20) \\
&= 16{,}000{,}000 + 300{,}000[(P/F, 6\%, 5) + (P/F, 6\%, 10) \\
&\quad + (P/F, 6\%, 15) + (P/F, 6\%, 20)] \\
&= 16{,}000{,}000 + 300{,}000(0.7473 + 0.5584 + 0.4173 + 0.3118) \\
&= 16{,}000{,}000 + 300{,}000 \times 2.0348 \\
&= \$16{,}610{,}440
\end{aligned}$$

Therefore[4],

$$\begin{aligned}
\text{BCR} &= \frac{\text{PW}_{\text{benefits}}}{\text{PW}_{\text{costs}}} \\
&= 19{,}174{,}500/16{,}610{,}440 \\
&= 1.15
\end{aligned}$$

EXAMPLE 10.4

To boost the city economy, the mayor and the council are considering building a four-lane road that will provide faster access to the nearby interstate highway. The construction cost of the road, which includes a tunnel, is $300 million, while its annual

[4]Whether the overhaul costs should be treated as part of the initial investment cost or as a disbenefit depends on how the cities budget for their projects. We are assuming here that the benefits are paid to the city-dwellers as relief on garbage collection fees and are not available to the cities to pay for overhaul costs. If the cities plan to pay the overhaul costs out of the savings, then these costs are disbenefits. In that case, the BCR = (19,174,500 − 300,000 × 2.0348)/16,000,000 = 1.16.

maintenance is estimated to cost \$10 million. The annual benefits have been projected to be additional commerce valued at \$50 million, billboard advertising fees of \$5 million, and reduction in accidents, saving the city \$500,000. The disbenefits are \$2 million yearly compensation to farmers for their land and businesses, and \$300,000 annual loss in tax from the businesses along the existing dirt road. Assuming a 50-year life for the proposed road, should the project be funded if the decision criterion is BCR? The city uses 6% as annual interest rate in its funding decisions.

Solution

Since most of the given data are annual, we are better off using AW in the BCR evaluation. The net annual benefit is

$$\begin{aligned} AW_{\text{benefits}} &= \text{Total annual benefit} - \text{Total annual disbenefit} \\ &= (50{,}000{,}000 + 5{,}000{,}000 + 500{,}000) - (2{,}000{,}000 + 300{,}000) \\ &= 53{,}200{,}000 \end{aligned}$$

The annual worth of the costs is

$$\begin{aligned} AW_{\text{costs}} &= \text{First cost's annual worth} + \text{Annual maintenance cost} \\ &= 300{,}000{,}000(A/P, 6\%, 50) + 10{,}000{,}000 \\ &= 300{,}000{,}000 \times 0.0634 + 10{,}000{,}000 \\ &= 29{,}020{,}000 \end{aligned}$$

Therefore,

$$\begin{aligned} BCR &= \frac{AW_{\text{benefits}}}{AW_{\text{costs}}} \\ &= 53{,}200{,}000/29{,}020{,}000 \\ &= 1.83 \end{aligned}$$

Since the BCR is greater than one, the project should be funded.

10.3 MULTIALTERNATIVE PROJECTS

In the previous sections we discussed no-alternative projects. For such projects, the application of the BCR criterion involves checking whether the project BCR is more than one. In multialternative projects, BCR analysis becomes relatively complex, as is evident in this section.

As was discussed earlier, engineering economics projects may be one of the three types:

1. Fixed input
2. Fixed output
3. Variable input, variable output

In *fixed-input* projects, the alternatives' costs are the same, while in *fixed-output* projects, their benefits are the same. Multialternative projects of these two types are analyzed the same way. In both, the alternatives are compared for their individual BCR value[5], and the one with the higher (if two alternatives) or highest (if more than two alternatives) BCR is selected. Example 10.5 offers an illustration.

For variable-input, variable-output projects *incremental analysis* is carried out. We have discussed such an analysis in Chapter 9 under the ROR method. Its application to the BCR method is presented in Section 10.4.

EXAMPLE 10.5

Either robot X or Y can carry out the loading and unloading tasks for a conveyor system, saving \$20,000 annually. X costs \$50,000 and has a useful life of 7 years and a salvage value of \$5,000. The corresponding data for Y are \$40,000, 5 years, and \$4,000. With the interest rate at 8% per year, compounded annually, which one should be selected on the basis of the BCR criterion?

Solution

This is a *fixed-output* problem, because whether we select X or Y the output of the investment is the same, namely, automation of loading and unloading tasks, resulting in annual savings of \$20,000. Since the output is fixed, the decision making can be based on a comparison of the individual BCRs of X and Y.

Which of the two approaches, PW or AW, should be used for BCR evaluation in this case? There is an important reason why we adopt the AW approach. That way we overcome the analytic difficulty due to the different useful lives of X and Y. In the AW method, the analysis period is simply the alternative's life. The PW approach would have required an LCM-based analysis period of 35 years, which is too long.

AW-based evaluation means converting for each (X and Y) the first cost and salvage value to their annual equivalents over the useful life, and subtracting the latter from the former. Thus, the robots' BCRs are

$$
\begin{aligned}
\mathrm{BCR}_X &= \frac{\mathrm{AW}_{\text{benefits}}}{\mathrm{AW}_{\text{costs}}} \\
&= \frac{20{,}000}{50{,}000(A/P, 8\%, 7) - 5{,}000(A/F, 8\%, 7)} \\
&= \frac{20{,}000}{50{,}000 \times 0.1921 - 5{,}000 \times 0.1121} \\
&= 2.21
\end{aligned}
$$

---

[5]In fixed-input projects, since the cost is the same, one can select the alternative on the basis of benefit only. Likewise, in fixed-output projects, one can do the selection on the basis of cost only.

$$\text{BCR}_Y = \frac{\text{AW}_{\text{benefits}}}{\text{AW}_{\text{costs}}}$$
$$= \frac{20{,}000}{40{,}000(A/P, 8\%, 5) - 4{,}000(A/F, 8\%, 5)}$$
$$= \frac{20{,}000}{40{,}000 \times 0.2505 - 4{,}000 \times 0.1705}$$
$$= 2.14$$

Since *X's* BCR is higher than that of *Y*, select[6] *X*.

In this example, the salvage value has been treated as cost recovery, and hence subtracted from the first cost. If it is considered a benefit, then it becomes a part of the numerator. In that case,

$$\text{BCR}_X = \frac{20{,}000 + 5{,}000(A/F, 8\%, 7)}{50{,}000(A/P, 8\%, 7)}$$
$$= \frac{20{,}000 + 5{,}000 \times 0.1121}{50{,}000 \times 0.1921}$$
$$= 2.14$$

$$\text{BCR}_Y = \frac{20{,}000 + 4{,}000(A/F, 8\%, 5)}{40{,}000(A/P, 8\%, 5)}$$
$$= \frac{20{,}000 + 4{,}000 \times 0.1705}{40{,}000 \times 0.2505}$$
$$= 2.06$$

As can be seen, these BCR values differ from those in Example 10.5, though the final decision in favor of *X* remains unchanged. In situations where there are doubts about whether a cash inflow, such as salvage value, should be treated as a benefit (part of the numerator) or as a cost recovery (part of the denominator reducing the first cost), it is better to use the modified version of the BCR method. In the modified version we look at the *benefit–cost difference* (BCD). For the case discussed, the BCDs for *X* and *Y* are

$$\text{BCD}_X = 20{,}000 + 5{,}000 \times 0.1121 - 50{,}000 \times 0.1921$$
$$= \$10{,}956$$

$$\text{BCD}_Y = 20{,}000 + 4{,}000 \times 0.1705 - 40{,}000 \times 0.2505$$
$$= \$10{,}662$$

---

[6]Due to the fixed output, the numerator in the BCR equation is the same for both *X* and *Y*. Since the numerator is the same, one can decide on the basis of the denominator alone, selecting the one whose denominator, i.e., PW_{costs}, is lower.

Robot X, with the higher BCD, is selected. Thus, the decision based on BCD is the same as that based on BCR. BCD's merit is that BCD values for X and Y are not affected by how we treat the salvage value—benefit or cost recovery. The BCD approach thus avoids the confusion the BCR approach can create. Note however that the BCD criterion is the same[7] as the PW or AW criterion, depending on which one has been used to evaluate the BCD. Here, $BCD_X = AW_X$ and $BCD_Y = AW_Y$, since BCD has been based on annual worth.

10.4 INCREMENTAL ANALYSIS

Engineering economics projects can be categorized into one of two groups: no-alternative or multialternative. In no-alternative projects, there is only one candidate, and hence no choice. We have discussed BCR analysis of such projects in the earlier sections, which involves determining the project BCR and approving it if the BCR ≥ 1.

As mentioned earlier, multialternative projects can be of three types: fixed input, fixed output, and variable input, variable output. We have discussed the first two types in Section 10.3. This section presents variable-input, variable-output problems that *must* be solved by the *incremental analysis* method.

10.4.1 Two Alternatives

In incremental BCR analysis, the two alternatives are compared for the differences in their costs and benefits. Since differential data are generated by subtracting the data of the smaller investment from the other, the difference in investment (first cost) is positive, that is, incremental—hence the name *incremental* analysis. Using the *incremental costs and benefit* data we determine the *incremental BCR*, denoted by ΔBCR. If $\Delta BCR \geq 1$, then the additional cost on the larger-investment alternative is justified. The concept behind incremental analysis can be explained as follows.

Suppose machine A costs \$4,000 and yields certain benefits. Consider that another machine B can also do the job, but it costs \$6,000 and will yield benefits that look superior. Is B really better than A from the BCR viewpoint? We want to know whether the additional cost of \$2,000 on B is worth the benefits expected of it. Our goal is to determine the ΔBCR—the *incremental benefit–cost ratio* (additional benefits divided by additional costs). If ΔBCR is greater than one, then machine B is selected; otherwise A is selected. If ΔBCR is exactly one, then either can be selected.

Conceptually, the alternative needing higher investment is a *challenger*. The incremental technique judges a challenger's investment worthiness not on its own but relative to that of the smaller-investment alternative (defender). In the illustrative discussion here, machine A, due to its lower first cost, is the obvious choice (defender), and B is the challenger. Example 10.6 illustrates the application of incremental analysis to a two-alternative project.

[7] While comparing two or more alternatives, the BCD evaluation should be LCM based as in PW analysis. This is not essential if BCD is AW based.

EXAMPLE 10.6

Two mutually exclusive alternatives A and B exist for a project. Given the following cash-flows, which one should be selected on the basis of BCR if $i = 10\%$ per year?

n	A	B
0	−\$4,500	−\$18,000
1	2,000	6,500
2	2,700	9,250
3	2,250	9,500

Solution

With different costs and benefits, alternatives A and B represent a variable-input, variable-output problem[8], and hence the solution *must* be based on incremental analysis.

Which of the two alternatives has lower first cost? With \$4,500, it is A, which becomes the obvious choice. We prefer to select A unless the \$13,500 additional cost on B (18,000−4,500) proves to be beneficial. We therefore analyze the $B-A$ incremental data by creating an additional column to post them:

n	A	B	$B-A$
0	−\$4,500	−\$18,000	−\$13,500
1	2,000	6,500	4,500
2	2,700	9,250	6,550
3	2,250	9,500	7,250

Next, we evaluate ΔBCR_{B-A} based on the incremental data. Between PW or AW as the basis of evaluation, here the AW approach is not simple, since the cash inflows are not uniform. We thus follow the PW approach, wherefrom

$$\begin{aligned}\Delta \text{BCR}_{B-A} &= \frac{\Delta \text{PW}_{\text{benefits}}}{\Delta \text{PW}_{\text{costs}}}\\ &= \frac{4{,}500(P/F, 10\%, 1) + 6{,}550(P/F, 10\%, 2) + 7{,}250(P/F, i, 3)}{13{,}500}\\ &= \frac{4{,}500 \times 0.9091 + 6{,}550 \times 0.8264 + 7{,}250 \times 0.7513}{13{,}500}\\ &= \frac{4{,}091 + 5{,}413 + 5{,}447}{13{,}500}\\ &= 1.11\end{aligned}$$

[8]The input of A, being \$4,500, differs from that of B, which is \$18,000. Their outputs (benefits) during their lives also differ.

Since ΔBCR_{B-A} is greater than one, the additional benefits from B are worth the additional cost on it. Hence, B is selected[9].

10.4.2 More Than Two Alternatives

BCR analysis of projects with more than two alternatives is basically an extension of that for a two-alternative project discussed above. Here too, the selection based on highest individual BCR can lead to an erroneous decision. Hence, incremental technique *must* be applied, as illustrated in Example 10.7

EXAMPLE 10.7

If each of the following five mutually exclusive alternatives has 8 years of useful life, which one should be selected based on the BCR criterion? Assume $i = 8\%$ per year.

	A	*B*	*C*	*D*	*E*
First cost	$600	$500	$965	$800	$750
Annual benefit	100	120	130	110	105
Salvage value	375	40	800	747	230

---

[9]The analysis based on individual BCR of A and B would have led to the wrong decision, as illustrated here.

$$\begin{aligned}BCR_A &= \frac{PW_{\text{benefits}}}{PW_{\text{costs}}}\\ &= \frac{2{,}000(P/F, 10\%, 1) + 2{,}700(P/F, 10\%, 2) + 2{,}250(P/F, i, 3)}{4{,}500}\\ &= \frac{2{,}000 \times 0.9091 + 2{,}700 \times 0.8264 + 2{,}250 \times 0.7513}{4{,}500}\\ &= \frac{1{,}818 + 2{,}231 + 1{,}690}{4{,}500}\\ &= 1.28\end{aligned}$$

$$\begin{aligned}BCR_B &= \frac{PW_{\text{benefits}}}{PW_{\text{costs}}}\\ &= \frac{6{,}500(P/F, 10\%, 1) + 9{,}250(P/F, 10\%, 2) + 9{,}500(P/F, i, 3)}{18{,}000}\\ &= \frac{6{,}500 \times 0.9091 + 9{,}250 \times 0.8264 + 9{,}500 \times 0.7513}{18{,}000}\\ &= \frac{5{,}909 + 7{,}644 + 7{,}137}{18{,}000}\\ &= 1.15\end{aligned}$$

On the basis of individual BCR, we would have selected A, since its BCR is greater. Note that this decision contradicts the one based on incremental analysis. Selecting A would have been a wrong decision, depriving the company of the opportunity to invest in B. The $13,500 additional investment in B, as shown in Example 10.6, is attractive.

Solution

For this problem incremental analysis is essential, since it is a multialternative project of the variable-input, variable-output type.

However, prior to the incremental analysis of such problems, it is essential to check the alternatives' individual BCRs. This helps eliminate alternatives with a BCR under one. The elimination reduces the number of alternatives to be incrementally analyzed. Let us therefore determine the individual BCRs. For project A we have

$$\begin{aligned}\text{BCR}_A &= \frac{100(P/A, 8\%, 8) + 375(P/F, 8\%, 8)}{600} \\ &= \frac{100 \times 5.747 + 375 \times 0.5403}{600} \\ &= \frac{574.7 + 202.6}{600} \\ &= \frac{777.3}{600} \\ &= 1.30\end{aligned}$$

In a similar way, we determine the individual BCRs of projects B, C, D, and E to be 1.42, 1.22, 1.29, and 0.97 respectively. Since alternative E's BCR is less than one, it is eliminated. The remaining four participate in the incremental analysis.

The incremental analysis procedure is similar to that in Example 10.6. We compare two alternatives at a time. Alternative B is the obvious first choice, since its investment of \$500 is the lowest. The next-higher-first-cost alternative is A with its \$600 investment. So we compare B with A to check whether the additional cost on A brings in enough additional benefits to yield $\Delta\text{BCR}_{A-B} \geq 1$. The winner between A and B will be compared with the next-higher-cost alternative, and so on. This process of comparing two alternatives at a time continues until all the alternatives have been examined for their worthiness.

For $A-B$ analysis, we tabulate the incremental data, as in the last column:

Year	B	A	$A-B$
0	−500	−600	−100
1–8	120	100	−20
8	40	375	335

We next analyze the incremental $A-B$ cashflows to determine ΔBCR. Basing the analysis on PW, we have[10]

$$\Delta\text{BCR}_{A-B} = \frac{335(P/F, 8\%, 8) - 20(P/A, 8\%, 8)}{100}$$

---

[10]The salvage value is being considered a benefit, since only the first cost is usually treated as investment. Also note that the incremental annual benefit can be negative, as here.

$$= \frac{335 \times 0.5403 - 20 \times 5.747}{100}$$

$$= \frac{181 - 115}{100}$$

$$= 0.66$$

Since ΔBCR_{A-B} is less than one, the additional cost of A is not justified. Hence we discard A; B, being the winner, continues to remain the choice.

B is now compared with the next higher-first-cost alternative, that is, D. The procedure is very similar to the above. Create a table to post $D-B$ data, if felt necessary. The incremental BCR for $D-B$ is found as

$$\Delta BCR_{D-B} = \frac{707(P/F, 8\%, 8) - 10(P/A, 8\%, 8)}{300}$$

$$= \frac{707 \times 0.5403 - 10 \times 5.747}{300}$$

$$= \frac{382 - 57.5}{300}$$

$$= 1.08$$

Since ΔBCR_{D-B} is greater than one, alternative D wins (B is discarded).

Alternative D is next compared with C, the last alternative. Following the same procedure,

$$\Delta BCR_{C-D} = \frac{53(P/F, 8\%, 8) + 20(P/A, 8\%, 8)}{165}$$

$$= \frac{53 \times 0.5403 - 20 \times 5.747}{165}$$

$$= \frac{28.64 + 114.94}{165}$$

$$= 0.87$$

Since ΔBCR_{C-D} is less than one, alternative C loses to D.

Thus, alternative D should be selected.

SUMMARY

The benefit–cost ratio (BCR) of a project represents its time-valued benefit per unit investment (first cost). The BCR is commonly used in economic analysis of government projects, which aim at benefitting as many people as possible. Present worth, future worth, or annual worth, whichever is easier to set up depending on the given cashflow pattern, is used as the basis of BCR evaluation. However, in projects with alternatives of unequal lives, AW-based BCR evaluation is preferred, since it does away with the need for a LCM-based analysis period. BCR analysis can at times be confusing, especially if it is difficult to decide whether the cash outflows other than the first cost should be treated as costs or disbenefits. Under the concept of disbenefit, all costs other than the first cost are paid out of project benefits. A similar confusion arises with salvage value, that is, whether to treat it as a benefit or a recovery of the first cost (investment). If such a confusion can't be resolved, then BCD analysis should be carried out instead. However, note that BCD analysis is similar to PW analysis. For fixed-output or fixed-input multialternative projects, alternatives' individual BCRs are ranked, and the alternative of the highest BCR is selected. For variable-input, variable-output multialternative projects, incremental analysis must be carried out. However, prior to the incremental analysis, individual BCRs are evaluated and examined, so that alternatives with BCR under one can be discarded.

EXERCISES

Discussion Questions

10.1 Why is BCR most popular in analyzing government projects for funding?

10.2 If there are controversies about whether a cash outflow is a cost or a disbenefit, what can be done in BCR analysis?

10.3 Explain the concept underlying *disbenefit*.

10.4 What role, if any, does the input–output concept play in BCR analysis?

10.5 How does the BCR method account for time value of money?

10.6 How do you decide whether PW or AW should be used in determining the BCR?

10.7 Does incremental BCR analysis conceptually differ from incremental ROR analysis? If yes, how? If no, why not?

Multiple-Choice Questions (Circle the *best* answer.)

10.8 If alternatives' useful lives are different, BCR analysis should preferably be based on
a. PW.
b. FW.
c. AW.
d. ROR.

10.9 For which multialternative project is incremental analysis essential?
a. Fixed input
b. Fixed output
c. Variable-input, variable-output
d. Fixed-input, fixed-output

10.10 In the BCR equation, the disbenefit data belong to the
a. numerator.
b. denominator.
c. both a and b, depending on interest rate.
d. either a or b, depending on salvage value.

10.11 When there is doubt about a data item—whether it is disbenefit or cost—we generally use a modified version of the BCR method, which is basically the
a. PW method.
b. AW method.
c. ROR method.
d. FW method.

10.12 In BCR analysis, if the salvage value is used to recover the first cost, it is being considered as a
a. cost.
b. disbenefit.
c. benefit.
d. negative cost.

10.13 While analyzing a multialternative project for decision making under the BCR criterion, the alternatives' individual BCRs are evaluated _____ the incremental analysis.
a. prior to
b. after
c. a or b, depending on the type of project
d. a or b, depending on the cashflow pattern

10.14 If ΔBCR_{C-D} is 1.25, alternative
a. C is selected.
b. D is selected.
c. a or b, depending on whose first cost is lower.
d. a or b, depending on whose annual benefit is higher.

Numerical Problems

10.15 If in Example 10.4 a court order decrees that the city pay the farmers an additional one-time compensation of \$35 million when the project begins, should the city still construct the road?

10.16 Mauritius asks the World Bank for a loan to finance the construction of a canal. The estimated construction cost is \$6.5 million. The estimated maintenance cost is \$200,000 per year. The annual benefit from irrigation fees is projected to be \$450,000. Assuming the canal will last for 50 years, should the World Bank fund the project if the decision criterion is BCR? The Bank uses a subsidized interest rate of 2% for third-world development.

10.17 What is the BCR for the project whose cashflow diagram is shown in Fig. 10.3? The first two cash outflows are investments, while the last two are overhaul costs. Assume $i = 8\%$ per year and project life $= 5$ years.

10.18 The new landfill is estimated to cost the City of Joy \$3 million and to last for 60 years. It will annually save the city \$375,000 in fees presently paid to a private contractor for garbage disposal outside the city limits. The environmental disbenefit of having a landfill within the city limits is estimated to be \$135,000 per year. If the city can borrow money at a federally subsidized annual rate of 3%, should the city fund the new landfill based on the BCR criterion?

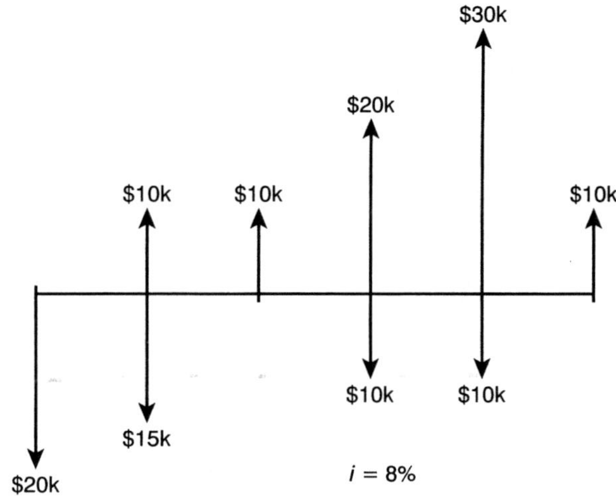

Figure 10.3 Diagram for Problem 10.17

10.19 An EPA-imposed water cleaning system is to be installed in a plant. There are two options: basic and deluxe. The basic model costs $2,500, is to be maintained at an annual cost of $200, and has no salvage value. The deluxe model costs $4,000, is maintenance-free, and has a salvage value of $500. Assuming that both are expected to last for 10 years, and $i = 12\%$ per year, which one should be selected based on the BCR criterion?

10.20 Two models (P and Q) of a machine are under consideration for procurement. Their cashflows are given in Fig. 10.4 which does not show the salvage values. If the salvage values are $5,000 for P and $15,000 for Q, which model is better based on the BCR criterion? Assume $i = 8\%$ per year.

10.21 Under a court decree, Universal Forgers has to erect a noise barrier to shield the nearby residential community from its noise. The barrier could be made of wood or plastic. Their costs are

	Plastic	Wood
First cost	$765,000	$550,000
Annual maintenance cost	50,000	120,000

As part of the court settlement the community will share the cost by annually paying Universal Forgers $50,000 for the plastic barrier or $40,000 for the wood barrier. On the basis of BCR, which material is better? Assume the barriers' lives to be 20 years and i to be 10% per year.

10.22 Which one of the following five mutually exclusive alternatives should be selected on the basis of the BCR criterion? The figures are in dollars.

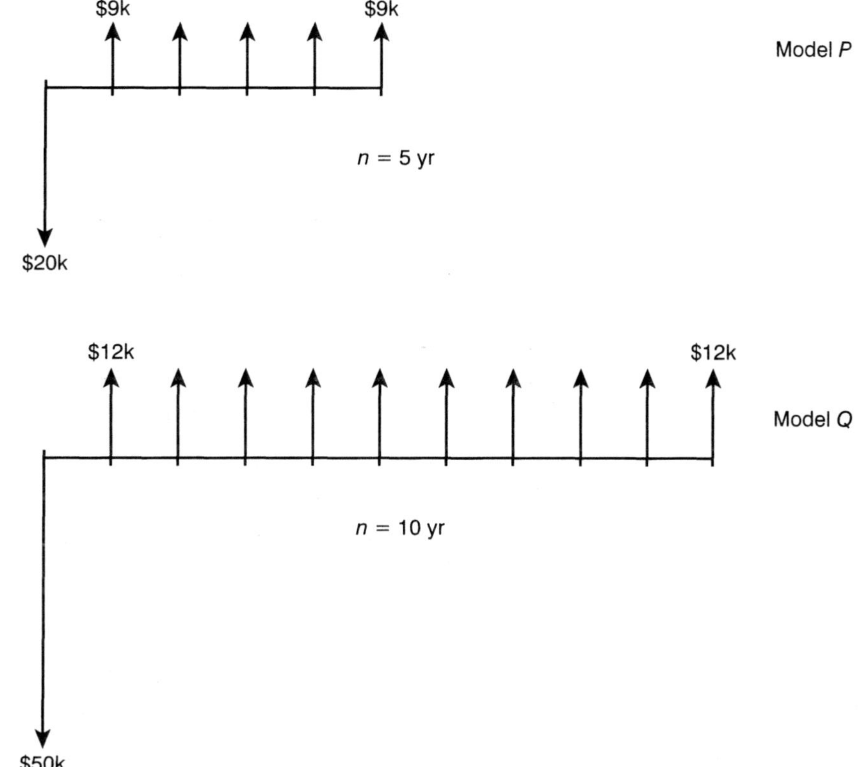

Figure 10.4 BCR Analysis of Two-Alternative Projects

	A	*B*	*C*	*D*	*E*
First cost	4,000	3,500	6,500	5,000	4,800
$PW_{benefits}$	7,500	6,000	5,750	6,500	8,000

10.23 A small city by the Mississippi river has occasionally experienced flooding. As a solution to the problem, the city is considering the following four options. Which one of these should be selected if the decision criterion is BCR?

	AW_{costs}	Annual damage
No flood control	0	$400,000
Construct levees	$50,000	250,000
Small reservoir	100,000	120,000
Large reservoir	200,000	20,000

(*Hint*: The annual benefit from an option is the saving from reduced damage.)

CHAPTER ELEVEN

Comparison

IN THIS CHAPTER YOU WILL LEARN ABOUT

- Comparison among the six methods
- Why and when to prefer a method
- Relative popularity of the methods
- Computer impact on method popularity
- Computer-aided analyses

In Chapters 5 through 10, we have extensively discussed the six methods commonly used in economic analysis of engineering projects. An obvious question is, Why so many methods? Others are, Which one should be preferred for analyzing a given problem and why? How can we ensure that the decision made on the basis of a particular criterion is reliable? Which method is commonly used in industry and why? In this last chapter of Part II we seek answers to such questions.

The discussions in this chapter are comparative in nature. Though the merits and limitations of the six methods may have been obvious in the respective chapters, a comparative study prepares engineers and technologists to apply the methods properly. It enables them to understand the methods' intricacies.

Engineering economics is basically decision making, as was pointed out in the very beginning of the text (Chapter 1). Decision making is much more than solving a numerical problem. An early and crucial step in the decision making process is to select the analysis method to use. In Chapters 5 through 10 we learned *how* to apply the methods in solving numerical problems; in here we discuss the *why* and *when* of these methods.

11.1 COMPARATIVE OVERVIEW

An important point to keep in mind while analyzing engineering economics problems is that the same problem may sometimes yield different answers, depending on the method used for analysis. In a multialternative project, for example, alternative *A* might be the best using one method, while alternative C using the other. How can engineering economists be sure of the decision in such a situation?

Different answers to the same problem beg the obvious question, which of the answers is correct, or are they all incorrect? We discuss this issue throughout Chapter 11 by comparatively reviewing what we have learned in Chapters 5 through 10.

Of the six, the payback period method is the simplest, most straightforward, and least confusing. This is also the only method that normally does not account for the time value of money, though it can be enhanced to do so, as was illustrated in the last section of Chapter 5.

The PW, FW, and AW methods discussed in Chapters 6, 7, and 8 are all based on the *worth* of the project. They differ only in the reference time to which the worth corresponds. In the PW method, the reference time is the present[1], while in the FW method it is the future. The future could be located anywhere beyond the present, though it is normally at the end of useful life. In the AW method, rather than a reference time period, the entire duration of the useful life or analysis period becomes the time window. These three analysis methods usually lead to the same decision and hence can be grouped under what may be called the *worth-based method*.

The other two analysis methods, namely ROR and BCR, discussed in Chapters 9 and 10, are complex[2] and may create confusion in decision making. They can yield different answers for the same problem, which may differ from those using payback-period or worth-based methods.

Thus, the six analysis methods form four different groups. After discussing each group in the following four sections (11.2 through 11.5), we study them comparatively in Section 11.6.

11.2 PAYBACK-PERIOD METHOD

The payback-period method sometimes leads to decisions that are incompatible with those by the other methods, as illustrated in Example 11.1. What can analysts do in such a situation? They should discuss the discrepancy with the project manager and apply one or more of the other methods to achieve more confidence in the final decision.

EXAMPLE 11.1

a. Which of the following two projects should be selected on the basis of payback period?

---

[1]In engineering economics, the sense of the *present* is relative. The *present* is prior to the *future* and can be located at any time period, though it is usually at $t = 0$.

[2]Although the BCR is based on PW, and therefore on FW or AW, BCR evaluation may at times be confusing, as discussed in Chapter 10.

Year	A	B
0	−$400	−$425
1	250	150
2	250	150
3	0	150
4	0	150

b. What's the decision under the PW criterion for $i = 12\%$ per year?
c. Explain any discrepancy in the results of (a) and (b).

Solution

a. The payback period for A is easily determined as 400/250 = 1.6 years, the time duration in which the $400 investment is recovered. For project B, it is 425/150 = 2.83 years. With shorter payback period as the decision criterion, project A is selected.

b. Next we analyze the problem under the PW criterion. The PWs of the two projects are

$$\begin{aligned} PW_A &= 250(P/F, 12\%, 1) + 250(P/F, 12\%, 2) - 400 \\ &= 250 \times 0.8929 + 250 \times 0.7972 - 400 \\ &= 223 + 199 - 400 \\ &= \$22 \end{aligned}$$

$$\begin{aligned} PW_B &= 150(P/A, 12\%, 4) - 425 \\ &= 150 \times 3.037 - 425 \\ &= 456 - 425 \\ &= \$31 \end{aligned}$$

With higher (positive) PW as decision criterion, project B is selected.

c. The PW-based decision in (b) is the opposite of that under payback period in (a). Payback period and PW can yield contradictory results if the project benefits are skewed to (concentrated at) the beginning or end of the analysis period. In this example, the $250 benefit for A is skewed to the beginning, being zero during the latter part of project life. For project B, the $150 benefit is uniformly spread over its four-year life.

In general, if the cashflows are fairly distributed throughout the project life, the payback and PW (and other worth-based) methods are likely to yield the same result. Let us consider for project A a cashflow pattern similar to that of B, such as

Year	A	B
0	−$400	−$425
1	125	150
2	125	150
3	125	150
4	125	150

Following the calculation procedure in Example 11.1, these data yield

$$\text{Payback period for } A = 400/125 = 3.2 \text{ years}$$
$$\text{Payback period for } B = 2.83 \text{ years (as in Example 11.1)}$$

$$\begin{aligned} \text{PW}_A &= 125(P/A, 12\%, 4) - 400 \\ &= 125 \times 3.037 - 400 \\ &= 380 - 400 \\ &= -\$20 \end{aligned}$$

$$\text{PW}_B = \$31 \qquad \text{(as in Example 11.1)}$$

We thus see that both methods lead to the same result (selection of alternative B) if cashflow patterns are similar.

11.3 WORTH-BASED METHODS

The three worth-based methods are very similar, and that is why they always lead to the same decision. Their similarities arise from the duality of functional equations:

$$\begin{aligned} \text{FW} &= \text{PW}(F/P, i, n) &\quad \text{or} \quad \text{PW} &= \text{FW}(P/F, i, n) \\ \text{FW} &= \text{AW}(F/A, i, n) &\quad \text{or} \quad \text{AW} &= \text{FW}(A/F, i, n) \\ \text{AW} &= \text{PW}(A/P, i, n) &\quad \text{or} \quad \text{PW} &= \text{AW}(P/A, i, n) \end{aligned}$$

From these equations it is clear that any of the three worths is simply a scaled version of the other two. The scaling is due to the multiplication with the (translating) functional factor, which is constant for a given i and n. It is this scaling that leads to the same conclusion by any of the worth methods. This is true for any cashflow pattern. Example 11.2 illustrates the compatibility among the three worth-based methods.

EXAMPLE 11.2

Given the following cashflows which alternative should be selected if decision criterion is (a) PW, (b) FW, (c) AW? Assume $i = 10\%$ per year.

	A	B	C
First cost	\$200	\$600	\$450
Annual benefit	110	150	140
Useful life, years	2	6	4

Solution
Since the alternatives' useful lives are different, a common analysis period is decided upon on the basis of their LCM. The LCM of 2, 6, and 4 is 12. We thus analyze the alternatives over 12 years[3].

[3]See the discussion later following Example 11.3. Under the AW method the analysis periods of the alternatives, being their useful lives, can differ.

a. *PW Analysis.* During the analysis period, alternative A is replaced five times, at periods 2, 4, 6, 8, and 10, as seen in the cashflow diagram Fig. 11.1(a). For B and C, replacement occurs once and twice, as seen in Figs. 11.1(b) and (c). Assuming that the costs and benefits for all replacements remain unchanged during the analysis period, the alternatives' PWs are

$$\begin{aligned} PW_A &= PW_{\text{benefits}} - PW_{\text{costs}} \\ &= 110(P/A, 10\%, 12) - [200 + 200(P/F, 10\%, 2) + 200(P/F, 10\%, 4) \\ &\quad + 200(P/F, 10\%, 6) + 200(P/F, 10\%, 8) + 200(P/F, 10\%, 10)] \\ &= 110 \times 6.814 - 200\,(1 + 0.8264 + 0.6830 + 0.5645 + 0.4665 + 0.3855) \\ &= 749.54 - 200 \times 3.9259 \\ &= -\$36 \end{aligned}$$

$$\begin{aligned} PW_B &= PW_{\text{benefits}} - PW_{\text{costs}} \\ &= 150(P/A, 10\%, 12) - [600 + 600(P/F, 10\%, 6)] \\ &= 150 \times 6.814 - 600(1 + 0.5645) \\ &= 1{,}022 - 600 \times 1.5645 \\ &= \$83 \end{aligned}$$

$$\begin{aligned} PW_C &= PW_{\text{benefits}} - PW_{\text{costs}} \\ &= 140(P/A, 10\%, 12) - [450 + 450(P/F, 10\%, 4) \\ &\quad + 450(P/F, 10\%, 8)] \\ &= 140 \times 6.814 - 450(1 + 0.6830 + 0.4665) \\ &= 954 - 450 \times 2.1495 \\ &= -\$13 \end{aligned}$$

With highest positive PW as the decision criterion, alternative B is selected.

b. *FW Analysis.* Based on the three cashflow diagrams in Fig. 11.1 and on the assumption that the costs and benefits for all replacements remain unchanged, the alternatives' FWs at the twelfth period are

$$\begin{aligned} FW_A &= FW_{\text{benefits}} - FW_{\text{costs}} \\ &= 110(F/A, 10\%, 12) - [200(F/P, 10\%, 12) + 200(F/P, 10\%, 10) \\ &\quad + 200(F/P, 10\%, 8) + 200(F/P, 10\%, 6) + 200(F/P, 10\%, 4) \\ &\quad + 200(F/P, 10\%, 2)] \\ &= 110 \times 21.384 - 200(3.138 + 2.594 + 2.144 + 1.772 + 1.464 + 1.210) \\ &= 2{,}352 - 200 \times 12.322 \\ &= -\$112 \end{aligned}$$

$$\begin{aligned} FW_B &= FW_{\text{benefits}} - FW_{\text{costs}} \\ &= 150(F/A, 10\%, 12) - [600(F/P, 10\%, 12) + 600(F/P, 10\%, 6)] \\ &= 150 \times 21.384 - 600(3.138 + 1.772) \\ &= 3{,}208 - 600 \times 4.910 \\ &= \$262 \end{aligned}$$

$$\begin{aligned} FW_C &= FW_{\text{benefits}} - FW_{\text{costs}} \\ &= 140(F/A, 10\%, 12) - [450(F/P, 10\%, 12) + 450(F/P, 10\%, 8) \\ &\quad + 450(F/P, 10\%, 4)] \\ &= 140 \times 21.384 - 450(3.138 + 2.144 + 1.464) \\ &= 2{,}994 - 450 \times 6.746 \\ &= -\$42 \end{aligned}$$

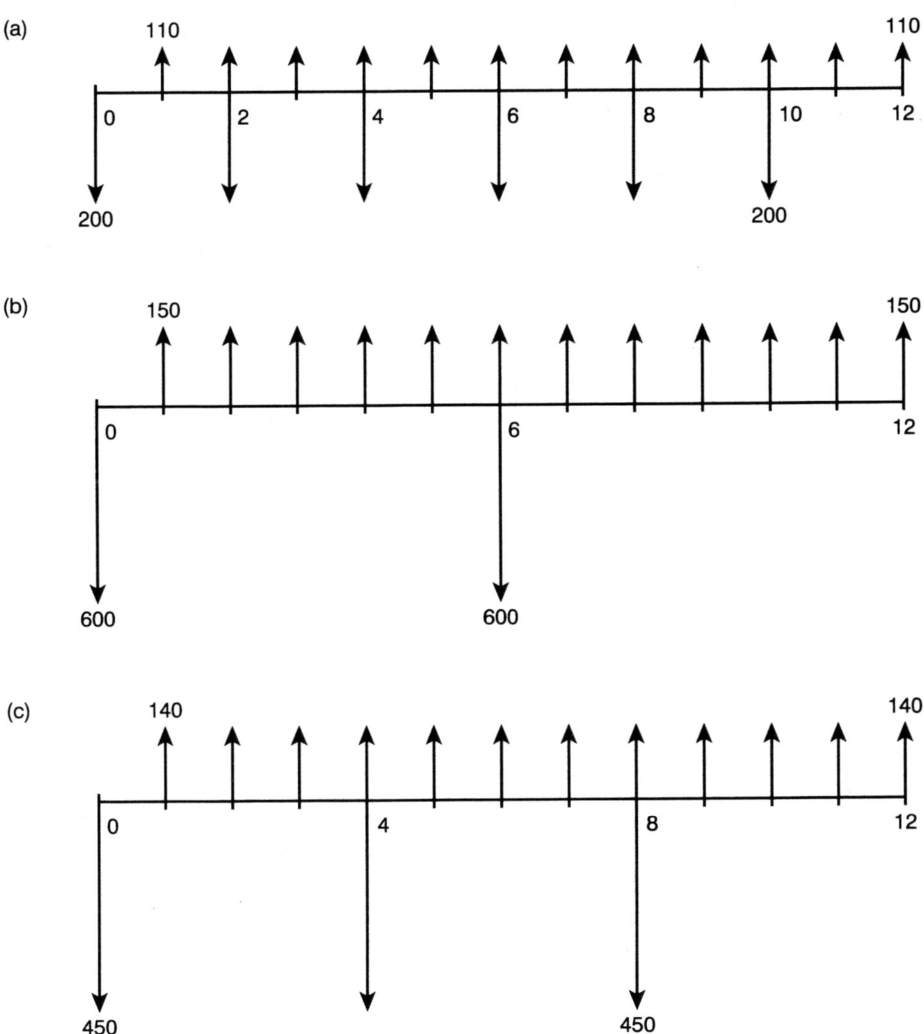

Figure 11.1 Diagram for Example 11.2

With highest positive FW as the criterion, alternative *B* is selected. Thus, the decision under FW criterion is the same as under the PW criterion.

c. *AW Analysis.* Based on the cashflow diagrams in Fig. 11.1 and on the assumption of unchanged costs and benefits for all replacements, the alternatives' AWs are[4]

[4]In the evaluation of AW_A, the factor $(P/F, 10\%, 2)$ in the third term on the right of the equal sign yields the *P* value at period 0 of the 200 cashflow at period 2, while the factor $(A/P, 10\%, 12)$ converts this *P* value into its equivalent annual worth over the 12-year analysis period.

$$\begin{aligned}
AW_A &= AW_{benefits} - AW_{costs} \\
&= 110 - [200(A/P, 10\%, 12) + 200(P/F, 10\%, 2)(A/P, 10\%, 12) \\
&\quad + 200(P/F, 10\%, 4)(A/P, 10\%, 12) + 200(P/F, 10\%, 6)(A/P, 10\%, 12) \\
&\quad + 200(P/F, 10\%, 8)(A/P, 10\%, 12) + 200(P/F, 10\%, 10)(A/P, 10\%, 12)] \\
&= 110 - 200(A/P, 10\%, 12)[1 + (P/F, 10\%, 2) + (P/F, 10\%, 4) \\
&\quad + (P/F, 10\%, 6) + (P/F, 10\%, 8) + (P/F, 10\%, 10)] \\
&= 110 - 200 \times 0.1468(1 + 0.8264 + 0.6830 + 0.5645 + 0.4665 + 0.3855) \\
&= 110 - 200 \times 0.1468 \times 3.9259 \\
&= 110 - 115 \\
&= -\$5
\end{aligned}$$

$$\begin{aligned}
AW_B &= AW_{benefits} - AW_{costs} \\
&= 150 - 600(A/P, 10\%, 12) - 600(P/F, 10\%, 6)(A/P, 10\%, 12) \\
&= 150 - 600 \times 0.1468 - 600 \times 0.5645 \times 0.1468 \\
&= 150 - 88 - 50 \\
&= \$12
\end{aligned}$$

$$\begin{aligned}
AW_C &= AW_{benefits} - AW_{costs} \\
&= 140 - 450(A/P, 10\%, 12) - 450(P/F, 10\%, 4)(A/P, 10\%, 12) \\
&\quad - 450(P/F, 10\%, 8)(A/P, 10\%, 12) \\
&= 140 - 450(A/P, 10\%, 12)[1 + (P/F, 10\%, 4) + (P/F, 10\%, 8)] \\
&= 140 - 450 \times 0.1468(1 + 0.6830 + 0.4665) \\
&= 140 - 142 \\
&= -\$2
\end{aligned}$$

With highest positive AW as the criterion, alternative B is selected. Thus, the decision under the AW criterion is the same as under the PW or FW criteria.

As seen in the above example, the decision remains the same irrespective of which of the three worth-based methods is used for analysis.

In Example 11.2, FWs and AWs were evaluated directly from the given cashflow diagrams, with no attempt to utilize the relationships that exist among the three worths. This resulted in lengthy calculations for parts (b) and (c). Had the relationships been utilized, the solutions would have been briefer and simpler, as illustrated in Example 11.3.

EXAMPLE 11.3

How can the solution of Example 11.2, especially of parts (b) and (c), be made concise?

Solution

To achieve brevity in the solution of Example 11.2 we should attempt to benefit from the relationship among the three worths.

a. *PW Analysis.* Since FW or AW is not known at this stage, evaluation of the PW can't be made any simpler than that in Example 11.2. However, parts (b) and (c) can be made concise because the values of PW, as determined in part (a), are by then known.

b. *FW Analysis.* There are basically two approaches to solve this part. The first one is based on the given cashflow diagram, as done in Example 11.2. The second approach utilizes the functional relationship between PW and FW, keeping the solution brief. Since PWs are already known from part (a) we simply convert them to their FWs:

$$\begin{aligned} \text{FW}_A &= \text{PW}_A(F/P, 10\%, 12) \\ &= -\$36 \times 3.138 \\ &= -\$113 \end{aligned}$$

$$\begin{aligned} \text{FW}_B &= \text{PW}_B(F/P, 10\%, 12) \\ &= \$83 \times 3.138 \\ &= \$260 \end{aligned}$$

$$\begin{aligned} \text{FW}_C &= \text{PW}_C(F/P, 10\%, 12) \\ &= -\$13 \times 3.138 \\ &= -\$41 \end{aligned}$$

The conciseness offered by the second approach is obvious when these FW evaluations are compared with those in part (b) of Example 11.2.

c. *AW Analysis.* Again, the first approach involves direct analysis of the given cashflow diagram, as in Example 11.2. The second approach, based on the functional relationship between PW[5] and AW, yields a brief solution. Since the PWs are already known, from part (a), simply convert them into their AWs, as below.

$$\begin{aligned} \text{AW}_A &= \text{PW}_A(A/P, 10\%, 12) \\ &= -\$36 \times 0.1468 \\ &= -\$5 \end{aligned}$$

$$\begin{aligned} \text{AW}_B &= \text{PW}_B(A/P, 10\%, 12) \\ &= \$83 \times 0.1468 \\ &= \$12 \end{aligned}$$

$$\begin{aligned} \text{AW}_C &= \text{PW}_C(A/P, 10\%, 12) \\ &= -\$13 \times 0.1468 \\ &= -\$2 \end{aligned}$$

A comparison of the above AW evaluations with those in part (c) of Example 11.2 illustrates that utilizing the functional relationship between the worths leads to briefer and simpler solutions.

---

[5]Known FWs determined in part (b) of Example 11.2 or 11.3 can be used instead.

There is still another way in the case of the AW method to solve Example 11.2. We can apply the basic concept of annual worth learned in Chapter 8. Since AW represents the net annual cashflow that is equivalent to all the costs and benefits, the analysis period in the AW method does not have to be the same for all the alternatives. That means we can consider just one complete cycle[6] of the cashflows for each alternative. In other words, the analysis period of an alternative is simply its useful life. *This is a unique advantage of the AW method,* and is nonexistent with the PW and FW methods.

Based on this explanation we can solve part (c) of Example 11.2 by considering just the first cycle of cashflows for each alternative. Referring to Fig. 11.1, from the first complete cycle of the alternative we get

$$\begin{aligned} \text{AW}_A &= 100 - 200(A/P, 10\%, 2) \\ &= 110 - 200 \times 0.5762 \\ &= 110 - 115 \\ &= -\$5 \\ \text{AW}_B &= 150 - 600(A/P, 10\%, 6) \\ &= 150 - 600 \times 0.2296 \\ &= 150 - 138 \\ &= \$12 \\ \text{AW}_C &= 140 - 450(A/P, 10\%, 4) \\ &= 140 - 450 \times 0.3155 \\ &= 140 - 142 \\ &= -\$2 \end{aligned}$$

These AW values are the same[7] as in part (c) of Examples 11.2 and 11.3. The AW method frees you from the inconvenience alternatives' unequal lives create in PW and FW analyses. AW is evaluated by considering just one complete cycle of cashflows for each alternative. This is quite a relief in multialternative projects, especially where the LCM-based analysis period becomes long.

We can thus say that the three worth-based methods lead to the same decision. Major attraction of these methods is that they do not require incremental analysis. Even in multialternative projects the selection is based simply on alternatives' individual worths. Of the three worth-based methods, FW is the least used in engineering economic analysis. This is primarily due to the nature of most engineering projects, which require decisions based on investment considerations as of now, or on annual opera-

[6]By complete cycle we mean the span of one useful life, which avoids any consideration of replacement. Thus, for alternative *A*, the cycle comprises the first \$200 cost followed by the two \$100 benefits. For *B*, it comprises the first \$600 cost followed by the six \$150 benefits, and for *C* the first \$450 cost followed by the four \$140 benefits.

[7]If a difference occurs in spite of correct calculations, it is due to the rounding off of functional factors' values in the tables.

tional advantage[8]. The FW method, on the other hand, focusing on the future, is applicable mostly in financial decisions. It is therefore basically nonengineering in nature. Engineers and technologists normally do not deal with financial investments which fall within the scope of business economics. Thus, PW and AW are directly applicable to engineering economics, while FW is "tangentially" so.

In the following comparative discussions we use only PW, as representative of the worth-based methods.

11.4 ROR METHOD

The attraction of the ROR method lies in the fact that ROR is measured in percentage, which is understood by most people. Moreover, this method does not need in its analysis the value of the interest rate, which is usually difficult to assess. In fact, the interest rate i is the unknown to be evaluated.

The ROR method's recent popularity in industry is attributed to the ubiquitous growth in computer technology. Computer systems have rendered ROR-based decisions manageable. As seen in Chapter 9, ROR analysis is lengthy and error-prone in comparison to the other five types of analyses. The enhanced user-friendliness of modern engineering economics software aids ROR analysis the most.

ROR-based decisions may at times contradict those by the other methods. This happens when alternatives' cashflow patterns differ significantly from each other. Skewed cashflows, for example heavy in the early years of the project, can give rise to discrepancy in results, as illustrated in Example 11.4. For projects whose cashflows are fairly distributed, the ROR method usually leads to the same decision as the other methods, as illustrated in Example 11.5.

EXAMPLE 11.4

a. Which of the following two project alternatives should be selected on the basis of payback period?

Year	*A*	*B*
0	−$400	−$400
1	250	150
2	250	150
3	0	150
4	0	150

[8]The PW method suits projects where decisions are based on economic consideration as of now. The AW method is suitable for projects where recurring cashflows are of paramount importance, as in maintaining equipment or operating a facility.

b. Does the decision in (a) change if ROR is the criterion?
c. Explain any contradiction in the results of (a) and (b).

Solution
a. The payback period for A is easily determined to be $400/250 = 1.6$ years, the duration in which the \$400 investment is recovered. For B, it is $400/150 = 2.67$ years. With shorter payback period as the decision criterion, alternative A is selected.

b. Since a two-alternative project is being analyzed under the ROR criterion, we first need to establish whether incremental analysis is necessary. Since the alternatives have the same input — an investment of \$400 — the project is of the fixed-input type, and hence an individual-ROR-ranking-based decision will suffice. In other words, there is no need for incremental analysis.

The alternatives' RORs are obtained by solving their PW functions for i. For A, the PW function is $250(P/A, i, 2) - 400 = 0$, whose solution[9] yields i to be greater than 15%, but less than 18%. Thus, $15\% < ROR_A < 18\%$. The PW function for B is $150(P/A, i, 4) - 400 = 0$, whose solution[10] yields $ROR_B > 18\%$.

Thus, based on the criterion of higher ROR, alternative B is selected. This decision contradicts the one in part (a) based on payback period.

c. The above inconsistency in the ROR and payback results can be explained by the difference in cashflow distribution for A and B. Alternative A is better under the payback criterion, since its \$250 annual benefit is large, returning the investment faster. Alternative B's \$150 annual benefit, being relatively smaller and spread out, takes longer to recover the investment. Such an inconsistency usually arises when alternatives' cashflow distributions are significantly different, as in this case.

Since A returns the investment faster while B earns a higher ROR, each has its own attraction. Companies with long-term investment strategies will prefer B to A. Capital-tight companies, on the other hand, may select A, since a shorter payback replenishes the limited capital for other investments.

---

[9]From the PW function, $(P/A, i, 2) = 400/250 = 1.6$. If the PW function comprises just one factor, as here, the range in which the ROR lies can be found simply by thumbing through the interest tables. Turn to any page of the table and look at the value of $(P/A, i, 2)$. Is it 1.6? If not, thumb through the pages until it is 1.6 or near about. The i corresponding to the page becomes ROR_A. In this case, $(P/A, 15\%, 2) = 1.626$, while $(P/A, 18\%, 2) = 1.566$. So $i_A > 15\%$, since $(P/A, i, 2) = 1.6$, but less than 18%. Thus, $ROR_A < 18\%$ but greater than 15%. Until we check where ROR_B lies, there is no need for exact determination of ROR_A (which by the way is 16.3% through interpolation).

As seen in the next footnote, ROR_B is greater than 18%. With ROR_A less than 18%, we select alternative B based on its higher rate of return. Note that we did not have to determine the exact values of RORs in this problem, where our interest was simply to know whose ROR is higher. Such judicious actions reduce time and effort in reaching a decision, and should be taken wherever appropriate.

[10]Here we have $(P/A, i, 4) = 400/150 = 2.67$. Thumb through the pages looking at the P/A column for $n = 4$. You will find that for $i = 18\%$, the entry is 2.69, which is greater than 2.67 we are interested in. Only a higher i will reduce this P/A value to 2.67. So we can say that $ROR_B > 18\%$. Sometimes, if you are lucky, the exact value being searched for is found on a page, yielding an exact ROR. The exact value of ROR_B, though not essential here, is through interpolation 18.46%.

EXAMPLE 11.5

Assuming an 8-year useful life for each of the following four mutually exclusive alternatives and an 8% MARR, which one should be selected under (a) the PW criterion, (b) the ROR criterion?

	A	*B*	*C*	*D*
First cost	\$600	\$500	\$965	\$800
Annual benefit	100	120	130	110
Salvage value	375	40	800	747

Solution

a. The selection is based on the largest positive PW corresponding to an interest rate equal to the MARR, that is, 8% per year. The PWs of the four alternatives are

$$\begin{aligned}PW_A &= 100(P/A, 8\%, 8) + 375(P/F, 8\%, 8) - 600\\ &= 100 \times 5.747 + 375 \times 0.5403 - 600\\ &= 574.70 + 202.61 - 600\\ &= \$177\end{aligned}$$

$$\begin{aligned}PW_B &= 120(P/A, 8\%, 8) + 40(P/F, 8\%, 8) - 500\\ &= 120 \times 5.747 + 40 \times 0.5403 - 500\\ &= 689.64 + 21.61 - 500\\ &= \$211\end{aligned}$$

$$\begin{aligned}PW_C &= 130(P/A, 8\%, 8) + 800(P/F, 8\%, 8) - 965\\ &= 130 \times 5.747 + 800 \times 0.5403 - 965\\ &= \$214\end{aligned}$$

$$\begin{aligned}PW_D &= 110(P/A, 8\%, 8) + 747(P/F, 8\%, 8) - 800\\ &= 110 \times 5.747 + 747 \times 0.5403 - 800\\ &= \$236\end{aligned}$$

With largest positive PW as the criterion, *D* is selected.

b. Since it is a multialternative problem, let us first check whether incremental analysis is essential. As explained in Chapter 9, incremental analysis is necessary in variable-input, variable-output problems. Since the alternatives' first costs (input) and benefits (output) are different, this problem is of the variable-input, variable-output type. So we need to carry out incremental analysis.

Under incremental analysis we first check the alternatives' individual RORs to ensure that they are above the MARR. Any alternative with the ROR less than the MARR is discarded from further (incremental) analysis. In this case, the alternatives' individual RORs are all greater than the MARR, since their PWs corresponding to the given MARR have been determined in part (a) to be positive. So all four alternatives participate in incremental analysis, which begins by comparing two alternatives at a time. Since alternative *B*'s first cost is the lowest, we prefer to select it. The next-higher-first-cost alternative is

A. So we compare B with A to determine whether the additional cost on A and its additional benefits yield a rate of return greater than the MARR. The winner between A and B is compared with the next-higher-first-cost alternative. This process continues until all the alternatives have been examined, one at a time.

For the first comparison generate the $A-B$ incremental data as in the last column:

Year	B	A	$A-B$
0	−500	−600	−100
1–8	120	100	−20
8	40	375	335

By applying[11] Tests 1 and 2 on the incremented data we can predict that only one positive ROR is likely, since there is one sign change in the $A-B$ cashflows. This incremental rate of return ΔROR_{A-B} is determined[12] from the PW function

$$-20(P/A, i, 8) + 335(P/F, i, 8) - 100 = 0$$

A trial and error solution yields $i \approx 5\%$, which equals ΔROR_{A-B}. Since ΔROR_{A-B} is less than the 8% MARR, the additional cost of A is not justified. Hence we decide to continue to prefer B.

Next, B is compared with the ensuing higher-first-cost alternative, that is D. The procedure is very similar. The PW function for $D-B$ incremental cashflows is

$$-10(P/A, i, 8) + 707(P/F, i, 8) - 300 = 0$$

---

[11] See Chapter 9, if necessary.

[12] Since the MARR is known, one can use the MARR-based shorter approach over the lengthier procedure of determining the exact ROR. One simply has to check the sign of the project's PW corresponding to i = MARR. If it is positive, then ROR > MARR; otherwise ROR < MARR. In this case, we have

$$\begin{aligned} \text{PW}_{A-B} &= 335(P/F, 8\%, 8) - 20(P/A, 8\%, 8) - 100 \\ &= 335 \times 0.5403 - 20 \times 5.747 - 100 \\ &= -\$33.9 \end{aligned}$$

Since PW_{A-B} is negative, ROR_{A-B} will be less than the MARR (8%). Thus, alternative A is not better than B. It is therefore discarded, and B continues to remain selected.

We adopt this approach to the next pair, yielding

$$\begin{aligned} \text{PW}_{D-B} &= 707(P/F, 8\%, 8) - 10(P/A, 8\%, 8) - 300 \\ &= 707 \times 0.5403 - 10 \times 5.747 - 300 \\ &= \$24.5 \end{aligned}$$

Since PW_{D-B} is positive, ROR_{D-B} will be greater than 8%. So, alternative D is better than B. Thus, discard B and select D.

Next, compare D with C. From the incremental $C-D$ data, we have

$$\begin{aligned} \text{PW}_{C-D} &= 53(P/F, 8\%, 8) + 20(P/A, 8\%, 8) - 165 \\ &= 53 \times 0.5403 + 20 \times 5.747 - 165 \\ &= -\$21.4 \end{aligned}$$

Since PW_{C-D} is negative, ROR_{C-D} will be less than 8%. Alternative C is not better than D. Thus, D is finally selected.

From trial and error, ΔROR_{D-B} is evaluated to be almost 9%. Since this is greater than the 8% MARR, alternative D wins.

Alternative D is finally compared with C, the last alternative. The PW function for $C-D$ incremental cashflows is

$$20(P/A, i, 8) + 53(P/F, i, 8) - 165 = 0$$

From trial and error, ΔROR_{C-D} is found to be 5%. Since this is less than the 8% MARR, alternative C loses to D.

Thus, alternative D is finally selected.

Note that it was also the choice on the basis of PW. Thus, the PW and ROR criteria both led in this case to the same decision, primarily because the alternatives' cashflow distributions are alike.

11.4.1 Common Error

A common error in the ROR method is *not to apply* incremental analysis in multi-alternative projects of the variable-input, variable-output type. Let us reconsider Example 11.5 and solve it without incremental analysis. This involves evaluation of the alternatives' individual RORs by solving their PW function for the unknown i. For alternative A, the PW function is

$$100(P/A, i, 8) + 375(P/F, i, 8) - 600 = 0$$

Its solution yields ROR_A as just under 14%. Solving the PW functions of other alternatives, their rates[13] of return are $ROR_B = 18\%$, $ROR_C \approx 12\%$, and $12\% < ROR_D < 15\%$. On the basis of largest positive ROR we would have selected alternative B—obviously a wrong decision compared to that in Example 11.5. Therefore, *under the ROR criterion one must carry out incremental analysis in selecting the best of two or more variable-input, variable-output alternatives.* This is also true for the BCR criterion.

Why does an individual-ROR-based decision sometimes differ from that based on incremental analysis? Let us review the preceding discussion along with the results of Example 11.5. Based on individual ROR alternative B is the best, but on the basis of incremental analysis D is the best. In general, engineers will prefer to choose the alternative whose first cost is the least, which in this case is alternative B. But the company may be looking for opportunities to invest capital as long as it can earn a return in excess of the MARR. Assuming this to be the case, the engineer has to check whether any

---

[13]We did not bother to determine the exact values of the RORs, where it involved interpolation, to save time and effort. Since we are interested only in knowing which ROR is the largest—rather than in their absolute values—so that the best alternative can be selected, we may save time by determining ROR values only as approximately as essential. For example, we saved time by avoiding interpolation for alternative D whose ROR was under 15%, less than the 18% already found for B. Carry out the interpolation(s) only if necessary.

of the other higher-first-cost alternatives (A, C, or D) offers such an opportunity. This is where incremental analysis becomes indispensable—in helping determine whether any of the other alternatives offers an investment opportunity. If so, the company seizes the opportunity. To keep the additional investment low, obviously, the engineer first tries the next-higher-first-cost alternative, A in this case. Since the additional \$100 cost on A over B did not yield a rate of return greater than the MARR, alternative A was discarded. Such a process of elimination continued until the best alternative "filtered out" to be D.

Had the incremental analysis been not applied, and therefore D not chosen, an opportunity to invest to earn a return in excess of the set MARR would have been lost. Obviously, no company would like that.

11.5 BCR METHOD

The benefit–cost ratio method is used primarily in evaluating government projects. As seen in Chapter 10, it is closely related[14] to the worth methods. Due to its "ratio" measure, the BCR method may at times yield misleading results. This arises from the ambiguity of whether cash outflows other than the investment should be treated as costs or disbenefits, or some as costs and others as benefits. As explained in Chapter 10, any such doubt affecting BCR-based results can be avoided by adopting the BCD method, which is based on the difference between benefits and costs, much like the PW method.

A point of contention in the BCR method is the interest rate to use in the analysis. Assuming benevolence as a hallmark of modern democratic governments, interest rates for public investments are usually subsidized, that is, below the cost of capital in the free (financial) market.

The BCR method also can lead to decisions that contradict those made by other methods, as illustrated in Example 11.6. The incompatibility in decisions is likely to arise when alternatives' cashflow distributions are not alike.

As in the ROR method, the BCR method also requires incremental analysis for variable-input, variable-output multialternative problems. The discussions in Subsection 11.4.1 apply to the BCR method as well. The other four methods (payback and the three worth-based methods) do not require incremental analysis at all.

---

[14]The BCR is a ratio, while the worths are differences. Each worth is related to the BCR as follows:

$$PW = PW_{benefits} - PW_{costs}$$

Dividing by PW_{costs},

$$PW/PW_{costs} = PW_{benefits}/PW_{costs} - 1$$
$$= BCR - 1$$

Therefore,

$$PW = PW_{costs}(BCR - 1)$$

Similarly,

$$FW = FW_{costs}(BCR - 1)$$
$$AW = AW_{costs}(BCR - 1)$$

Note that with BCR = 1, PW = 0, FW = 0, and AW = 0, as expected.

EXAMPLE 11.6

Assuming $i = 12\%$ per year, a six-year useful life, and no salvage value, which of the following four alternatives should be selected based on payback period? Does the decision change if the criterion is BCR?

	A	*B*	*C*	*D*
First cost	\$80	\$50	\$15	\$90
Annual benefit	18	15	5	25

Solution
For the first part of the problem, we evaluate the payback period for each alternative and select the one with the shortest period.

$$\begin{aligned}\text{Payback period for } A &= 80/18 = 4.44 \text{ years}\\ \text{Payback period for } B &= 50/15 = 3.34 \text{ years}\\ \text{Payback period for } C &= 15/5 = 3 \text{ years}\\ \text{Payback period for } D &= 90/25 = 3.6 \text{ years}\end{aligned}$$

On the basis of shortest payback period, alternative C should be selected.

For the second part of the problem, first note that the alternatives have different inputs (costs) and outputs (benefits). Thus, the solution of this variable-input, variable-output problem requires incremental analysis. Before we carry out this analysis, we should check whether any of the alternatives has a BCR less than one. Such alternatives are eliminated from further analysis. Noting that BCR $= \text{PW}_{\text{benefits}}/\text{PW}_{\text{costs}}$,

$$\text{BCR}_A = \frac{18(P/A, 12\%, 6)}{80} = \frac{18 \times 4.111}{80} = 0.92$$

$$\text{BCR}_B = \frac{15(P/A, 12\%, 6)}{50} = \frac{15 \times 4.111}{50} = 1.23$$

$$\text{BCR}_C = \frac{5(P/A, 12\%, 6)}{15} = \frac{5 \times 4.111}{15} = 1.37$$

$$\text{BCR}_D = \frac{25(P/A, 12\%, 6)}{90} = \frac{25 \times 4.111}{90} = 1.14$$

Based on these results, alternative A with a BCR less than one is discarded. The incremental analysis is carried out with the remaining three alternatives—B, C, and D.

We begin the incremental analysis by choosing alternative C as our favorite, since its first cost is the lowest. Since the next-higher-first-cost alternative is B, analyze the incremental $B-C$ data as

$$\Delta\text{BCR}_{B-C} = \frac{\Delta\text{PW}_{\text{benefits}}}{\Delta\text{PW}_{\text{costs}}} = \frac{(15 - 5)(\text{P/A}, 12\%, 6)}{(50 - 15)} = \frac{10 \times 4.111}{35} = 1.17$$

Since ΔBCR_{B-C} is higher than one, alternative *B* wins, that is, *B* is selected and *C* discarded. We now carry out the next incremental analysis, by comparing alternative *B* with *D*.

$$\Delta BCR_{D-B} = \frac{\Delta PW_{benefits}}{\Delta PW_{costs}} = \frac{(25 - 15)(P/A, 12\%, 6)}{(90 - 50)} = \frac{10 \times 4.111}{40} = 1.03$$

Since ΔBCR_{D-B} is higher than one, alternative *D* is preferred to *B*.

Thus, on the basis of BCR alternative *D* should be selected. Note that the BCR-based decision of selecting *D* is different from that based on the payback criterion.

11.6 COMPREHENSIVE COMPARISON

An overall comparison of the six analysis methods is provided in this section through a table and one example. Table 11.1 presents a comprehensive summary of these methods in four groups. Example 11.7 offers a comprehensive illustration of their applications. It shows that for similar cashflow distributions the decision is the same irrespective of the method used. However, where the distributions are not reasonably alike, the decisions may be different.

Table 11.1 Summary of the Analysis Methods

Comparison Basis	Payback	Worth[15]-based	ROR	BCR
Analysis complexity	Low	Moderate	High	Moderate
Need for computers	Low	Moderate	High	Low
Incremental analysis	No	No	Yes[16]	Yes[17]
Industrial[18] popularity	High	Moderate	High	Little
Usage by Governments[19]	Little	Little	Little	High
Multianswers[20]	No	No	Yes[21]	Yes[22]
Value of i as input	Yes	Yes	No	Yes
External investment	No	No	Yes[23]	No

The BCR method is almost exclusively used for analyzing government and not-for-profit projects. The other five are practiced mostly in for-profit industries.

---

[15]Includes PW, FW, and AW methods

[16]Only in the case of variable-input, variable-output multialternative projects

[17]Only in the case of variable-input, variable-output multialternative projects

[18]Profit-making industries

[19]As well as not-for-profit organizations

[20]The same problem has more than one answer

[21]Apply the four tests to determine IROR, see Chapter 9

[22]Depends on whether future cash outflows are treated as disbenefits or costs

[23]The available surplus cash is invested external to the project, yielding IROR.

EXAMPLE 11.7

Which of the following three alternatives should be selected as per the six analysis methods discussed in this text? Explain any contradiction(s) in the selection. Assume MARR = 10% per year, and zero salvage value for each.

	A	*B*	*C*
First cost	\$50	\$100	\$61
Annual benefit	30	40	37
Useful life, years	2	3	2

Solution
We consider the six methods in the order presented in the text.

a. *Payback Period.* Under this method we evaluate the payback period for each alternative and select the one with the shortest period.

$$\begin{aligned}\text{Payback period for } A &= 50/30 = 1.67 \text{ years}\\ \text{Payback period for } B &= 100/40 = 2.5 \text{ years}\\ \text{Payback period for } C &= 61/37 = 1.65 \text{ years}\end{aligned}$$

On the basis of shortest payback period alternative C should be selected.

b. *PW.* Since the alternatives' useful lives are different, a common analysis period based on their LCM should be used. The LCM of their lives (2, 3, and 2 years) is 6. So we analyze the alternatives over a 6-year time window.

During the 6-year analysis period, alternatives A and C will be replaced twice, at periods 2 and 4, as shown in Figs. 11.2(a) and (c). Alternative B will be replaced only once, at period 3, as seen in Fig. 11.2(b). Assuming that the replacement cost(s) and the resulting benefits remain unchanged during the analysis period, the alternatives' PWs are

$$\begin{aligned}\text{PW}_A &= 30(P/A, 10\%, 6) - [50 + 50(P/F, 10\%, 2) + 50(P/F, 10\%, 4)]\\ &= 30(P/A, 10\%, 6) - 50[1 + (P/F, 10\%, 2) + (P/F, 10\%, 4)]\\ &= 30 \times 4.355 - 50(1 + 0.8264 + 0.6830)\\ &= 130.7 - 125.5\\ &= \$5.2\end{aligned}$$

$$\begin{aligned}\text{PW}_B &= 40(P/A, 10\%, 6) - [100 + 100(P/F, 10\%, 3)]\\ &= 40(P/A, 10\%, 6) - 100[1 + (P/F, 10\%, 3)]\\ &= 40 \times 4.355 - 100(1 + 0.7513)\\ &= 174.2 - 175.1\\ &= -\$0.9\end{aligned}$$

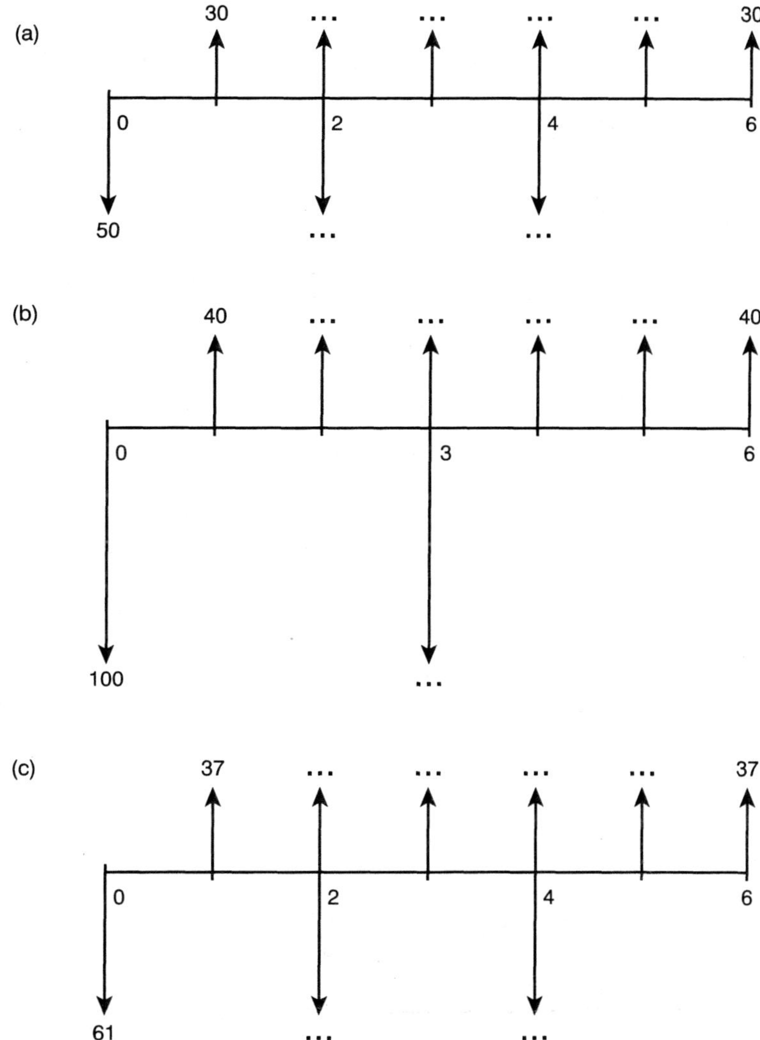

Figure 11.2 Diagram for Example 11.7

$$\begin{aligned}
\mathrm{PW}_C &= 37(P/A, 10\%, 6) - [61 + 61(P/F, 10\%, 2) + 61(P/F, 10\%, 4)] \\
&= 37(P/A, 10\%, 6) - 61[1 + (P/F, 10\%, 2) + (P/F, 10\%, 4)] \\
&= 37 \times 4.355 - 61(1 + 0.8264 + 0.6830) \\
&= 161.1 - 153.1 \\
&= \$8
\end{aligned}$$

With largest positive PW as the decision criterion, alternative C is selected. This decision is the same as under the payback criterion in (a). The compatibility in these two results is attributable to the similarity in alternatives' cashflow distributions.

c. *FW*. The future is selected to be at the end of the analysis period. Based on the three cashflow diagrams in Fig. 11.2 and on the assumption that the costs and benefits for all replacements remain unchanged, the alternatives' FWs at the sixth period are[24]

$$\begin{aligned}
\mathrm{FW}_A &= 30(F/A, 10\%, 6) - [50(F/P, 10\%, 6) + 50(F/P, 10\%, 4) \\
&\qquad + 50(F/P, 10\%, 2)] \\
&= 30(F/A, 10\%, 6) - 50[(F/P, 10\%, 6) + (F/P, 10\%, 4) \\
&\qquad + (F/P, 10\%, 2)] \\
&= 30 \times 7.716 - 50(1.772 + 1.464 + 1.210) \\
&= 231.5 - 222.3 \\
&= \$9.2
\end{aligned}$$

$$\begin{aligned}
\mathrm{FW}_B &= 40(F/A, 10\%, 6) - [100(F/P, 10\%, 6) + 100(F/P, 10\%, 3)] \\
&= 40 \times 7.716 - 100(1.772 + 1.331) \\
&= 308.6 - 310.3 \\
&= -\$1.7
\end{aligned}$$

$$\begin{aligned}
\mathrm{FW}_C &= 37(F/A, 10\%, 6) - [61(F/P, 10\%, 6) + 60(F/P, 10\%, 4) \\
&\qquad + 60(F/P, 10\%, 2)] \\
&= 37(F/A, 10\%, 6) - 61[(F/P, 10\%, 6) + (F/P, 10\%, 4) \\
&\qquad + (F/P, 10\%, 2)] \\
&= 37 \times 7.716 - 61(1.772 + 1.464 + 1.210) \\
&= 285.5 - 271.2 \\
&= \$14.3
\end{aligned}$$

On the basis of largest positive FW, alternative C is selected. As expected, this decision is compatible with that based on PW.

d. *AW*. As we learned in Chapter 8 and elsewhere in the text, there is no need for the analysis period to be the same for each alternative when annual worth is the decision criterion. One can simply use the alternative's individual life as its analysis period, resulting in concise calculations. This is a strong point in favor of AW analysis when alternatives' useful lives are different. Referring to the first complete cycle of each cashflow diagram in Fig. 11.2, the alternatives' AWs are evaluated[25] as follows.

---

[24]We could have followed the shorter approach illustrated in Example 11.3. By multiplying the PWs determined in part (b) with $(F/P, 10\%, 6)$ we could have determined the FWs as

$$\begin{aligned}
\mathrm{FW}_A &= \mathrm{PW}_A(F/P, 10\%, 6) = 5.2 \times 1.772 = 9.2 \\
\mathrm{FW}_B &= \mathrm{PW}_B(F/P, 10\%, 6) = -0.9 \times 1.772 = -1.6 \\
\mathrm{FW}_C &= \mathrm{PW}_C(F/P, 10\%, 6) = 8 \times 1.772 = 14.2
\end{aligned}$$

Note that these FWs are approximately the same as those based on the diagrams.

[25]We could have used the shorter approach illustrated in Example 11.3. By multiplying the PWs determined in part (b) with $(A/P, 10\%, 6)$ we could have determined the AWs as

$$\begin{aligned}
\mathrm{AW}_A &= \mathrm{PW}_A(A/P, 10\%, 6) = 5.2 \times 0.2296 = 1.2 \\
\mathrm{AW}_B &= \mathrm{PW}_B(A/P, 10\%, 6) = -0.9 \times 0.2296 = -0.2 \\
\mathrm{AW}_C &= \mathrm{PW}_C(A/P, 10\%, 6) = 8 \times 0.2296 = 1.8
\end{aligned}$$

For A, the relevant diagram comprises the \$50 first cost, followed by the two \$30 benefits. Thus,

$$\begin{aligned} AW_A &= 30 - 50(A/P, 10\%, 2) \\ &= 30 - 50 \times 0.5762 \\ &= 30 - 28.8 \\ &= \$1.2 \end{aligned}$$

For B, the relevant diagram comprises the \$100 first cost, followed by the three \$40 benefits. Thus,

$$\begin{aligned} AW_B &= 40 - 100(A/P, 10\%, 3) \\ &= 40 - 100 \times 0.4021 \\ &= 40 - 40.2 \\ &= -\$0.2 \end{aligned}$$

For C, the relevant diagram comprises the \$61 first cost, followed by the two \$37 benefits. Thus,

$$\begin{aligned} AW_C &= 37 - 61(A/P, 10\%, 2) \\ &= 37 - 61 \times 0.5762 \\ &= 37 - 35.1 \\ &= \$1.9 \end{aligned}$$

With largest positive AW as the criterion, alternative C is selected. As expected, this decision is compatible with that by the PW or FW methods in parts (b) and (c). It is also compatible with the payback period method for the reason explained in part (b).

e. *ROR*. With the three alternatives' inputs (costs) and outputs (benefits) being different, the given problem is of the variable-input, variable-output type, and hence its solution must be based on incremental analysis. Prior to this analysis, however, we need to determine whether any of the alternatives has an ROR less than the MARR so that it can be eliminated from incremental analysis. There are two ways to proceed: (i) Since the MARR is given, one can check the sign of the alternatives' PW (if positive, then ROR > MARR), or (ii) evaluate each ROR exactly by setting PW = 0. We follow the first approach, which is usually simpler.

Since the PWs evaluated in part (b) are based on the MARR, we do not need to carry out any new calculations. Of the three present worths there, PW_B is negative. Hence B is discarded. Thus, the three-alternative problem has reduced to a two-alternative problem. In the incremental analysis we compare alternatives A and C. Since A's first cost is lower, we prefer to select it. We thus evaluate C over A to check whether the additional cost of C and the resulting additional benefits yield a rate of return ΔROR_{C-A} greater than the MARR.

For determining ΔROR_{C-A}, generate the C–A incremental data and carry out the calculations. Since the useful life of these two alternatives is the same at two years, the analysis period is two years. From the given data, we get

Year	A	C	$C-A$
0	$-\$50$	$-\$61$	$-\$11$
1	30	37	7
2	30	37	7

By applying[26] Tests 1 and 2 on these $C-A$ cashflows we predict a positive rate of return, since there is only one sign change. Next, we evaluate PW_{C-A} to check its sign.

$$\begin{aligned}\text{PW}_{C-A} &= 7(P/A, 10\%, 2) - 11\\ &= 7 \times 1.736 - 11\\ &= 12.152 - 11\\ &= 1.152\end{aligned}$$

Since PW_{C-A} is positive, investment-wise C is better than A. Hence C should be selected. Note that it was also the choice based on all the previous methods. No contradiction in results has occurred so far, since the cashflow distributions for the three alternatives are alike.

f. *BCR*. Since this is a multialternative problem, we need to first check whether incremental analysis is essential. Looking at the cashflows, the alternatives are of the variable-input, variable-output type. Hence the solution must be based on incremental analysis. Prior to the analysis, however, we need to check for any alternative with a BCR of less than one. Such alternatives are excluded from incremental analysis. Considering one complete cycle of the cashflows, and noting that $\text{BCR} = \text{PW}_{\text{benefits}}/\text{PW}_{\text{costs}}$, we get

$$\text{BCR}_A = \frac{30(P/A, 10\%, 2)}{50} = \frac{30 \times 1.736}{50} = 1.04$$

$$\text{BCR}_B = \frac{40(P/A, 10\%, 3)}{100} = \frac{40 \times 2.487}{100} = 0.99$$

$$\text{BCR}_C = \frac{37(P/A, 10\%, 2)}{61} = \frac{37 \times 1.736}{61} = 1.05$$

Based on these results, alternative B, with a BCR less[27] than one, is discarded from any further consideration. In other words, incremental analysis encompasses alternatives A and C only. With lower first cost, alternative A is our preferred choice. We therefore analyze the $C-A$ data to evaluate ΔBCR_{C-A}. The calculations should be simpler, since both A and C have the same useful life. Had it been different, we would have needed to analyze over an LCM-based analysis period.

$$\Delta\text{BCR}_{C-A} = \frac{\Delta\text{PW}_{\text{benefits}}}{\Delta\text{PW}_{\text{costs}}} = \frac{(37 - 30)(P/A, 10\%, 2)}{(61 - 50)} = \frac{7 \times 1.736}{11} = 1.10$$

[26]See Chapter 9 if necessary.

[27]Since B's BCR is almost equal to one, the engineer should check its data once more to ensure that they are accurate. Even a slight variation in the data can render B acceptable.

Since ΔBCR_{C-A} is higher than one, alternative C wins. Thus, on the basis of BCR too, alternative C should be selected. This decision is compatible with those in (a) through (e).

In the above example, all six methods led to the same decision, since the alternatives' cashflows were alike. Where the cashflows are significantly different, the decisions may be contradictory.

11.7 INDUSTRIAL USAGE

Companies use one or more of the six methods discussed in the text for analyzing engineering economics projects. The analysis method is decided upon by the company management considering several financial factors. Engineers working in small companies, with fewer people knowledgeable in financial matters, or in their own business may select the method by themselves.

The choice of a method suitable for economic analysis of engineering projects is influenced by several factors, such as

1. Level of investment
2. Simplicity of analysis sought
3. Time horizon of the project: short-term, long-term, etc.
4. Financial health of the company
5. Economic environment and market stability
6. Government incentives for engineering investment
7. Type of business (computer chips or potato chips!)

According to fragmented surveys on their popularity, the ROR method is the most widely used in industry, as shown in Fig. 11.3. It is closely followed by PW and AW, the two engineering-oriented worth-based methods. The payback-period method is also common in industry in spite of its nonconsideration of the time value of money. In companies with a limited capital pool, it may be the only one in use. The BCR method is most prevalent in government and not-for-profit organizations.

Most companies use only one of the six methods, as recommended by the upper management. If this method gives rise to any doubt about the result of the analysis, then other methods may be used to enhance the reliability of the final decision. Several companies use more than one method, especially payback period for initial "screening" of the alternatives.

SUMMARY

A comparison among the six methods—payback period, PW, FW, AW, ROR, and BCR—commonly used in analyzing engineering economics projects has been presented in Chapter 11. We discussed why a method may yield results that contradict those obtained by the other methods. The three worth-based methods, namely PW, AW, and FW, always lead to the same decision. This is not true when other methods are involved. In general, for projects with heavily skewed cash inflows (more benefits either in the beginning or at the end of the project life), the payback-period, ROR, and BCR methods may yield re-

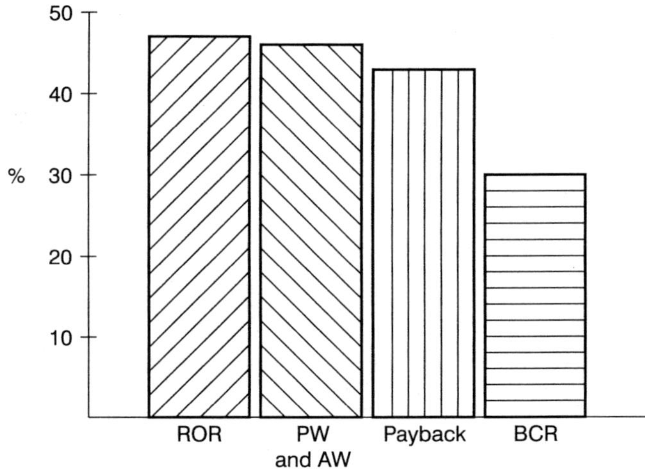

Figure 11.3 Industrial Usage of the Methods

sults that differ from those by the worth-based methods. If the cash inflows are fairly distributed, all the methods lead to the same decision. The payback-period method favors projects with heavy cash inflows in the beginning of the project. Of the six, the ROR and BCR methods may involve incremental analysis. The ROR method is widely used in industry, especially in financially healthy companies with long-term investment strategies. The advent of computer systems has rendered the otherwise complex ROR method popular. The simpler payback-period method, which does not account for the time value of money, is commonly used by companies struggling to survive, that is, with a short-term investment strategy. This method is sometimes used to "filter" out some of the alternatives, prior to detailed economic analyses of the more promising ones.

EXERCISES

Discussion Questions

11.1 Why do engineers sometimes apply more than one method to analyze an investment project?

11.2 Explain the concept of incremental analysis to a manager who has a college degree but did not study engineering economics.

11.3 Discuss why engineering economics is much more than solving a given numerical problem.

11.4 Describe an engineering project for which FW analysis might be more appropriate. (*Hint*: Consider expansion of an existing manufacturing facility that will take 3 years to complete.)

11.5 If the payback-period method is both simple and popular in industry, why do we need the other methods, especially ROR, which is complex and lengthy?

11.6 Can there be a cashflow distribution for which decisions based on the three worth-based methods may differ? If yes, when? If no, why not?

11.7 When is incremental analysis applied? Explain through an example.

Multiple-Choice Questions (Circle the *best* answer.)

11.8 Of the six methods for analyzing engineering economics problems, the most widely used is
a. PW.
b. AW.
c. ROR.
d. payback period.

11.9 The PW method always leads to the same decision as the
a. ROR method.
b. BCR method.
c. payback period method.
d. AW method.

11.10 Incremental analysis may have to be carried out while analyzing a problem under the ______ criterion.
a. PW
b. BCR
c. payback-period
d. AW

11.11 The least likely method for analyzing an engineering project is
a. FW.
b. PW.
c. ROR.
d. payback period.

11.12 Computer technology has popularized the ______ method in industry.
a. PW
b. BCR
c. ROR
d. payback-period

11.13 If the MARR is given, ROR analysis can be simplified by evaluating the
a. BCR.
b. PW.
c. ROR.
d. payback period.

11.14 ROR is evaluated by setting
a. payback period = 3 years.
b. MARR = prevalent interest rate.
c. BCR > 1.
d. PW = 0.

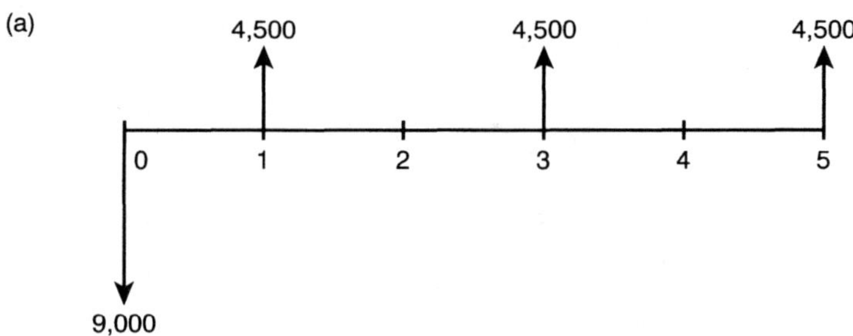

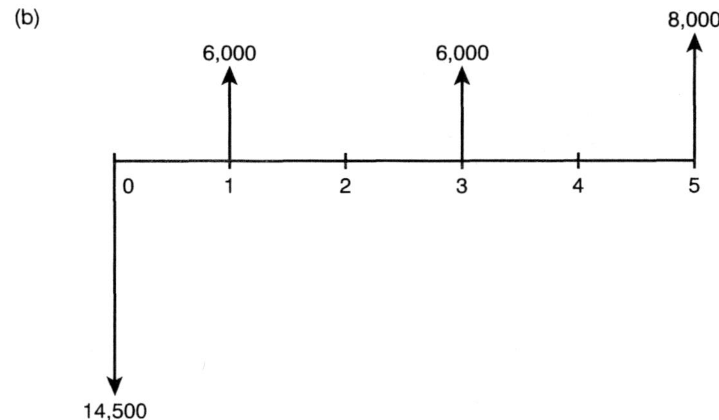

Figure 11.4 Diagram for Problem 11.15

Numerical Problems

11.15 Which of the two projects whose cashflow diagrams are given in Fig. 11.4 should be selected on the basis of the PW criterion if the annual interest rate is 9% compounded yearly? Does the decision change under FW and AW criteria?

11.16 If the criterion in Problem 11.15 is payback period, does the decision change?

11.17 There are three alternatives in a productivity project. Alternative *A* involves the use of robots, *B* of modular fixtures, and *C* of an upgraded machine. The alternatives' first costs in dollars are 3,600, 5,000, and 8,400 respectively. If their net annual benefits are \$1,890, \$2,680, and \$3,200, and their useful lives are 2, 2, and 3 years, which alternative should be selected on the basis of payback period? Does the decision change if the ROR criterion is used? Explain any discrepancy in the decision. Assume annual compounding and 12% MARR.

11.18 Which of the two projects, *A* or *B*, whose cashflow diagrams are given in Fig. 11.5 should be selected on the basis of payback criterion? Does the decision change if the criterion is BCR? Assume a 9% interest rate compounded annually.

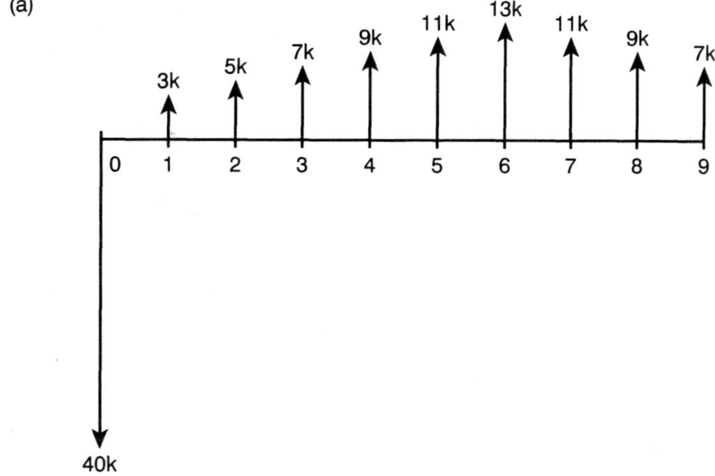

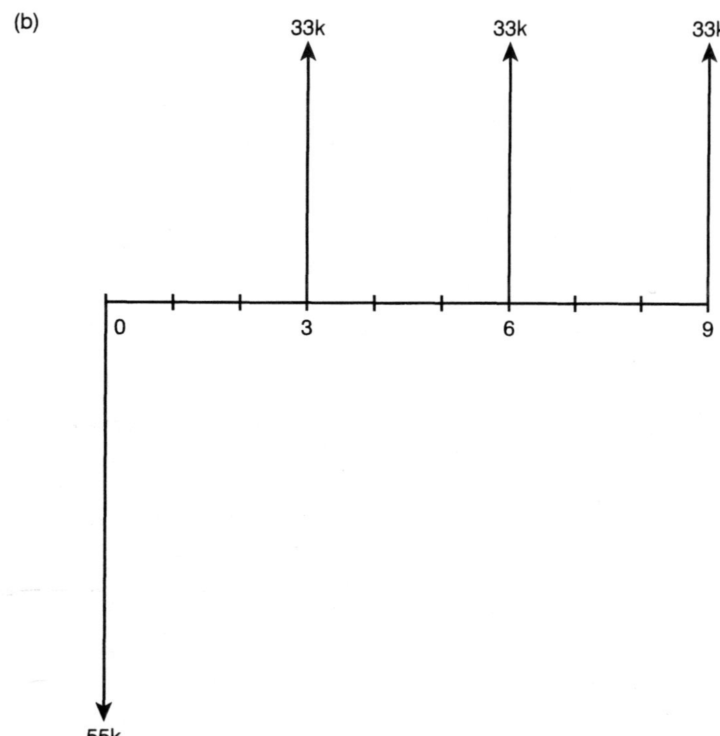

Figure 11.5 Diagram for Problem 11.18

11.19 For the following cashflows which alternative should be selected on PW basis if $i = 20\%$ per year? Is the decision affected if the criterion is ROR?

Year	*A*	*B*
0	−\$80	−\$150
1–5	40	70

11.20 For the following cashflows, which alternative should be selected on the basis of the BCR? Is the decision affected if ROR is the criterion? Assume the MARR = 12%.

Year	*A*	*B*
0	−\$900	−\$900
1	50	600
2	350	300
3	700	200
4	400	150

11.21 Which of the two projects whose cashflow diagrams are given in Fig. 11.6 should be selected on the basis of PW if the annually compounded interest rate is 5% per year? Does the decision change under the BCR criterion?

11.22 Which of the two projects whose cashflow diagrams are given in Fig. 11.7 should be selected on the basis of PW if the MARR = 15%? Does the decision change under ROR or BCR criteria?

11.23 If in Example 11.7, alternative *B*'s cash inflows are \$10, \$40, and \$90 respectively for the three years of its life, instead of \$40 annually, how are the various results affected? Discuss any contradictions.

11.24 Which of the following alternatives should be selected by the six analysis methods discussed in the text? Explain any contradiction(s) in the results. Assume MARR = 10%, and zero salvage value. Note that the useful lives of alternatives *A* and *C* are two years, and that of *B* three years. If necessary, assume an external interest rate of 8% per year.

Year	*A*	*B*	*C*
0	−\$50	−\$100	−\$61
1	30	90	37
2	30	20	37
3		10	

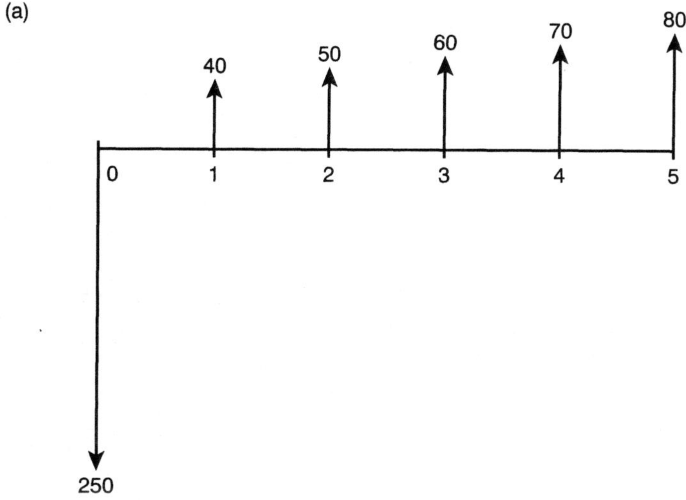

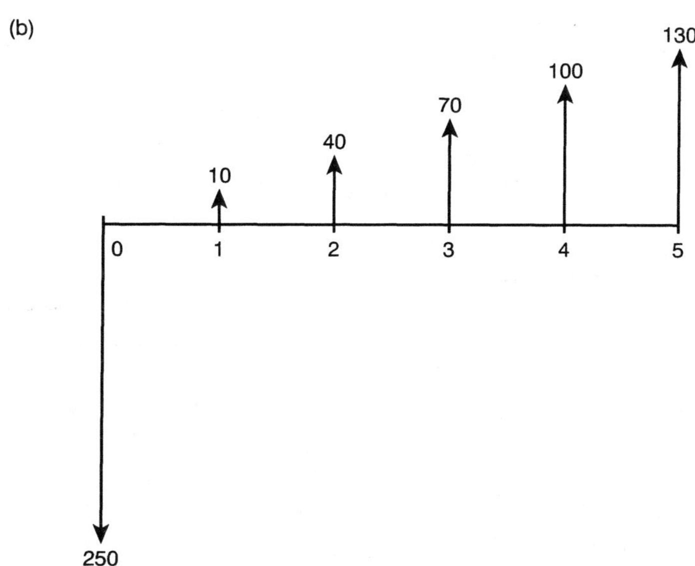

Figure 11.6 Diagram for Problem 11.21

(a)

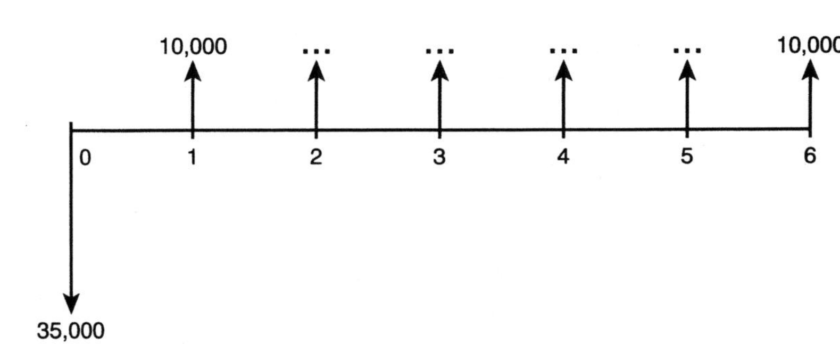

(b)

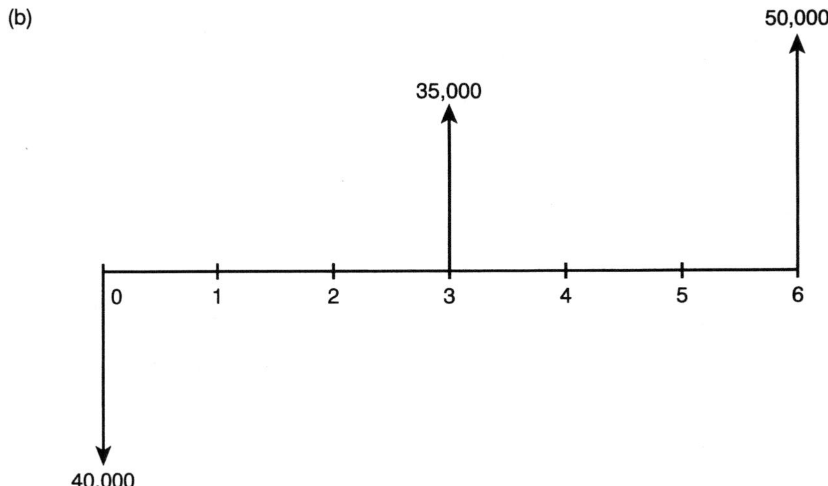

Figure 11.7 Diagram for Problem 11.22

PART III

Realism

In Part II we discussed the application of six commonly used analysis methods. However, we did not include any consideration of two important facts of the engineering industry, namely depreciation and income tax. We cover them in Part III along with replacement analysis.

The organization of this text in three parts, as mentioned earlier, is metaphorically similar to the construction of a house. Part I was the *foundation,* while Part II the *structure,* comprising floors, walls, and ceilings. Part III represents the *furnishings* such as drapes, carpets, and kitchen appliances. We discuss here some essential refinements to the analyses covered in the text.

Part III therefore adds realism to what we have learned so far, by incorporating "real-world" considerations. In Chapter 12, we learn to account for depreciation of resources. One of the ways governments guide national economies is through taxation, whose impact on economic analyses is discussed in Chapter 13. An important and distinct area of engineering economics is resource replacement—the subject matter of Chapter 14. Finally, in Chapter 15 we discuss the effect of inflation on analyses and also summarize advanced topics relevant to engineering economics.

CHAPTER TWELVE

Depreciation

IN THIS CHAPTER YOU WILL LEARN ABOUT

- The meaning of depreciation
- Net cashflows to account for depreciation
- An asset's book value
- How to cost depreciation
- A depreciation schedule
- The straight-line method
- The SOYD method
- The declining-balance method
- The double-declining-balance method
- The unit-of-production method
- MACRS as a government-approved method
- The composite depreciation method
- Analysis of depreciation-modified net cashflows

Machines, equipment, and other resources or assets in which companies invest capital deteriorate with use, resulting in a decrease in the value. The loss in value is described by the term *depreciation* and expressed as *depreciation cost*.

Depreciation costs are also called *depreciation charges* because they are deducted from, or charged to, the profit realized from the invested resource. As a financial incentive, governments allow companies to deduct depreciation cost from their profit before it is taxed. In this chapter we study three interrelated aspects of depreciation:

1. Determination of depreciation costs
2. The effect of depreciation costs on cashflows, and
3. Economic analysis with depreciation accounted for.

12.1 MEANING

Depreciation measures the *decrease in value* of an asset. It is the opposite of *appreciation*, which means an increase in value. In periods of peace and tranquility, engineering assets generally depreciate, since better ones are continually developed, and marketed, through technical innovations. In periods of war and uncertainty, assets may appreciate due to the imminent danger of destruction and slowdown of the economy. In the 20th century, we experienced appreciation during World War II. Since then, the world economy has operated with depreciation as a fact.

Consider that XYZ company procured a robot for $20,000 to improve its assembly line productivity. At the end of the first year of use, the robot's value due to wear, tear, and obsolescence is $16,000. This decrease of $4,000 in the robot's value is its depreciation for the first year. Let us say that the robot saved $9,000 during the first year of its use. This saving adds to the company profit, which is taxable. The investment in the robot thus generated a $9,000 gross profit, but at a loss of $4,000 in the robot's value. The company therefore accounts this $4,000 as a cost, resulting in a *net profit* of $5,000 ($9,000 − $4,000). The income tax is due on the net profit of $5,000, not on the gross profit of $9,000. The provision of charging depreciation to the income from an invested resource thus reduces the tax burden.[1]

Capital equipment depreciates due to obsolescence or deterioration from use, or both. A machine tool, for example, depreciates due to wear and tear from its use. The same is true for an automobile. Some equipment, such as personal computers, depreciates primarily due to obsolescence created by new and better products. With the exception of land, almost every asset needed by the engineering industry is depreciable and subject to depreciation costs. With an infinite life, land is nondepreciable even if its market value for the year has decreased.

For depreciation purposes, an asset's useful life is considered finite. The asset's initial value reduces to its salvage value at the end of useful life. We can illustrate this through a graph as in Fig. 12.1. The values are plotted along the y-axis and the useful life along the x-axis. The equipment costs P at period 0 when its use begins. Over time, at the end of its useful life N its value drops to the salvage value S. Thus, total depreciation over the useful life is $P-S$, the difference between the initial cost and salvage value. If this total depreciation is charged uniformly, then the depreciation cost is $(P-S)/N$ per year.

12.2 TERMINOLOGY

In the course of depreciation costing we come across several terms that seem familiar. The important terms are discussed here because we need to understand their meanings precisely to avoid any error in costing.

[1] What is company's gain through lower tax is the government's loss through reduced collection. Governments therefore limit the amount of depreciation that can be charged in a year.

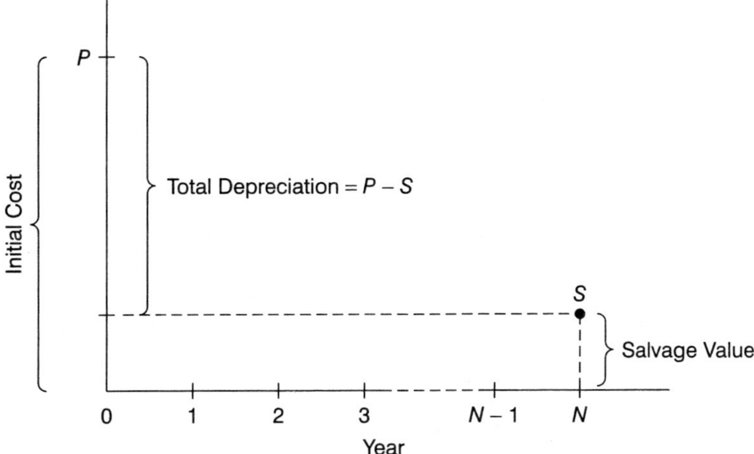

Figure 12.1 Total Depreciation Equals $P-S$

Asset An asset is any resource used in business. It may be tangible or intangible. Equipment, machines, computers, and buildings are examples of tangible assets. Goodwill, prompt after-sale service, quality reputation, and a community-caring image are examples of intangible assets. Tangible assets can easily be quantified in monetary terms, while intangibles can't.

Useful Life Useful life is the duration of time equipment or resources are expected to be in use. Its value is usually known, but not with 100% certainty, since it relates to the future. It is only an estimate, usually based on past data. Insurance companies normally keep track of the useful lives of a variety of equipment.

For a given first cost, shorter useful lives allow larger annual depreciation charges, thus reducing the income tax more. While faster and larger depreciation charges are desirable for the company, governments lose, since they collect less in income tax. To avoid any controversy on the amount of allowable depreciation, largely due to the difference in *expected* useful life as judged by the company and the taxing authority, government may fix the value of the allowable useful life for an asset. We discuss this further in Subsection 12.3.5.

The term *useful life* differs from similar terms such as *service life*, *economic life*, *market life*, and *shelf life*. We discuss the difference between useful life and economic life later in Chapter 14.

Market Value We have earlier defined depreciation as a loss in value, without specifying the type of value. An asset has *market value* and *book value*. The market value is what the asset can be disposed of at the time of sale. It is what others, especially prospective buyers, are prepared to pay for the asset. It is therefore determined by the marketplace, irrespective of how much the owner thinks the asset is worth. It also depends on the timing of asset disposal. While the market value of a capital asset may not change as abruptly as the value of a company stock, it nevertheless fluctuates.

Book Value The term *book* is derived from the age-old practice of maintaining company books to record asset costs. An asset's first cost is its initial book value. With use, the value decreases[2] each year, thus reducing the asset's book value. The book value is what the owner needs to sell the asset for to avoid a loss. However, depending on the buyer's need for the used asset, its market value may be above or below the book value.

The market value and book value of an asset usually differ. Several factors influence this difference. For products under a continual onslaught of evolving technologies, the market value of a used asset is usually lower than its book value. Personal computers and electronic products are good examples. For most engineering assets, however, such as machines, concrete mixers, electric motors, market value and book value are about the same. For an antique, market value is much higher than its book value, which is usually zero.

Depreciation Schedule A depreciation schedule is simply a listing of the annual[3] depreciation costs (or charges) for an asset. A depreciation schedule looks like this:

Year	Depreciation Cost
1	$9,000
2	9,000
3	9,000
4	9,000
5	9,000

If the depreciation cost is constant during the asset life, the entries in the schedule are the same, as here. Note that the tabulated depreciation schedule looks like a cash-flow table.

The book value is a function of depreciation costs. An asset's cost is usually entered in appropriate company books at the time of procurement. This entry is known as first cost, initial cost, or procurement cost. Every year certain depreciation is charged to the asset. By subtracting the depreciation charge for the year from the book value (at the beginning of the year), we get the year-end book value. The year-end book value becomes the book value at the beginning of the next year, as illustrated in the following table for an asset whose first cost is $10,000.

[2]Depreciation is a charge to compensate for this.

[3]Could a depreciation charge be other than annual? In theory yes, but in practice no. Since taxes are paid annually, depreciation costs are also charged on an annual basis.

Year	Beginning Book Value	Depreciation Cost	Year-End Book Value
1	$10,000	$2,000	$8,000
2	8,000	1,500	6,500
3	6,500	1,000	5,500
4	5,500	750	4,750
⋮	⋮	⋮	⋮

As can be seen, the book value at the end of the year is equal to the beginning book value minus the depreciation cost; for example, for year 3 it is $6,500 − $1,000 = $5,500. Note that the first and third columns together form the depreciation schedule.

Often we keep track of the *cumulative depreciation* cost to know the depreciation to date. This is done by adding another column to the table:

Year	Beginning Book Value	Depreciation Cost	Cumulative Depreciation	Year-End Book Value
1	$10,000	$2,000	$2,000	$8,000
2	8,000	1,500	3,500	$6,500
3	6,500	1,000	4,500	5,500
4	5,500	750	5,250	4,750
⋮	⋮	⋮	⋮	⋮

The book value at any time during the asset's life is the difference between the first cost and cumulative depreciation up to that time. Thus,

$$\text{Book value} = \text{First cost} - \text{Cumulative depreciation}$$

For example, as posted in the preceding table, the year-end book value for year 3 is

$$\begin{aligned}\text{First cost} - \text{Cumulative depreciation up to year 3} &= \$10{,}000 - \$4{,}500 \\ &= \$5{,}500\end{aligned}$$

EXAMPLE 12.1

A machine costs $11,500. At the end of its five-year useful life its salvage value is estimated to be $1,500. Prepare its depreciation schedule if annual depreciation is (a) 20% of the first cost, and (b) 20% of the book value. The depreciation charge for the last (fifth) year must be adjusted so that the total does not exceed the maximum allowable.

Solution

Total[4] allowable depreciation for the machine, being the difference between the first cost ($11,500) and salvage value ($1,500) is $10,000 over five years of its useful life.

a. The annual depreciation, being 20% of the first cost, is 0.2 × $11,500 = $2,300. A complete depreciation schedule is shown[5] in the following table. Note, however, that the depreciation for the fifth year is $800—the remainder[6] of the total allowable. Since by the end of the fourth year cumulative depreciation is already $9,200, the fifth-year depreciation = $10,000 − $9,200 = $800.

Year	First Cost	Depreciation Cost	Cumulative Depreciation
0	$11,500		
1	11,500	$2,300	$2,300
2	11,500	2,300	4,600
3	11,500	2,300	6,900
4	11,500	2,300	9,200
5	11,500	800	10,000

b. In this case depreciation is 20% of the book value. The complete schedule is

Year	Depreciation Cost	Cumulative Depreciation	Book Value
0			$11,500
1	$2,300	$2,300	9,200
2	1,840	4,140	7,360
3	1,472	5,612	5,888
4	1,178	6,790	4,710
5	3,210	10,000	1,500

Note that the depreciation cost for each of the first four years is 20% of the book value at the end of the previous[7] year. For example, for year 3, it is 20% of $7,360 = $1,472. The fifth-year depreciation of $3,210 is the remainder that could be charged, that is, total allowable depreciation minus the cumulative depreciation up to the fourth year ($10,000 − $6,790 = $3,210).

---

[4]The word *total* or *maximum* as an adjective to *allowable depreciation* means the same.

[5]The data in depreciation schedule tables usually correspond to period ends, as in cashflow tables.

[6]The cumulative depreciation cannot exceed $10,000—the total allowable.

[7]The end of the previous year marks the beginning of the current year, so the book value at the end of the previous year is the book value at the beginning of the current year.

If you find the entries in the table confusing, an alternative table format may be the following, in which book value is posted in the second column.

Year	Book Value	Depreciation Cost	Cumulative Depreciation
0	$11,500		
		$2,300	
1	9,200		$2,300
		1,840	
2	7,360		4,140
		1,472	
3	5,888		5,612
		1,178	
4	4,710		6,790
		3,210	
5	1,150		10,000

In this format the depreciation costs have been posted in between the years, since they relate to the duration rather than to a particular point in time. The entries are perhaps more comprehensible in this format. For example, the depreciation cost of $1,472 for the duration in between years 2 and 3 is 20% of $7,360—the book value at the beginning of year 2. As another example, the $4,710 book value at the beginning of year 4 is simply the book value at the beginning of year 3 ($5,888) minus the depreciation cost for the duration between years 3 and 4 ($1,178). Note that for any year the total of the book value and cumulative depreciation equals the first cost ($11,500). For example, for year 3, it is $5,888 + $5,612 = $11,500.

12.3 METHODS

There are several methods to determine the annual depreciation charge for an asset. They differ primarily in how the total allowable depreciation, $P-S$, is distributed over the asset's useful life N; in other words, referring to Fig. 12.1, in how to arrive from point P to point S. Obviously, there can be several paths connecting P to S, the simplest one being a straight line.

12.3.1 Straight Line

The straight-line method (SL method) is commonly used in depreciation costing. This method distributes the total depreciation $P-S$ *uniformly* over the useful life N. In other words,

$$\text{Annual depreciation} = \frac{1}{N}(P-S) \tag{12.1}$$

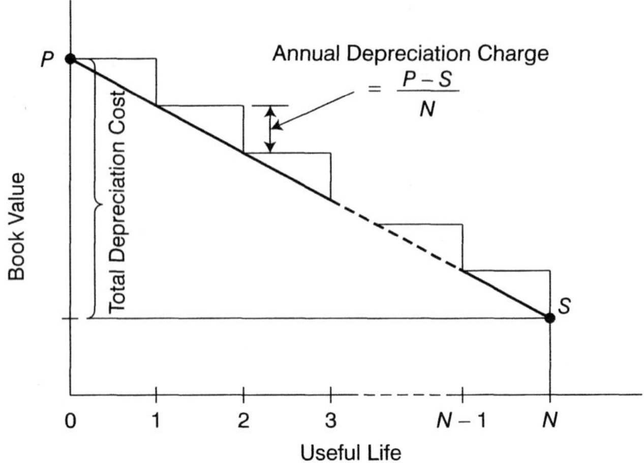

Figure 12.2 Straight-Line Depreciation

The factor $1/N$ is called the *SL depreciation rate*, since the annual depreciation is $1/N$ times the maximum allowable depreciation $P-S$. For example, for an asset with a 5-year useful life, this rate is $1/5 = 0.20 = 20\%$. Thus, the annual depreciation is 20% of the total allowable.

Why is it called the straight-line method? Refer to Fig. 12.2, which shows the annual depreciation charges according to the straight-line method. Note that the charges are constant and equal to $(P-S)/N$, as represented by the vertical lines of the stair pattern. Since this pattern rests on a straight line connecting P to S, it is called the straight-line method.

The merit of the straight-line method is its simplicity—an equal charge each year. This simplicity is its limitation too. It does not allow for higher charges in the beginning of the useful life when an asset loses its value more rapidly. Higher charges in the beginning would have enabled faster recovery of the investment. The straight-line method thus delays recovery of the investment. The other methods, discussed in the following subsections, do not suffer from this limitation.

EXAMPLE 12.2

Universal Arm (a robot) costs $25,000. At the end of its four-year useful life its salvage value is estimated to be $5,000. Prepare its depreciation schedule under the straight-line method.

Solution

Here, $P = \$25{,}000$, $S = \$5{,}000$, and $N = 4$ years. From Equation[8] (12.1),

[8]Straight-line depreciation problems can be solved without needing the equation, as illustrated in Example 12.1.

$$\text{Annual depreciation} = \frac{P-S}{N} = \frac{\$25{,}000 - \$5{,}000}{4} = \$5{,}000$$

The depreciation schedule for Universal Arm can be prepared in the way explained in Example 12.1, and is tabulated here.

Year[9]	Beginning Book Value	Depreciation	Ending Book Value
1	$25,000	$5,000	$20,000
2	20,000	5,000	15,000
3	15,000	5,000	10,000
4	10,000	5,000	5,000
		Total: $20,000	

Note that the total depreciation charged over the four years of useful life is $20,000. The resulting book value of $5,000 at the end of the fourth year (useful life), being equal to the salvage value, confirms the accuracy of the results.

We are often interested in an asset's book value at the end of a specific year. One can read this value off a complete depreciation schedule table if it exists. The alternative is to use an equation, which is derived here.

We intend to determine the book value at the end of jth year. Since the annual depreciation is $(P-S)/N$,

$$\text{Cumulative depreciation up to } j\text{th year} = \frac{J \times (P-S)}{N}$$

Since the asset's first cost is P, of which $J(P-S)/N$ has been recovered through depreciation charges, its remaining value, or the book value, must be the difference between the two. Thus,

$$\text{Book value at } j\text{th year end} = P - \frac{J \times (P-S)}{N} \qquad (12.2)$$

---

[9]Year 0 is irrelevant and hence left out. Note that year 1 begins when the resource is put in service. Hence, the beginning book value for year 1 is the first cost.

Substituting from Equation (12.1), this can be expressed also as

$$\text{Book value at } j\text{th year end} = P - J \times \text{Annual depreciation}$$

We can apply Equation (12.2) to Example 12.2. For example, the book value at the end of the third year ($J = 3$) from this equation[10] is

$$\begin{aligned} & 25{,}000 - \frac{3 \times (25{,}000 - 5{,}000)}{4} \\ &= 25{,}000 - 15{,}000 \\ &= \$10{,}000 \end{aligned}$$

As expected, this book value is the same as that in the last column of Example 12.2's depreciation schedule for year 3.

12.3.2 Sum-of-Years-Digits (SOYD)

As mentioned earlier, the SL method results in constant annual depreciation during the asset life. Since most assets depreciate faster in the beginning of their lives, companies would prefer to charge depreciation accordingly. Methods that enable higher depreciation charges in the early years of the asset life are therefore more realistic. The sum-of-years-digits (SOYD) method is one of them. It charges higher depreciation in the beginning by factoring in the asset's remaining life.

The SOYD is simply the sum of all the digits up to the useful life. For example, if an asset's useful life is five years, then its SOYD = 1 + 2 + 3 + 4 + 5 = 15 years. Similarly, for an eight-year asset the SOYD is 36 years (1 + 2 + ⋯ + 8). An asset's SOYD can be determined by either of the following two approaches.

1. *Short* Useful Life (N small).
   If the useful life N is short, then simply add[11] all the digits up to N. For example, if the useful life is 4 years, then its SOYD = 1 + 2 + 3 + 4 = 10 years. For most engineering economics problems, this approach suffices, since N is usually small.

---

[10]Rather than use the equation, we can follow first principles. The annual depreciation is (25,000 − 5,000)/4 = $5,000. The total depreciation over three years will thus be $15,000, resulting in a book value of $25,000 − $15,000 = $10,000. The first-principles approach works well only for the SL method. For other depreciation methods, the equation-based approach is preferred.

[11]The SOYD calculation for a long-life asset is simpler if the SOYD for a shorter life is known. Let us say that we need to determine the SOYD of a 10-year asset and that the SOYD of an 8-year asset is known to be 36 years. Since

$$\text{SOYD}_{10\text{ years}} = 1 + 2 + 3 + 4 + 5 + 6 + 7 + 8 + 9 + 10$$

we can say that

$$\begin{aligned} \text{SOYD}_{10\text{ years}} &= (1 + 2 + 3 + 4 + 5 + 6 + 7 + 8) + 9 + 10 \\ &= \text{SOYD}_{8\text{ years}} + 9 + 10 \\ &= 36 + 19 \\ &= 55\text{ years} \end{aligned}$$

2. *Long* Useful Life (N large).
   If the useful life is long, then adding the digits may be lengthy and prone to error. In that case, use the equation derived here.
   For an asset of useful life N years,

$$\text{SOYD} = 1 + 2 + 3 + \cdots + (N-1) + N$$

Rewriting the right side in reverse order,

$$\text{SOYD} = N + (N-1) + (N-2) + \cdots + 2 + 1$$

Adding these two expressions, we get

$$\begin{aligned} \text{SOYD} + \text{SOYD} &= (N + 1) + (N + 1) + (N + 1) + \cdots + (N + 1) + (N + 1) \\ 2(\text{SOYD}) &= N \text{ times } (N + 1) \\ &= N(N + 1) \end{aligned}$$

Therefore,

$$\text{SOYD} = N(N + 1)/2 \tag{12.3}$$

For example, for a 15-year asset this equation yields

$$\begin{aligned} \text{SOYD}_{15} &= 15 \times 16/2 \\ &= 120 \text{ years} \end{aligned}$$

Under SOYD, the asset's depreciation for the year is obtained by multiplying the total depreciation $P-S$ by a factor that equals *Remaining useful life*/SOYD. The remaining useful life must be reckoned from the beginning of the year for which depreciation is being calculated. For the first year, the remaining life equals the useful life N. For the third year, the remaining life equals $N-2$, since the asset has already been used for two years.

Consider an asset with first cost P and salvage value S, that is, with a total depreciation of $P-S$. If N is the useful life in years, then

$$\begin{aligned} \text{First-year depreciation} &= \frac{\text{Remaining useful life}}{\text{SOYD}} \times (\text{P}-\text{S}) \\ &= \frac{N}{\text{SOYD}} \times (P-S) \\ \text{Second-year depreciation} &= \frac{\text{Remaining useful life}}{\text{SOYD}} \times (P-S) \\ &= \frac{N-1}{\text{SOYD}} \times (P-S) \end{aligned}$$

Similarly,

$$\text{Third-year depreciation} = \frac{N-2}{\text{SOYD}} \times (P-S)$$

$$\vdots$$

$$N\text{th-year depreciation} = \frac{N-(N-1)}{\text{SOYD}} \times (P-S)$$

$$= \frac{1}{\text{SOYD}} \times (P-S)$$

Note that the multiplying numerator in these relations has decreased from N to 1. We illustrate the application of the SOYD method through Example 12.3.

EXAMPLE 12.3

Universal Arm (a robot) costs \$25,000. At the end of its four-year useful life, its salvage value is estimated to be \$5,000. Prepare its depreciation schedule under the SOYD method.

Solution

Here $P = \$25{,}000$, $S = \$5{,}000$, and $N = 4$ years. We first determine the SOYD. Since the useful life is short, we can do without the equation. From first principles,

$$\text{SOYD} = 1 + 2 + 3 + 4 = 10 \text{ years}$$

$$\text{Total depreciation} = P - S$$

$$= \$25{,}000 - \$5{,}000$$

$$= \$20{,}000$$

Using the relationship,

$$\text{Year's depreciation} = \frac{\text{Remaining useful life}}{\text{SOYD}} \times (P-S)$$

we get

$$\text{First-year depreciation} = \frac{N}{\text{SOYD}} \times (P-S)$$

$$= \frac{4}{10} \times \$20{,}000$$

$$= \$8{,}000$$

$$\text{Second-year depreciation} = \frac{N-1}{\text{SOYD}} \times (P-S)$$

$$= \frac{3}{10} \times \$20{,}000$$
$$= \$6{,}000$$

$$\text{Third-year depreciation} = \frac{N-2}{\text{SOYD}} \times (P - S)$$
$$= \frac{2}{10} \times \$20{,}000$$
$$= \$4{,}000$$

$$\text{Fourth-year depreciation} = \frac{N-3}{\text{SOYD}} \times (P - S)$$
$$= \frac{1}{10} \times \$20{,}000$$
$$= \$2{,}000$$

The SOYD-based depreciation[12] schedule for Universal Arm can be tabulated in the way explained in Examples 12.1 and 12.2, and is

Year	Beginning Book Value	Depreciation	Ending Book Value
1	$25,000	$8,000	$17,000
2	17,000	6,000	11,000
3	11,000	4,000	7,000
4	7,000	2,000	5,000
		20,000	

The ending book value of $5,000, being equal to the salvage value, confirms the accuracy of the results. This can also be done by adding the depreciation charges to check their sum, which should equal $P - S$, as in this table ($20,000 total).

In comparison to the schedule under the SL method (Example 12.2), note that

1. The yearly SOYD depreciation charge is not constant, and
2. The depreciation charge is higher in the beginning.

---

[12]We could have shortened the calculations by first determining the value of $(P - S)/\text{SOYD}$ as $20{,}000/10 = 2{,}000$. The depreciations would then have been simply $(N)2{,}000$ for the first year, $(N - 1)2{,}000$ for the second year, and so on. This might have simplified the calculations as

First-year depreciation $= 4 \times 2{,}000 = \$8{,}000$
Second-year depreciation $= 3 \times 2{,}000 = \$6{,}000$
Third-year depreciation $= 2 \times 2{,}000 = \$4{,}000$
Fourth-year depreciation $= 1 \times 2{,}000 = \$2{,}000$

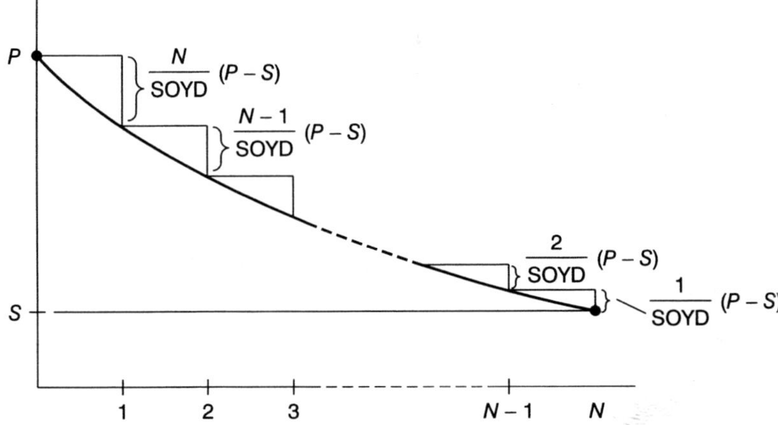

Figure 12.3 SOYD Depreciation

The characteristic of SOYD-based depreciation costing can be illustrated graphically, as in Fig. 12.3. Note how the depreciation charge reduces with time and thus is not constant as in Fig. 12.2 (SL method). The base of the staircase pattern in Fig. 12.3 is nonlinear, and hence the SOYD method can be called a *nonlinear* method. The nonlinearity is also obvious in the yearly depreciation equation, where the remaining life as a multiplier reduces each year.

12.3.3 Declining Balance

The declining balance method charges a fixed (constant) percentage of the asset's book value as depreciation for the year. Since the book value decreases each year, the annual depreciation charge[13] also decreases, being highest for the first year. For this reason the declining balance method is also nonlinear, like the SOYD method.

The percentage is decided by the taxing authority, usually a department of the federal or central government. In the United States two rates are allowed: 150% or 200% of the straight-line rate. The 200%-based approach is often called the *double-declining-balance (DDB) method*, the term *double* signifying the fact that the rate is twice that of the SL method. Since DDB enables faster depreciation than the 150% rate, DDB is preferred.

Declining-balance depreciation may be determined by first principles, as illustrated in Example 12.4. For the mathematically inclined readers who prefer equations, the following derivations should be of interest. Equation-based analysis has been illustrated in Example 12.5 by redoing Example 12.4.

Consider an asset of initial cost P, salvage value S, and useful life N. The straight line rate is $1/N$, since, as explained earlier in conjunction with Equation (12.1), the annual SL depreciation is obtained by multiplying the total depreciation $(P-S)$ with $1/N$.

[13]*Charge* and *cost* are synonymous in the context of depreciation.

In the declining-balance method the rates are 150% and 200% of the straight-line rate, that is, $1.5/N$ or $2/N$. However these rates are multiplied with the *book value*, not with the *total depreciation* $(P - S)$ as in the straight-line method. The procedure for determining declining-balance depreciation for the 200% rate[14] is as follows.

Year 1

Book value at the *beginning* of the first year
= First (initial) cost
$= P$

First-year depreciation $= (2/N) \times$ first year book value
$= (2/N) \times P$
$= 2P/N$

Book value at the *end* of the first year = Book value at the *beginning* of the first year − First-year depreciation
$= P - 2P/N$
$= P(1 - 2/N)$

Year 2

Book value at the *beginning* of the second year
= Book value at the *end of the first year*
$= P(1 - 2/N)$

Second-year depreciation $= (2/N) \times$ second year book value
$= (2/N) \times P(1 - 2/N)$
$= (2P/N)(1 - 2/N)$

Book value at the *end* of the second year = Book value at the *beginning* of the second year − Second-year depreciation
$= P(1 - 2/N) - (2P/N)(1 - 2/N)$
$= P[1 - 2/N - 2/N + (2/N)^2]$
$= P[1 - 4/N + (2/N)^2]$
$= P(1 - 2/N)^2$

Year 3

Book value at the *beginning* of the third year
= Book value at the *end of the second year*
$= P(1 - 2/N)^2$

Third-year depreciation $= (2/N) \times$ third year book value
$= (2/N) \times P(1 - 2/N)^2$
$= (2P/N)(1 - 2/N)^2$

---

[14]For the 150% rate, simply replace $2/N$ throughout the derivation with $1.5/N$.

Continuing the procedure for other years,

$$\vdots$$

$$j\text{th-year depreciation} = (2P/N)\,(1 - 2/N)^{j-1} \qquad \textit{(12.4)}$$

$$\vdots$$

$$N\text{th-year depreciation} = (2P/N)\,(1 - 2/N)^{N-1}$$

Note that Equation (12.4) has been derived from first principles and that it is valid for the double-declining balance method. For a declining-balance rate of 150%, $2/N$ should be replaced in this equation and elsewhere by $1.5/N$.

There are two approaches to solve a declining-balance problem:

1. Follow the first principles, or
2. Use Equation (12.4), or its variation for the 150% rate.

In the first-principles approach, we basically follow the steps that led to Equation (12.4), but using the given data. This is illustrated in Example 12.4. The depreciation calculations can also be based on the equation approach, as illustrated in Example 12.5.

Where a complete depreciation schedule is to be prepared, the first-principles and equation approaches are equally suitable. However, when depreciation charges for only a specific year, or a few intermediate years, are to be determined, the equation approach is better, as illustrated in Example 12.6, since it does away with the intervening steps essential in the first-principles approach.

EXAMPLE 12.4

Universal Arm (a robot) costs $25,000. At the end of its four-year useful life its salvage value is estimated to be $5,000.

a. Prepare its DDB depreciation schedule.

b. Discuss any discrepancy in the schedule, especially between book value at the end of useful life and salvage value.

Solution

The given data are $P = \$25{,}000$, $S = \$5{,}000$, and $N = 4$ years.

a. For DDB depreciation the rate is $2/N$. Since the entire depreciation schedule is being asked for, we can either follow the first principles or use Equation (12.4). Let us follow the first principles, especially since N is small[15]. This involves following the steps used in deriving Equation (12.4), but with the given data. Thus,

Year 1

$$\text{Book value at the } \textit{beginning} \text{ of the first year} = \text{First (initial) cost}$$
$$= \$25{,}000$$

[15]For large values of N, the first-principles approach becomes lengthy.

$$
\begin{aligned}
\text{First-year depreciation}^{16} &= (2/N) \times \text{First-year book value} \\
&= (2/4) \times \$25{,}000 \\
&= \$12{,}500
\end{aligned}
$$

$$
\begin{aligned}
\text{Book value at the } \textit{end} \text{ of the first year} &= \text{Book value at the beginning of the first year} \\
&\quad - \text{First-year depreciation} \\
&= \$25{,}000 - \$12{,}500 \\
&= \$12{,}500
\end{aligned}
$$

Year 2

Book value at the *beginning* of the second year

$$
\begin{aligned}
&= \text{Book value at the } \textit{end of the first year} \\
&= \$12{,}500
\end{aligned}
$$

$$
\begin{aligned}
\text{Second-year depreciation} &= (2/N) \times \text{Second-year book value} \\
&= (2/4) \times \$12{,}500 \\
&= \$6{,}250
\end{aligned}
$$

Book value[17] at the *end* of the second year

$$
\begin{aligned}
&= \text{Book value at the } \textit{beginning} \text{ of the second year} \\
&\quad - \text{Second-year depreciation} \\
&= \$12{,}500 - \$6{,}250 \\
&= \$6{,}250
\end{aligned}
$$

Year 3

Book value at the *beginning* of the third year

$$
\begin{aligned}
&= \text{Book value at the } \textit{end of the second year} \\
&= \$6{,}250
\end{aligned}
$$

$$
\begin{aligned}
\text{Third-year depreciation} &= (2/N) \times \text{Third-year book value} \\
&= (2/4) \times \$6{,}250 \\
&= \$3{,}125
\end{aligned}
$$

$$
\begin{aligned}
\text{Book value at the } \textit{end} \text{ of the third year} &= \text{Book value at the } \textit{beginning} \text{ of the third year} \\
&\quad - \text{Third-year depreciation} \\
&= \$6{,}250 - \$3{,}125 \\
&= \$3{,}125
\end{aligned}
$$

---

[16]One can evaluate $2/N$ as $2/4 = 0.5$ at this stage and use this 0.5, instead of $2/N$, throughout the calculations to achieve simplicity.

[17]Since the given data, rather than symbols, are being used as the calculation proceeds, the results get simplified; we do not face lengthy terms as we did while deriving Equation (12.4). This illustrates that the first-principles approach can be simpler.

Year 4

Book value at the *beginning* of the fourth year
= Book value at the *end of the third year*
= \$3,125

Fourth-year depreciation = $(2/N) \times$ Fourth-year book value
= $(2/4) \times$ \$3,125
= \$1,563

Book value at the *end* of the fourth year
= Book value at the *beginning* of the fourth year
− Fourth-year depreciation
= \$3,125 − \$1,563
= \$1,562

The DDB depreciation schedule can be summarized in a table as

Year	Beginning Book Value	Depreciation	Ending Book Value
1	\$25,000	\$12,500	\$12,500
2	12,500	6,250	6,250
3	6,250	3,125	3,125
4	3,125	1,563	1,562
	Total depreciation =	\$23,438	

b. We notice a discrepancy between the expected salvage value \$5,000 and the book value \$1,562 at the end of useful life. These two should have been the same. This discrepancy is also reflected in the total depreciation cost \$23,438 being greater than the maximum allowable of \$20,000 ($P - S$ = \$25,000 − \$5,000). Thus, there is something wrong with these results.

The discrepancy can be explained through Fig. 12.4. In any depreciation method the book value should reduce from its initial value P to the salvage value S by the end of useful life. Conceptually, a "depreciation journey" begins at point P and should end at S. The SL method leads from P to S through the shortest path—a straight line. The SOYD method also leads to the destination S, but via a nonlinear path. The declining-balance method also follows a nonlinear path, but may miss the destination S. In the present case, the depreciation journey ends at point T, which is below point S. In other words, the book value at the end of useful life (point T) is less than the salvage value (point S). In some cases the declining balance method may lead to a point above S (book value greater than salvage value). In rare cases (if the analyst is lucky!), this method leads exactly to the destination S (book value equal to salvage value).

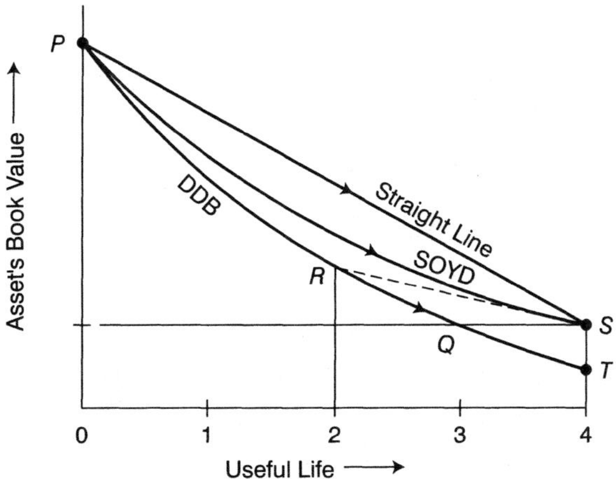

Figure 12.4 Linear and Nonlinear Depreciation

What can be done about this likely discrepancy with the declining balance method? The choices are

Point *T* below *S*	Stop charging depreciation once the salvage value has been reached (point *Q* in Fig. 12.4). At the appropriate year charge just enough depreciation to have the book value match the salvage value
Point *T* above *S*	If the resource is used beyond the useful life, keep charging depreciation until the salvage value is reached.

A third choice, mostly practiced, is to switch over to the SL method just in time to reach the destination point S by the end of useful life. In Fig. 12.4 this happens at point *R*, wherefrom *RS* is the linear path. We discuss this switching over later in Section 12.4. Since it involves two methods, one before and the other after the switchover, the procedure is called a *composite method*.

For the example under discussion, with point *T* below *S*, the depreciation schedule of Example 12.4 is modified by reducing the total depreciation to the maximum allowed. This necessitates charging none for the fourth year and only $1,250 for the third year, as explained below.

We need to reduce the total charged from $23,438 to $20,000. Our attempt at reduction should begin backward from the end of useful life, that is, from year 4. Assuming no charge for the fourth year, the total charged up to the third year is $23,438 − $1,563 = $21,875. Since this is still higher than the maximum allowed ($20,000), we next reduce the charge for the third year. This reduction must equal $21,875 − $20,000 = $1,875. So the third-year depreciation is $3,125 − $1,875 = $1,250.

The resulting schedule is as follows, where we have added another column for the cumulative depreciation for the sake of completeness.

Year	Beginning Book Value	Depreciation	Ending Book Value	Cumulative Depreciation
1	\$25,000	\$12,500	\$12,500	\$12,500
2	12,500	6,250	6,250	18,750
3	6,250	1,250	5,000	20,000
4	5,000	0	5,000	20,000

EXAMPLE 12.5

Solve Example 12.4 using the equation approach.

Solution

The given data are $P = \$25{,}000$, $S = \$5{,}000$, and $N = 4$ years.

a. Since DDB depreciation is required, the rate to use is $2/N$. The relevant equation is Equation (12.4),

$$j\text{th-year depreciation} = (2P/N)(1 - 2/N)^{j-1}$$

By substituting the given values of P (\$25,000) and N (4 years), and the appropriate value of j (1, 2, 3, and 4), depreciation for each year is determined as

$$\begin{aligned}\text{First-year depreciation} &= (2 \times \$25{,}000/4)\,(1 - 2/4)^{1-1} \\ &= (\$12{,}500) \times (1 - 0.5)^0 \\ &= (\$12{,}500) \times 0.5^0 \\ &= \$12{,}500 \times 1 \\ &= \$12{,}500\end{aligned}$$

$$\begin{aligned}\text{Second-year depreciation} &= (2 \times \$25{,}000/4)\,(1 - 2/4)^{2-1} \\ &= (\$12{,}500) \times (1 - 0.5)^1 \\ &= (\$12{,}500) \times 0.5^1 \\ &= \$12{,}500 \times 0.5 \\ &= \$6{,}250\end{aligned}$$

$$\begin{aligned}\text{Third-year depreciation} &= (2 \times \$25{,}000/4)\,(1 - 2/4)^{3-1} \\ &= (\$12{,}500) \times (1 - 0.5)^2 \\ &= (\$12{,}500) \times 0.5^2 \\ &= \$12{,}500 \times 0.25 \\ &= \$3{,}125\end{aligned}$$

$$\begin{aligned}\text{Fourth-year depreciation} &= (2 \times \$25{,}000/4)(1 - 2/4)^{4-1} \\ &= (\$12{,}500) \times (1 - 0.5)^3 \\ &= (\$12{,}500) \times 0.5^3 \\ &= \$12{,}500 \times 0.125 \\ &= \$1{,}563\end{aligned}$$

The discussions in part (b) of Example 12.4 requiring modification in the third and fourth year depreciation charges remain applicable. Hence, the final depreciation schedule is

Year	Beginning Book Value	Depreciation	Ending Book Value	Cumulative Depreciation
1	$25,000	$12,500	$12,500	$12,500
2	12,500	6,250	6,250	18,750
3	6,250	1,250	5,000	20,000
4	5,000	0	5,000	20,000

Note that the use of equations rendered the solution in Example 12.5 more concise than that in Example 12.4. In addition, the yearly depreciation computation was "decoupled," that is, could be calculated without needing the previous year's results. These make the equation approach better where the depreciation charge for a specific year is to be determined, as illustrated in Example 12.6. This approach is also desirable when a few intermediate years are involved.

EXAMPLE 12.6

What is the fifth-year DDB depreciation charge for a machine whose first cost is $50,000 and estimated salvage value at the end of its eight-year useful life is $2,000?

Solution

Since we need to determine the depreciation charge for a specific (fifth) year only, the use of equations will be simpler than the first principles. The latter would have required determining depreciation for the previous years too to get the book value for year 5. The relevant formula, Equation (12.4), is

$$j\text{th-year depreciation} = (2P/N)(1 - 2/N)^{j-1}$$

For $P = \$50{,}000$, $N = 8$ years, and $j = 5$, we get

$$\begin{aligned}\text{Fifth-year depreciation} &= (2 \times \$50{,}000/8)(1 - 2/8)^{5-1}\\ &= (\$12{,}500) \times (1 - 0.25)^4\\ &= \$12{,}500 \times 0.75^4\\ &= \$12{,}500 \times 0.3164\\ &= \$3{,}955\end{aligned}$$

Using Equation (12.4), two more equations are derived in the following two subsections. These equations are handy in determining the specific year-end cumulative declining balance and the asset book value.

12.3.3.1 Cumulative Declining Balance

The cumulative DDB depreciation at the end of the jth year is the total of all the annual depreciation charges up to the jth year. From Equation (12.4),

$$\begin{aligned}
\text{First-year DDB depreciation} &= (2P/N)\,(1 - 2/N)^{1-1} \\
&= (2P/N)\,(1 - 2/N)^{0} \\
&= (2P/N) \times 1 \\
&= 2P/N \\
\text{Second-year DDB depreciation} &= (2P/N)\,(1 - 2/N)^{2-1} \\
&= (2P/N)\,(1 - 2/N)^{1} \\
&= (2P/N)\,(1 - 2/N) \\
\text{Third-year DDB depreciation} &= (2P/N)\,(1 - 2/N)^{3-1} \\
&= (2P/N)\,(1 - 2/N)^{2} \\
&\vdots \\
j\text{th-year DDB depreciation} &= (2P/N)\,(1 - 2/N)^{j-1}
\end{aligned}$$

By adding the above, we get

$$\begin{aligned}
&\text{Cumulative DDB depreciation up to year } j \\
&\quad = 2P/N + (2P/N)(1 - 2/N) + (2P/N)(1 - 2/N)^2 + \cdots + (2P/N)(1 - 2/N)^{j-1} \\
&\quad = (2P/N)[1 + (1 - 2/N) + (1 - 2/N)^2 + \cdots + (1 - 2/N)^{j-1}]
\end{aligned}$$

Assuming the cumulative DDB depreciation up to year j to be Y,

$$Y = (2P/N)[1 + (1 - 2/N) + (1 - 2/N)^2 + \cdots + (1 - 2/N)^{j-1}]$$

Multiplying both sides by $(1 - 2/N)$

$$Y(1 - 2/N) = (2P/N)[(1 - 2/N) + (1 - 2/N)^2 + \cdots + (1 - 2/N)^{j}]$$

Subtracting the expression for Y from that for $Y(1 - 2/N)$, we get[18]

$$\begin{aligned}
Y(1 - 2/N) - Y &= (2P/N)[(1 - 2/N)^j - 1] \\
Y - 2Y/N - Y &= (2P/N)[(1 - 2/N)^j - 1] \\
-2Y/N &= (2P/N)[(1 - 2/N)^j - 1] \\
-Y &= P[(1 - 2/N)^j - 1] \\
Y &= P[1 - (1 - 2/N)^j]
\end{aligned}$$

that is,

$$\text{Cumulative DDB depreciation} = P[1 - (1 - 2/N)^j] \qquad \textit{(12.5)}$$

For a declining balance rate of 150%, in the preceding equation replace 2 by 1.5.

---

[18]All the terms on the right-hand side of the two expressions, except $(1 - 2/N)^j$ and 1, cancel with each other.

12.3.3.2 Book Value

The book value of an asset at the end of a specific year j will simply be its initial cost P minus the cumulative depreciation up to the jth year (Equation (12.5)). For a double declining balance, therefore,

$$\begin{aligned}\text{Book value at the end of year } j &= P - P[1 - (1 - 2/N)^j] \\ &= P - P + P(1 - 2/N)^j \\ &= P(1 - 2/N)^j\end{aligned}$$

Thus,

$$\text{Book value at } j\text{th year end} = P(1 - 2/N)^j \qquad (12.6)$$

EXAMPLE 12.7

Using the appropriate equations, determine the cumulative DDB depreciation and book value in Example 12.4 at the end of the third year.

Solution

The equations derived in the preceding two subsections are applicable. The cumulative DDB depreciation at the end of the third year is obtained by substituting $j = 3$ in Equation (12.5). For the given values of P ($25,000) and N (4 years), this yields

$$\begin{aligned}\text{Cumulative DDB depreciation at third year end} &= P[1 - (1 - 2/N)^3] \\ &= 25{,}000[1 - (1 - 2/4)^3] \\ &= 25{,}000[1 - 1/8] \\ &= 25{,}000 \times 7/8 \\ &= \$21{,}875\end{aligned}$$

The asset's book value at the end of the third year is obtained from Equation (12.6) as

$$\begin{aligned}\text{Book value at third year end} &= \text{P}(1 - 2/N)^j \\ &= 25{,}000(1 - 2/4)^3 \\ &= 25{,}000 \times 1/8 \\ &= \$3{,}125\end{aligned}$$

The book value could have been determined easily from first principles (without needing the equation), since the cumulative DDB depreciation at the third year end is already known[19]. Merely subtracting the cumulative DDB depreciation, $21,875, from the asset's first cost, $25,000, would have yielded $3,125.

[19]The use of Equation (12.6) is warranted if the corresponding cumulative DDB depreciation is unknown.

12.3.4 Unit of Production

Straight-line annual depreciation is based on the fraction a year is of the asset's useful life. The SL method, as well as the others, works well if the asset is in regular use. For assets that are used irregularly, wear and tear do not relate to the physical time duration. In such cases we prorate depreciation on the basis of actual use, that is, on actual units of production (UOP) for the year.

Consider an automobile, for example, whose first cost is \$16,000. It is expected to last seven years, during which it will be driven 100,000 miles. If its salvage value is \$2,000, then the annual straight-line depreciation, as one seventh of the total depreciation \$14,000 ($P - S = 16{,}000 - 2{,}000$), will be \$2,000. This annual charging assumes that the automobile is driven 100,000/7 = 14,286 miles per year. Consider, however, that the automobile belongs to a retiree who uses it for local driving of 5,000 miles a year. In that case, rather than the time-based depreciation, we should use mileage-based depreciation, that is, the UOP method. Based on the actual usage, the depreciation for 5,000 miles of driving will be one twentieth (5,000/100,000 = 1/20) of \$14,000, that is, \$700 per year. If during a particular year the retiree takes up a traveling salesperson's job that increases driving to, say, 20,000 miles, then for that year the depreciation charge will be one fifth (20,000/100,000 = 1/5) of \$14,000, that is, \$2,800.

The UOP method is suitable for depreciating natural resources based on annual extraction or harvesting as a proportion of the total available. It is basically similar to the straight-line method, except that the charge for the year is prorated by the ratio *Production or use for the year/Expected lifetime production or use*. For an asset of initial cost P and salvage value S, the prorating of total depreciation $P-S$ yields

$$\text{UOP depreciation for the year} = \frac{\text{Production for the year}}{\text{Expected lifetime production}} \times (P-S)$$

Example 12.8 illustrates the application of UOP method.

EXAMPLE 12.8

To support a road construction project, management decides to install at the site a concrete mixer for an initial cost of \$35,000. At the end of the construction period of five years the machine is likely to be sold for \$3,000. The annual concrete needs during the five years are 300, 400, 800, 200, and 100 tons. Prepare the mixer's depreciation schedule based on the UOP method, as well as its book values.

Solution

It is given that

$$\begin{aligned} P-S &= \$35{,}000 - \$3{,}000 \\ &= \$32{,}000 \end{aligned}$$

We need to know the lifetime use of the mixer to prorate $P-S$ on the basis of annual use. Adding the needs of the five years,

$$\text{Total usage} = 300 + 400 + 800 + 200 + 100 = 1{,}800 \text{ tons}$$

Thus,

$$\text{First-year depreciation} = \frac{\text{First-year use}}{\text{Total usage}} \times (P-S) = \frac{300}{1{,}800} \times 32{,}000 = \$5{,}333$$

$$\text{Second-year depreciation} = \frac{\text{Second-year use}}{\text{Total usage}} \times (P-S) = \frac{400}{1{,}800} \times 32{,}000 = \$7{,}111$$

$$\text{Third-year depreciation} = \frac{\text{Third-year use}}{\text{Total usage}} \times (P-S) = \frac{800}{1{,}800} \times 32{,}000 = \$14{,}222$$

$$\text{Fourth-year depreciation} = \frac{\text{Fourth-year use}}{\text{Total usage}} \times (P-S) = \frac{200}{1{,}800} \times 32{,}000 = \$3{,}556$$

$$\text{Fifth-year depreciation} = \frac{\text{Fifth-year use}}{\text{Total usage}} \times (P-S) = \frac{100}{1{,}800} \times 32{,}000 = \$1{,}778$$

The depreciation schedule for the mixer is a summary of these charges in a table. Check that the total depreciation equals $P-S$ ($32,000), confirming the accuracy of the results.

Year	Depreciation
1	$5,333
2	7,111
3	14,222
4	3,556
5	1,778
Total depreciation =	$32,000

The year-end book values can be determined by subtracting the yearly depreciation from the beginning book values, as shown in the following table. The last (fifth) year's book value of $3,000 being equal to the salvage value offers another check on the accuracy of the results.

Year	Beginning Book Value	Depreciation	Ending Book Value
1	$35,000	$5,333	$29,667
2	29,667	7,111	22,556
3	22,556	14,222	8,334
4	8,334	3,556	4,778
5	4,778	1,778	3,000
	Total depreciation =	$32,000	

12.3.5 Accelerated Cost Recovery System (ACRS)

We have discussed several methods of depreciation costing in the preceding sections. In general, companies are free to choose the method they use, but for tax purposes they must follow the one recommended by the taxing authority. In the United States, the federal government's Internal Revenue Service (IRS) collects income tax. Prior to 1954 the IRS required companies to follow the straight-line method. Beginning in 1954 the accelerated methods of DDB and SOYD were required. In 1981 these methods were superseded by a simpler procedure called Accelerated Cost Recovery System (ACRS), which was modified later under the Tax Reform Act 1986 as MACRS, where M stands for *Modified*. It may be noted that not all countries practice systems such as ACRS or MACRS, and where they do the systems may be different. Non-U.S. readers should refer to the cost recovery systems, if any, prevalent in their countries, and to the associated documents obtainable from the taxing arm of their governments.

12.3.5.1 MACRS

The Modified Accelerated Cost Recovery System (MACRS, pronounced "makers") simplifies depreciation costing for both the government and the companies. The simplification has been achieved by grouping all the depreciable assets into certain property classes of definite lives, called *recovery periods*. Altogether there are eight classes[20], with recovery periods of 3, 5, 7, 10, 15, 20, 27.5, and 39 years. For each property class the IRS has established fixed depreciation allowance percentages, which yield, when multiplied with the first cost, the depreciation charge for each year of the asset's useful life.

Some common items and their property class are given in Table 12.1, along with the *recovery periods*—the term used in the MACRS method for useful life.

[20]For complete information refer to the latest edition of IRS Publication 534. *Depreciation*. U.S. Government Printing Office: Washington, D.C.

Table 12.1 MACRS Property Classes

Property Type	Recovery Period, years
Fabricated metal products, special tools for the manufacture of plastic parts	3
High-tech equipment, R&D equipment, automobiles, light trucks	5
Office furniture, manufacturing equipment, fixtures	7
Railroad cars, barges, vessels, tugs	10
Telephone distribution plants and similar utility property, wastewater plants	15
Electrical power plants, municipal sewers	20
Residential rental property	27.5
Nonresidential property, elevators, escalators	39

The recovery percentages for different property classes are given in Table 12.2. For percentages for the real-estate property classes (last two in Table 12.1) refer to the original IRS document.

Table 12.2 MACRS Recovery[21] Percentages

Class Year	3-year	5-year	7-year	10-year	15-year	20-year
1	33.33	20.00	14.29	10.00	5.00	3.750
2	44.45	32.00	24.49	18.00	9.50	7.219
3	14.81*	19.20	17.49	14.40	8.55	6.677
4	7.41	11.52*	12.49	11.52	7.70	6.177
5		11.52	8.93*	9.22	6.93	5.713
6		5.76	8.92	7.37	6.23	5.285
7			8.93	6.55*	5.90*	4.888
8			4.46	6.55	5.90	4.522
9				6.56	5.91	4.462*
10				6.55	5.90	4.461
11				3.28	5.91	4.462
12					5.90	4.461
13					5.91	4.462
14					5.90	4.461
15					5.91	4.462
16					2.95	4.461
17						4.462
18						4.461
19						4.462
20						4.461
21						2.231

[21]The percentages begin as declining balance. The 15-year and 20-year properties are based on 150%, the other four on 200% (DDB). The percentages change to straight line at the asterisked point.

Example 12.9 illustrates the application of the MACRS method. Its simplicity—a depreciation method that is confusion-free and easy to use—has been the goal of the IRS.

EXAMPLE 12.9

XYZ company puts in use a computer workstation that cost $20,000. Determine the depreciation schedule as per MACRS.

Solution

First refer to Table 12.1 to determine the class the computer workstation belongs to; from there it is a five-year property. Next refer to Table 12.2 for recovery percentages for a five-year property which are:

Recovery Year	Allowable %
1	20.00
2	32.00
3	19.20
4	11.52
5	11.52
6	5.76

For the workstation costing $20,000, the first-year depreciation will thus be 20% of $20,000 = $4,000. For the second year, it will be 32% of $20,000 = $6,400. Similarly, for the subsequent years depreciation will be $3,840, $2,304, $2,304, and $1,152. The resulting depreciation schedule is

Year	Depreciation
1	$4,000
2	6,400
3	3,840
4	2,304
5	2,304
6	1,152
Total:	$20,000

Note that the total depreciation charged is $20,000, exactly equal to the investment cost to be recovered. Checking the total depreciation to ensure the accuracy of results is always a good practice.

The recovery percentages in Table 12.2 are based[22] on DDB depreciation with timely switchover to the SL method. Another fact underlying MACRS is that the depreciation for the first year is half of that by the DDB method. This is because MACRS assumes the asset to have been put in service at midyear. The effect of charging only half for the first year is a spillover of depreciation into the year following the end of useful life. For example, a five-year property has, as illustrated in Example 12.9, a depreciation charge in the sixth year. This is true for other property classes too, as seen in Table 12.2.

12.4 THE COMPOSITE METHOD

A depreciation method that enables faster recovery of the invested capital is always desirable. The SL method offers constant recovery; that is why it has been superseded by nonlinear methods that yield faster recovery during the early years of asset life. In general, it is acceptable to switch from one method to another during the asset life. Such a combination of two or more methods is termed a *composite depreciation method.*

---

[22]To understand how the recovery percentages have been arrived at, consider a five-year property as illustration. We begin charging depreciation by DDB and switch over to straight-line when beneficial. The salvage value in MACRS is always considered zero, and hence the total depreciation equals the first cost.

To calculate the DDB depreciation rate we need to determine first the straight-line rate. The straight-line rate for a five-year property is 1/5 = 0.20. The DDB rate, being twice of the straight-line rate, will be 0.40, or 40%. MACRS allows half depreciation in the first year on the assumption that the asset is put to use in the middle of the year. In this case, therefore, first-year depreciation is half of 40%, i.e., 20% as seen in Table 12.2 for the five-year property. For the subsequent years we determine both DDB and SL depreciations to check which one is greater, so that we can make timely switch over. The relevant calculations are illustrated below. Note that for SL depreciation during the second year we use a ratio of 1/4.5 because the first year depreciation corresponded to half a year only.

Year	Calculations	MACRS %
1	Half-year DDB depreciation = 0.50 × 40%	<u>20</u>
2	DDB depreciation = 0.40 (100% − 20%)	<u>32</u>
	SL depreciation = (1/4.5) (100% − 20%)	17.78
3	DDB depreciation = 0.40 (100% − 52%)	<u>19.20</u>
	SL depreciation = (1/3.5) (100% − 52%)	13.71
4	DDB depreciation = 0.40 (100% − 71.20%)	11.52
	SL depreciation = (1/2.5) (100% − 71.20%)	<u>11.52</u>
5	SL depreciation = (1/1.5) (100% − 82.72%)	<u>11.52</u>
6	SL depreciation = remaining 100% − 94.24%	<u>5.76</u>

The underlined percentage being greater of the two for the year becomes the MACRS recovery percentage for the year, as seen in Table 12.2 for the five-year property class. The recovery percentages for other property classes have been determined in the same way. Note that for all property classes the percentage for the last year is half of that for the previous year.

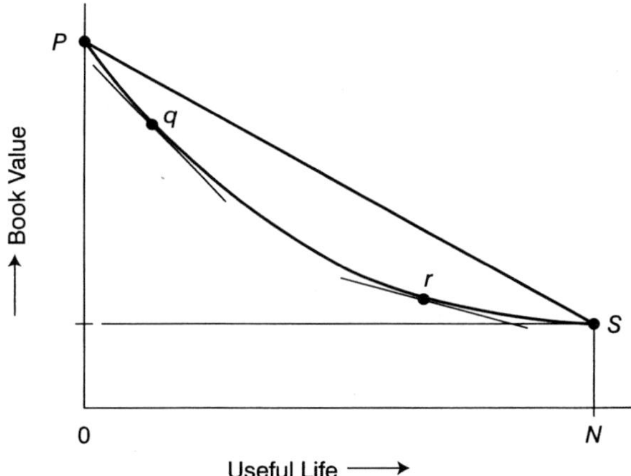

Figure 12.5 Concept of Composite Depreciation

Referring to Fig. 12.5, the SL method represented by line *PS* yields a constant annual charge. For nonlinear methods, such as SOYD or declining balance, the depreciation follows a curve like *PqrS*. The depreciation charge depends on the slope of the line or the curve, depending on the method used. In the case of straight line this slope is fixed, while it varies along the curve. In the early years, depreciation is large if the curve is followed. For example, the curve's slope at point *q* is greater than the fixed slope of the straight line *PS*. During the latter years depreciation charges along the curve are lower (slope milder) than that with the SL method. For example, at point *r* the curve's slope is milder than that of the straight line *PS*.

Thus, a nonlinear method (curve) is better than the SL method in the early years and worse in the latter years. To keep the depreciation charge high, it is thus essential to begin with the nonlinear method and switch to the SL method somewhere along the curve, for example in between *q* and *r* in Fig. 12.5. Obviously, the appropriate time to switch corresponds to the point where the curve's slope equals that of the line *PS*. This is the concept underlying composite depreciation, as illustrated in Example 12.10.

EXAMPLE 12.10

The first cost of a battery-operated machine is \$2,500, and its salvage value at the end of its five-year useful life is estimated to be \$100. Determine its depreciation schedule based on the DDB–SL[23] composite method.

[23]Begins with the DDB method and switches to the SL method.

Solution

From the discussions in Sections 12.4 and 12.5 as well as in the earlier sections, and also from Fig. 12.5, depreciation will be most rapid under the DDB–SL composite method.

It is given that the total depreciation = $P - S$ = \$2,500 − \$100 = \$2,400. With the SL method, the depreciation rate for a five-year asset is 1/5, and therefore the yearly SL charge[24] will be (1/5) × \$2,400 = \$480.

The DDB depreciation rate is twice that of the SL rate, that is, 2 × (1/5) = 2/5, and therefore the charges for the five years will be:

$$\begin{aligned}\text{First-year depreciation} &= (2/N) \times \text{Book value}\\ &= (2/5)(2{,}500)\\ &= \$1{,}000\end{aligned}$$

$$\begin{aligned}\text{Second-year depreciation} &= (2/N) \times \text{Book value}\\ &= (2/5)(2{,}500 - 1{,}000)\\ &= \$600\end{aligned}$$

$$\begin{aligned}\text{Third-year depreciation} &= (2/N) \times \text{Book value}\\ &= (2/5)(2{,}500 - 1{,}600)\\ &= \$360\end{aligned}$$

$$\begin{aligned}\text{Fourth-year depreciation} &= (2/N) \times \text{Book value}\\ &= (2/5)(2{,}500 - 1{,}960)\\ &= \$216\end{aligned}$$

$$\begin{aligned}\text{Fifth-year depreciation} &= (2/N) \times \text{Book value}\\ &= (2/5)\,(2{,}500 - 2{,}176)\\ &= \$130\end{aligned}$$

The DDB depreciation schedule is thus

Year	DDB Depreciation
1	\$1,000
2	600
3	360
4	216
5	130
	Total: \$2,306

[24]This annual charge will change depending on when we switch over to the SL method, as discussed later.

Since the total DDB depreciation of $2,306 is lower than the maximum allowed ($2,400), a switchover to the SL-method is called for[25]. This will enable the total depreciation charged to exactly equal the maximum allowable.

To do this we compare for each year the DDB charge with the SL charge and choose the greater of the two. For year 1, the DDB charge at $1,000 is more than the $480 SL charge. So we opt for DDB depreciation of $1,000 for year 1. For year 2, the DDB charge is $600, while the SL charge[26] will be

$$\begin{aligned}(1/4)(\text{total depreciation remaining to be charged}) &= 0.25 \times (2{,}400 - 1{,}000)\\ &= \$350\end{aligned}$$

So for year 2 too, we choose DDB since its $600 charge is greater than the $350 SL charge.

We now check the third year to see whether it makes sense to switch. From the table, the DDB charge for year 3 is $360. The cumulative depreciation charged so far is $1,000 + $600 = $1,600. So the SL method yields a depreciation charge of (1/3) × (2,400 − 1,600) = $267, which is less than $360. Thus, switching to the SL method is not yet desirable. We thus "stick with" the DDB for the third year too.

Next check for year 4. The SL charge for the fourth year is (1/2)(2,400 − 1,960) = $220, which is greater than the DDB charge of $216 for year 4. Hence, switching to the SL method at year 4 makes sense. The switching means charging $220 for year 4, and also[27] for year 5. The composite DDB-SL depreciation schedule will thus be

Year	Composite Depreciation
1	$1,000
2	600
3	360
4	220
5	220
	Total: 2,400

[25]Had this total been greater than $2,400, we would have reduced the charges for the latter years, since tax laws do not allow total depreciation to exceed $P - S$.

[26]If we switch to the SL method now at the end of year 1, then N is 4, and therefore the depreciation rate is 1/4. Since $1,000 has already been charged for the first year by the DDB method, the depreciation that remains to be charged is $2,400 (maximum allowed) minus $1,000.

[27]Once we switch to the SL method, the depreciation for the subsequent years remains the same due to the linear nature of the method.

The previous total of \$2,400, being equal to $P-S$, confirms the accuracy of the results.

Example 12.10 illustrated the combining of DDB with the SL method. In a similar way one can combine SOYD with the SL method, or any two or more methods, as long as only one method is used to determine the depreciation for any year. The ultimate goal of any compositing is to depreciate the asset as rapidly as possible without violating any tax law.

12.5 COMPARISON

All the methods other than the SL method allow for accelerated depreciation in the early part of the asset life. Of the two nonlinear methods (DDB and SOYD), DDB is in general more accelerating, that is, yields higher charges in the beginning. This is shown in Fig. 12.6, where the DDB curve has steeper slopes[28] in the beginning than the other two. Note, as explained earlier, that the DDB method may not always lead to point S, as shown in this figure, representing the salvage value. In such a case, a composite method that switches to SL, as illustrated in Example 12.10, is used to ensure that the total depreciation does not exceed $P - S$, the maximum allowable.

In most cases, the taxing-authority-sponsored depreciation method, such as MACRS, is practiced. Companies may be able to choose other methods in certain cases, especially for older assets under "grandfather" clauses in the tax law. For example, the DDB based on 200% of the SL rate is used for new assets, while the 150% rate is applicable to used assets or in special cases. Non-MACRS methods may eventually disappear in the United States, and in other countries as well, which may have already adopted, or are planning to adopt, methods similar to the MACRS. Meanwhile, industries continue to practice most of the methods discussed in this chapter. We covered the non-MACRS methods too because

1. In the United States, properties acquired before December 31, 1980, must continue to be depreciated according to pre-1981 methods.
2. Countries other than the United States may not be using MACRS or a similar method. Moreover, U.S. companies operating overseas are generally required to follow the depreciation method and tax laws of the host country.

---

[28]The plural form is used to signify that several values are possible depending on the point selected on the curve.

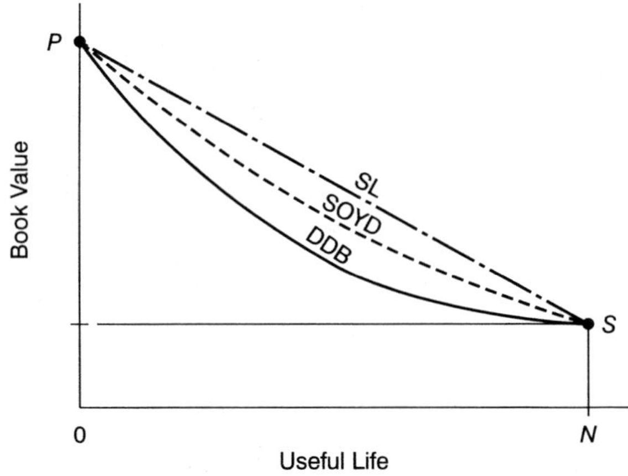

Figure 12.6 Comparison of Depreciation Methods

SUMMARY

Depreciation is a loss in the value of an asset. Almost all capital assets used by companies, with the exception of land, are depreciable. Depreciation is basically a means of recovering the cost of capital invested in the assets that generate income. Assets are allowed annual depreciation deductions from profits, reducing the income tax payable. Depreciation charges modify the gross incomes from the resource. A listing of the yearly charges is called a depreciation schedule. The maximum allowable depreciation during the asset life is usually $P-S$, the loss in value, where P is the first cost and S the salvage value.

Depreciation methods commonly practiced in industry have been discussed in this chapter. These are straight-line (SL), SOYD, declining balance such as DDB, unit of production, and MACRS. The SL method is the simplest of all, charging the same depreciation for each year of the asset life. Its weakness is its inability to charge higher depreciations during the early years of the asset life; DDB and SOYD do this. The declining balance method may not ensure that the life-end book value equals the salvage value. This means that the total depreciation charged could be different from the maximum allowable $P-S$. The analyst must be aware of this likely situation. By opting for a composite depreciation method, it is possible to ensure that, under a declining-balance method such as DDB, the total depreciation charge equals $P-S$. Sponsored by the U.S. Internal Revenue Service, MACRS is a composite DDB–SL method. Its tabular approach to depreciation costing streamlines the calculations. In the United States, the SL method is applicable only to pre-1981 assets; MACRS is used for depreciating recently acquired assets.

EXERCISES

Discussion Questions

12.1 Why are there so many methods for depreciation costing?

12.2 Compare the straight-line depreciation method with all the others discussed in Chapter 12 for constancy of depreciation charges and complexity of their costing.

12.3 Explain the concept behind the DDB depreciation method.

12.4 Companies account for depreciation because they pay taxes. Companies pay taxes because they account for depreciation. Which of these two statements is correct and why?

12.5 In a composite depreciation method why do we switch to the SL method where the slope of the declining balance curve equals that of the straight line?

12.6 Give an example from your own experience where the unit-of- production method will be appropriate.

12.7 Under MACRS, why do we charge depreciation during the sixth year for an asset with a five-year useful life?

Multiple-Choice Questions (Circle the *best* answer.)

12.8 Depreciation charges for equipment that cost $2,000 have been $600, $500, and $300. The book value in dollars is
a. 600.
b. 3,400.
c. 1,200.
d. 1,400.

12.9 Calculation-wise the simplest depreciation method is
a. SOYD.
b. straight-line.
c. declining balance.
d. DDB.

12.10 If a machine's useful life is six years, its SOYD value in years is
a. 15.
b. 6.
c. 21.
d. none of the above

12.11 In SOYD depreciation, the annual charge
a. remains the same.
b. decreases every year.
c. increases every year.
d. remains constant during the first half of the useful life.

12.12 The annual depreciation under the SL method is
a. $(S - P)/N$.
b. $(P + S)/N$.
c. $(P - S)/N$.
d. $(S + P)/N$.

12.13 The depreciation method that is primarily tabular is
a. SOYD.
b. declining balance.
c. DDB.
d. MACRS.

12.14 In a composite method, usually the
- a. SL method follows the DDB method.
- b. DDB method follows the SL method.
- c. SOYD method follows the SL method.
- d. DDB method follows the SOYD method.

Numerical Problems

12.15 A robot has been acquired for $50,000. It is expected to remain useful for the next five years, at the end of which its salvage value is likely to be $1,500. What is its uniform annual depreciation charge?

12.16 The first cost of a machining center is $120,000, and its salvage value is estimated to be $10,000. During its 10-year useful life it is expected to generate a net annual income of $20,000. Prepare the machining center's cashflow table assuming constant depreciation.

12.17 Beautiful Roses buys a truck for $14,000 to deliver flowers locally. It believes that the truck could be sold for $4,000 at the end of four years. If the business practices straight-line depreciation, how much could it deduct each year from the profit as a depreciation charge on the truck?

12.18 Premium Manufacturers acquires a fax machine for $1,350 whose estimated useful life and salvage value are five years and $100. Prepare the machine's depreciation schedule under the SOYD method.

12.19 Jack-of-all-Trades, a one-person home repair business, purchases a computer system to manage its accounting and customer data. The system costs $8,000 and is expected to have a useful life of five years. Assuming an estimated salvage value of $500, prepare the depreciation schedule using the DDB method. Do not switch to another depreciation method, but ensure that the total depreciation does not exceed the maximum allowable.

12.20 For a networked computer system costing $45,000, determine the DDB depreciation charges and book values over 8 years of useful life, assuming a salvage value of $9,000. Without switching to another method, ensure that the total depreciation does not exceed the maximum allowable.

12.21 A tool-room lathe has been acquired by Perfect Productions for $9,000. It is likely to be useful for the next five years, at the end of which its salvage value is estimated to be $300. Prepare a composite depreciation schedule that begins with DDB and switches over to the SL method at the most opportune time.

12.22 Aluminum International buys a truck for $50,000 for use in hauling bauxite. The estimated salvage value at the end of an expected 200,000-mile useful life is $2,000. The truck's use during the first two years was 40,000 and 45,000 miles. What is the third-year depreciation if its use during the year was 50,000 miles?

12.23 The engineering department of a company acquires a $12,000 workstation for designers. Prepare the workstation's depreciation schedule under MACRS.

12.24 Just-in-Time Janitors purchases an automatic floor cleaning and polishing machine for $9,000. The machine is expected to last for 10 years and to have a salvage value of $1,000. Prepare its straight-line, SOYD, DDB, and MACRS depreciation schedules.

CHAPTER THIRTEEN

Income Tax

IN THIS CHAPTER YOU WILL LEARN ABOUT

- How to account for income tax
- Income tax calculations
- *Incremental* income tax rates
- Tax-modified cashflows
- The effect of income tax through depreciation
- After-tax ROR and other analyses
- Combined federal and state income taxes
- Other tax implications in economic analyses

To add further realism to the economic analyses discussed so far, we consider in this chapter another fact of engineering business, namely income tax obligations. The past discussions ignored the tax for the sake of simplicity. For example, we excluded in Chapter 9 the effects of income tax on rate of return, conducting what may be called *before-tax* ROR analysis. To render any investment decision realistic the analyst must include tax obligations in the ROR and other analyses.

When taxes are taken into account, the procedure is appropriately called *after-tax analysis*. Such analyses reveal the effect of taxes on alternatives' attractiveness. Since taxes are a fact[1] of life in every business throughout the world, engineers invariably include tax considerations in investment decisions.

13.1 TAXES

Most governments consider it their duty to promote an economic "climate" conducive to wealth creation. Even in free market economies such as that of the United States, the

[1]The reality of taxes has nowhere been expressed as philosophically as by Benjamin Franklin, who said, "Two things are inevitable: death and taxes."

government is an active partner—albeit indirectly—of business. Governments[2] participate in economic activities in various ways, one of which is through taxation. In good times when companies make profits, governments charge taxes of various types, and in bad times they help them through "soft" credits, loans, tax write-offs, and similar incentives. They sometimes set up special agencies to implement their policies of participation. For example, the U.S. federal government has created the Small Business Administration (SBA) to encourage small businesses to become key players in the economy.

Though various types of taxes are collected from businesses, the discussions in this chapter are limited to income tax, since it has direct bearing on investment decisions. Additionally, we consider capital gain/loss and investment credit in Section 13.8.

13.2 INCOME TAX

Income tax is payable by both individuals and corporations. The taxes are calculated using rates fixed by the governments. In the United States, the federal government collects most of the income tax. Several states as well as city governments also levy income tax. Each government reviews its budget provisions once a year and decides the tax rates for the next fiscal (financial) year.

Income tax rates are usually variable or *graduated*, which means that the rate depends on the income level. In graduated tax systems, those with higher incomes pay taxes at higher rates. Most governments of the world practice such a system. In the United States, a move to a *flat* tax rate whereby all incomes are taxed at the same rate has been under discussion for some time.

Consider taxpayers A and B. Assume that A's taxable income for the year is $50,000, while B's is $80,000. Under a flat tax rate of 10%, A will pay an income tax of $5,000, B of $8,000. Next, consider graduated taxation under the following assumed schedule:

Taxable Income	Rate
$0–$60,000	10%
$60,000–$150,000	15%

These rates mean that incomes above $60,000 are taxed at 15%. Since A's income is $50,000, all of which falls in the $0–$60,000 range, the applicable tax rate is 10%; thus, A's tax is 10% of $50,000, that is, $5,000. In the case of B, with an income of $80,000, the tax calculations are slightly complex. Referring to the schedule, B's income of $80,000 should be considered divided in two components: $60,000 plus $20,000. For the $60,000 component the rate will be 10%, while for the $20,000 it will be 15%. Thus, B's tax will be 10% of $60,000 plus 15% of $20,000. This works out to be $6,000 + $3,000 = $9,000. Had the tax rate been flat at 10%, B would have paid only $8,000 (10% of $80,000). Thus, graduated taxation charged tax at a higher rate from B, whose income was higher. The justification for graduated taxation is that those who can afford to pay should pay more in tax.

[2]The plural form is more appropriate, since government participation could be at various levels: federal or central, state or province, county, and city.

Income tax laws are complicated, and the rates change each year. Almost all companies, and many individuals, hire accountants or tax attorneys to handle their income tax obligations and the associated investment opportunities. The purpose of this chapter is not to learn about the intricacies of taxes, but merely to acquire the skills necessary for assessing the implications of income taxes for engineering economic analyses. Such skills enable technical professionals to make more realistic and judicious engineering investment decisions.

EXAMPLE 13.1

On her graduation with a BS in manufacturing engineering technology, Brenda set up a job shop called AnyMachining Inc. Since her recent use of the Internet to advertise her business through a home page, the orders have increased, resulting in a taxable income of $95,000. How much income tax is payable by AnyMachining if the applicable tax rate schedule is as follows?

Taxable Income	Rate
Less than $50,000	10%
$50,000–$75,000	15%
$75,000–$100,000	20%
Above $100,000	25%

Solution

The given schedule comprises four tax brackets of 10%, 15%, 20%, and 25%. The taxable income of $95,000 falls in the third bracket of $75,000–$100,000. Thus, $95,000 will have three components. Since the upper cut-off for the first bracket is $50,000, the first component will be $50,000, which is taxable at 10%. This leaves $45,000 ($95,000 − $50,000) to be taxed. The upper cut-off of the second bracket is $75,000. Therefore, the second component will be $75,000 − $50,000 = $25,000, which is taxed at 15%. The remainder of the income, that is, $95,000 − $75,000 = $20,000, as the third component, is taxable at 20%.

On the first $50,000 component, at a 10% rate, the tax is $5,000. On the second component, the tax is 15% of $25,000 = $3,750. Since the remaining $20,000 of the income is taxed at 20%, the tax on this component is $4,000. Thus, the total income tax[3] payable by AnyMachining is

$$\$5{,}000 + \$3{,}750 + \$4{,}000 = \$12{,}750$$

---

[3]A table may be used to summarize the calculations and results.

Income Component	Rate	Tax
$50,000	10%	$5,000
25,000	15%	3,750
20,000	20%	4,000
Total: $95,000		$12,750

13.3 TAXABLE INCOME

Income tax is payable on *taxable* income. In the preceding example, the taxable income was given; in most cases, it is determined. Taxable income is the difference between gross income and the expenses incurred in generating the income.

In this section, we discuss how to determine taxable income, and therefrom tax, for a business. Most businesses are either proprietorships or corporations. If you own a business as a proprietor, alone or in partnership, the tax payable is at individual rates. Corporations, on the other hand, pay[4] taxes at corporate rates. The rates and the other related information are available in appropriate documents prepared by the taxing authorities. The discussions in this section are limited to taxation by the U.S. federal government[5].

13.3.1 Individual

For individual taxpayers, first determine the *total*[6] *income*, which is the sum of incomes from all sources. For a sole proprietor or business partner,

$$\textit{Total income} = \text{Net income from business} + \text{Other incomes}$$

For an employed person,

$$\textit{Total income} = \text{Taxable salary} + \text{Other incomes}$$

Taxable salary[7] is usually lower than gross salary by an amount invested in tax-deferred retirement and similar plans. The other incomes include interest earnings from saving accounts, dividends from investments, capital gains, and so on.

The total income is next adjusted to allow for certain deductions such as moving expenses and alimony paid. Thus,

$$\textit{Adjusted gross income} = \text{Total income} - \text{Allowable deductions}$$

---

[4]A company can be formed as an S corporation, which is better in "infancy" when it is likely to incur losses. S corporation owners can compensate business losses by deducting them from nonbusiness profitable incomes. However, they run the risk of personal liability under S corporation laws. Once such a company begins to earn profit, it changes itself to the usual corporation to avoid liability.

[5]U.S state governments as well as governments in other countries have similar income tax provisions. Non-U.S. readers should contact the nearest office of their income taxing authorities to obtain current information on tax rates and associated matters.

[6]The terminology used here is that in the Internal Revenue Service (IRS) document 1040.

[7]This information is provided by the employer to each employee at the time taxes are to be filed.

The taxing authority may allow certain personal exemptions, and a standard or itemized[8] deduction from the adjusted gross income. This yields taxable income as

$$\textit{Taxable income} = \text{Adjusted gross income} - \text{Personal and standard (or itemized) deductions}$$

The various deductions reflect government-sanctioned expenses likely to be incurred by an individual in earning the income, as well as other tax reliefs. They include a personal exemption for the employee, exemptions for dependents, and a lump-sum standard deduction, or itemized deductions in lieu. The itemized deductions[9] include expenses such as state and local income taxes, excessive medical costs, home mortgage interest, donations to charities, and so on. The taxpayer has the choice of claiming either a standard deduction or itemized deductions, whichever reduces the tax. The amounts of personal exemption and standard deduction may increase each year to account for inflation.

The taxable income is taxed at rates fixed by the government for the year. The rate schedules for the U.S. federal income tax are printed in various tables and are also available at the *www.irs.ustreas.gov* website. One of the tables, for example, is for individuals filing as singles; for the 1998 tax year, it is shown in Table 13.1.

The table has been formatted[10] to simplify the tax calculations. For example, the dollar figures in the table's second column (under Tax) are simply the taxes on the upper cut-off income of the previous bracket. The cut-off incomes are the income limits at which the tax rate changes; for example, at \$61,400 the tax rate increases from 28% to 31%. As an illustration, the figure \$13,896.50 in the second column is the tax on a \$61,400 income. Any income in excess of \$61,400 but less than \$128,100 falls in the third bracket and, as per the rate in the table's second column, is taxed at 31%; by adding this tax to \$13,896.50 the total tax is determined. The figure \$13,896.50 has been obtained by charging 15% tax on \$25,350 (the upper cut-off in the first bracket) and 28% on \$36,050 (the income range of the second bracket = \$61,400 − \$25,350).

[8]Taxpayers can claim either a standard or an itemized deduction. For the latter, the taxpayer lists the claims against what is allowed. Itemized deduction is obviously claimed only if it is beneficial, that is, if it is more than the standard deduction.

[9]The type of item and the maximum allowable deduction may vary from year to year.

[10]Table 13.1 is simply an extension of the following schedule.

Taxable Income	Tax Rate, %
Up to \$25,350	15
25,350–61,400	28
61,400–128,100	31
128,100–278,450	36
Above 278,450	39.6

One can determine the tax using this schedule, as explained in Example 13.1.

Table 13.1 Year 1998 Tax Rate Schedule for Singles

Taxable Income	Tax	Of the Amount Over
$0–$25,350	15%	$0
25,350–61,400	$3,802.50 + 28%	25,350
61,400–128,100	13,896.50 + 31%	61,400
128,100–278,450	34,573.50 + 36%	128,100
Above 278,450	88,699.50 + 39.6%	278,450

In other words, $13,896.50 is the result of 0.15 × $25,350 + 0.28 × $36,050. Other dollar figures in the table's second column have been obtained in the same way, that is, based on cut-off incomes and corresponding tax rates.

If you find the concise schedule under footnote 10 easier to use than Table 13.1, then do so following the procedure explained in Example 13.1. Example 13.2 illustrates how the income tax of a small company whose proprietor pays tax as single individual is calculated.

EXAMPLE 13.2

On finishing her BS in computer engineering, Jasmin set up a computer repair and system configuration business as a sole proprietor. Her business income in 1998 was $180,000. The rent, utilities, and other direct expenses of the business added to $93,000. Jasmin also had interest and dividend earnings of $5,000. How much income tax is payable if Jasmin is single with no dependents? Refer to IRS Form 1040 for any relevant data.

Solution
Jasmin's net income from the business is

$$\begin{aligned}\text{Business income} - \text{Expenses to earn the income} &= \$180{,}000 - \$93{,}000\\ &= \$87{,}000\end{aligned}$$

To this we add her other incomes. Thus, her

$$\begin{aligned}\text{Total income} &= \text{Business income} + \text{Any other income}\\ &= \$87{,}000 + \$5{,}000\\ &= \$92{,}000\end{aligned}$$

Assuming that this total income does not need to be modified any further as per IRS Form 1040, Jasmin's

$$\begin{aligned}\text{Adjusted gross income} &= \text{Total income}\\ &= \$92{,}000\end{aligned}$$

Jasmin is entitled to only one personal exemption of \$2,700 (referring to the 1998 IRS Form 1040) for herself, since she has no dependents. She is also entitled to the standard deduction or itemized deductions. Assuming that she is better off[11] claiming the standard deduction, Jasmin opts for it. For 1998, the standard deduction for single filers was \$4,250 (IRS Form 1040). Thus, Jasmin's

$$\begin{aligned}\text{Taxable income} &= \text{Adjusted gross income} \\ &\quad - \text{Personal exemption and standard deduction} \\ &= \$92{,}000 - (\$2{,}700 + \$4{,}250) \\ &= \$85{,}050\end{aligned}$$

Noting in Table 13.1 that a taxable income of \$85,050 falls in the third tax bracket of \$61,400–\$128,100, Jasmin's tax[12] is[13]

$$\begin{aligned}\text{Income tax} &= \$13{,}896.50 + 31\% \text{ of income over } \$61{,}400 \\ &= \$13{,}896.50 + 31\% \text{ of } (\$85{,}050 - \$61{,}400) \\ &= \$13{,}896.50 + 0.31 \times \$23{,}650 \\ &= \$13{,}896.50 + \$7{,}331.50 \\ &= \$21{,}228\end{aligned}$$

Another group of income taxpayers is married couples filing tax jointly. For them, the 1998 U.S. tax schedule was as shown in Table 13.2. Table 13.2 too has been formatted to simplify the tax calculations. It is basically an extension of the following schedule, which can be used to calculate income tax in a way explained in Example 13.1.

Taxable Income	Tax Rate, %
Up to \$42,350	15
42,350–102,300	28
102,300–155,950	31
155,950–278,450	36
Above 278,450	39.6

Example 13.3 illustrates the calculations for the income tax obligation of a married couple filing jointly.

---

[11]She had few expenses, resulting in itemized deductions totaling less than the standard deduction.

[12]According to Table 13.1, for the third bracket, the income tax is \$13,896.50 plus 31% of the taxable income over \$61,400.

[13]You should get this answer also by using the schedule at footnote 10 and following the procedure of Example 13.1.

Table 13.2 Married Couples Filing Jointly

Taxable Income	Tax	Of the Amount Over
\$0–\$42,350	15%	\$0
42,350–102,300	\$6,352.50 + 28%	42,350
102,300–155,950	23,138.50 + 31%	102,300
155,950–278,450	39,770.00 + 36%	155,950
278,450	83,870.00 + 39.6%	278,450

EXAMPLE 13.3

In Example 13.2 assume that Jasmin is married to a person who, as an associate in the local video rental store, earns \$18,000 as wages and \$5,000 as commission. If they filed tax jointly, how much income tax would be due? Refer to IRS Form 1040 for 1998 for any relevant data.

Solution
Since Jasmin is married and files jointly, she is entitled to more exemptions and deductions. However, now the

$$\begin{aligned}\text{Total income} &= \text{Jasmin's business income} + \text{Spouse's income}\\ &= \$92{,}000 + (\$18{,}000 + \$5{,}000)\\ &= \$115{,}000\end{aligned}$$

Assuming that this total income does not need to be modified any further as per IRS Form 1040, their

$$\begin{aligned}\text{Adjusted gross income} &= \text{Total income}\\ &= \$115{,}000\end{aligned}$$

Jasmin is now entitled to two personal exemptions of \$2,700 each, one for herself and one for her spouse. They are also entitled to a standard deduction or itemized deductions. Assuming that they have summed up the itemized deductions and found the total to be lower than the standard deduction for couples, they claim the latter. For 1998, the standard deduction for married couples filing jointly was \$7,100 (IRS Form 1040). Thus, their

$$\begin{aligned}\text{Taxable income} &= \text{Adjusted gross income}\\ &\quad - \text{Personal exemptions and standard deduction}\\ &= \$115{,}000 - (2 \times \$2{,}700 + \$7{,}100)\\ &= \$115{,}000 - (\$5{,}400 + \$7{,}100)\\ &= \$102{,}500\end{aligned}$$

From Table 13.2, a taxable income of \$102,500 falls in the third tax bracket of \$102,300–\$155,950; hence

$$
\begin{aligned}
\text{Income tax} &= \$23{,}138.50 + 31\% \text{ of income over } \$102{,}300 \\
&= \$23{,}138.50 + 31\% \text{ of } (\$102{,}500 - \$102{,}300) \\
&= \$23{,}138.50 + 0.31 \times \$200 \\
&= \$23{,}138.50 + \$62 \\
&= \$23{,}200.50
\end{aligned}
$$

Each year the U.S. federal government decides upon the type and extent of *allowable deductions*. For complete details, readers should refer to the IRS document Form 1040 and its attachments, as well as the accompanying instructions. They are available free from the nearest IRS office or at their website[14]. If you already pay, or have paid, U.S. income tax as an individual, you may have found the treatment in this subsection easier to follow.

13.3.2 Corporations

Gross income is the difference between the revenue from the sale of goods and services and the operating cost incurred in their production and marketing. Thus,

$$\textit{Gross income} = \text{Sales revenue} - \text{Operating cost}$$

The operating cost comprises both direct and indirect costs, which include the costs of material, labor, rent, utilities, maintenance, marketing, and others. Also included are costs on nonconsumable items lasting less than a year.

For incorporated businesses, taxable income is determined by deducting from the gross income expenditures on depreciable assets used in generating the sales revenue. Thus,

$$\textit{Taxable income} = \text{Gross income} - \text{Depreciation expenditures}$$

Depreciation expenditures pertain to resources (equipment and machines) that last more than a year. For tax purposes, the costs of depreciable assets are not allowed as deductions from the profits in the year the assets are put to service. They must, instead, be deducted as depreciation charges over the assets' useful lives. We have discussed the procedures of depreciation charging extensively in Chapter 12.

The taxing authorities usually provide companies information on the various tax incentives offered by the governments. Tax attorneys and accountants keep track of these matters, and are the ones businesses consult with to benefit from such incentives.

EXAMPLE 13.4

A small jobshop's sales revenue for the last year was $782,552, while its operating costs were $458,760. The shop acquired a new CAM system for its numerically controlled machines at a cost of $125,000. The CAM system's useful life is expected to be three years, at the end of which it is likely to be salvaged at $20,000. For straight-line depreciation, how much income tax is due under the following schedule?

---

[14] *www.irs.ustreas.gov*

Taxable Income	Rate
Below \$50,000	10%
\$50,000–\$75,000	15%
\$75,000–\$100,000	20%
Above \$100,000	25%

Solution
First, we determine the gross income as

$$\begin{aligned}\text{Gross income} &= \text{Sales revenue} - \text{Operating cost}\\ &= \$782{,}552 - \$458{,}760\\ &= \$323{,}792\end{aligned}$$

Next, we calculate the taxable income from

$$\text{Taxable income} = \text{Gross income} - \text{Depreciation expense}$$

For the given data, the straight-line annual depreciation charge[15] is

$$\begin{aligned}&= (P - S)/N\\ &= (\$125{,}000 - \$20{,}000)/3\\ &= \$35{,}000\end{aligned}$$

Thus,

$$\begin{aligned}\text{Taxable income for the year} &= \$323{,}792 - \$35{,}000\\ &= \$288{,}792\end{aligned}$$

This taxable income can be considered divided into its components to match the cut-off limits of the various tax brackets, as given in the tax schedule. Thus, \$288,792 can be considered to be the result of[16] \$50,000 + \$25,000 + \$25,000 + \$188,792.

Tax on the \$50,000 of the income at 10%	= \$5,000
Tax on the (first) \$25,000 of the income at 15%	= 3,750
Tax on the (second) \$25,000 of the income at 20%	= 5,000
Tax on the remaining \$188,792 of the income at 25%	= 47,198

Thus, the income tax due is

$$\$5{,}000 + \$3{,}750 + \$5{,}000 + \$47{,}198 = \$60{,}948$$

[15]Refer to Chapter 12 if you have difficulty following it.

[16]See Example 13.1 if necessary.

In the previous two sections we discussed how to calculate income tax. While these skills are useful to engineers and technologists, the tasks of figuring out income taxes and timing investments (whether to invest this year or next year) to reduce tax obligations are so complex that specialists, such as tax attorneys and chartered public accountants (CPAs), are hired or retained for the purpose. Engineers and technologists are usually concerned only with analyzing the impact of taxes on economic decisions.

13.4 INCREMENTAL TAX RATE

The concept of *incremental tax rate* facilitates the analysis of tax implications. An incremental tax rate is simply the tax rate that is applicable to an *incremental (additional)* income. Consider the following tax schedule.

Taxable Income	Rate
Up to $50,000	10%
$50,000–$75,000	20%
$75,000–$100,000	30%
Beyond $100,000	40%

As per this schedule, for a company with taxable income of $60,000, the tax rate for the first $50,000 is 10%, and for the remaining $10,000 it is 20%. Assume now that this company can increase its gross income by $8,000 by investing in new equipment. While carrying out an economic analysis, what is important is not the amount of tax, but the tax implication of the *additional* $8,000 gross income on the decision. Assume that this additional income yields a $6,000 profit after considering the operating costs. Due to depreciation charge this profit reduces to, let us say, a taxable income of $5,000. The total taxable income will thus be $60,000 + $5,000 = $65,000, which is still in the 20% bracket. Therefore, the additional taxable income of $5,000 resulting from investment[17] in the equipment will be taxed at 20%. This effect of the investment on income tax will be stated thus: The *incremental tax rate* is 20%. It is this rate that is relevant in economic analyses.

Now, consider this: What if the current $60,000 taxable income could be increased by a taxable income of $20,000 by investing in a deluxe model of the equipment? With the total taxable income becoming $80,000, the company is now in the 30% tax bracket, as per the given schedule. Since the lower cut-off income for 30% is $75,000, the $20,000 increment in taxable income could be considered as the total of the two components $15,000 and $5,000. The $15,000 component will raise the current $60,000 taxable income to $75,000—the upper cut-off limit of the 20% tax bracket, and hence this $15,000, being in the same bracket, will be taxed at 20%. The remaining $5,000 will, however,

[17]Tax obligation = 0.20 × $5,000 = $1,000. The after-tax return from the investment is thus $5,000 − $1,000 = $4,000. Note that in this illustration a gross income of $8,000 was finally reduced by the operating costs, depreciation, and the tax obligation to a $4,000 after-tax cashflow.

be taxed at the next higher rate of 30%, since this component lies in that bracket. Thus, of the $20,000 additional taxable income from investment in the deluxe model, for the first $15,000 the *incremental tax rate* is 20%, and for the remaining $5,000 it is 30%. Note that two or more incremental tax rates arise only when the additional income pushes the total income to higher tax brackets.

Engineers and technologists need to know the value of the incremental tax rate(s) applicable to a project. The concept explained in this section helps them in selecting the appropriate rate(s). Example 13.5 further illustrates how the incremental tax rate is assessed.

EXAMPLE 13.5

Kristi owns and operates a jobshop that is likely to generate this year a taxable income of $95,000. She is considering investing in a personal computer that is expected to increase the taxable annual income by $15,000. What incremental tax rate or rates are appropriate for use in decision making pertaining to this investment under the following schedule? How much additional income tax will be due on the earnings from the investment?

Taxable Income	Rate
Up to $50,000	10%
$50,000–$75,000	20%
$75,000–$100,000	30%
Beyond $100,000	40%

Solution

With the expected $95,000 taxable income being in the $75k–$100k[18] range, the jobshop's tax rate in the absence of investment will be 30%. With the additional $15,000 taxable income from investment in the PC, its taxable income will be $95,000 + $15,000 = $110,000. With this income, the tax liability shifts into the next higher bracket of 40%.

Not all of the additional income will be taxed at 40%, however. Of the $15,000, the first $5,000 shifts the current $95k income to $100k—the upper limit of the 30% bracket. Hence this $5,000 is taxable still at 30%. The remaining $10,000 of the additional income falls in the higher bracket of 40%. Thus, an incremental tax rate of 30% is applicable to the first $5,000, while 40% is applied to the remaining $10,000.

$$\begin{aligned}\text{Additional income tax due} &= 30\% \text{ of } \$5{,}000 + 40\% \text{ of } \$10{,}000\\ &= \$1{,}500 + \$4{,}000\\ &= \$5{,}500\end{aligned}$$

[18]k is an abbreviation for kilo, meaning 1,000. Thus, $100k means $100,000.

EXAMPLE 13.6

Sheri's business is expected to generate a taxable income of $60,000. Her choices for income tax purposes are

a. Run the business as a sole proprietor and pay income tax jointly with her spouse at a 40% incremental rate, or
b. Incorporate the business and pay income tax as a C corporation for which the tax rates are

Taxable Income	Rate, %
Under $50,000	15
Next $25,000	24
Next $25,000	34
Next $235,000	39
Over $335,000	34

From the tax obligation viewpoint, which choice is better?

Solution

a. As a joint income her $60,000 earnings will be taxed at the 40% incremental rate. Thus, Sheri's portion of the tax = 40% of $60,000 = $24,000.
b. As a C corporation, referring to the tax schedule given, her $60,000 income can be considered divided into two components of $50,000 and $10,000. She would pay 15% on the $50,000 component and 24% on the $10,000. Thus,

$$\begin{aligned}\text{Tax} &= 15\% \text{ of } \$50{,}000 + 24\% \text{ of } \$10{,}000\\ &= \$7{,}500 + \$2{,}400\\ &= \$9{,}900\end{aligned}$$

She is better off incorporating, saving $24,000 − $9,900 = $14,100 in income tax.

13.5 AFTER-TAX ANALYSES

Accounting for the obligation of income tax arising from engineering investment is called *after-tax* economic analysis. Such analyses can be carried out for any of the six methods discussed in Chapters 5 through 10, and involve depreciation considerations discussed in Chapter 12. The tax implications arising out of depreciation make the analysis complex, as discussed later in Subsection 13.5.2.

13.5.1 Nondepreciable Case

For simplicity we first consider the case that does not involve depreciation. The analyst should modify the given before-tax cashflows as per the applicable incremental rate(s).

The resulting after-tax cashflows are then analyzed following any of the procedures discussed in Part II of the text. In general, the following five steps are involved:

1. Gather the given (before-tax) data in a cashflow table.
2. Find out what the applicable incremental tax rates are.
3. Determine the tax payable each year.
4. Deduct the yearly tax from the before-tax (gross) cashflow to obtain after-tax (net) cashflows.
5. Analyze the after-tax cashflows under the criterion.

To minimize confusion, it is always better to implement steps 3, 4, and 5 by adding columns to the original (before-tax) cashflow table, as illustrated in Example 13.7.

EXAMPLE 13.7

A family-owned speciality store operates ten-to-six, five days of the week. Due to rising healthcare costs, the owner is in financial difficulty. One way out is to augment the income by keeping the store open on Saturdays too, which is expected to result in the following gross incomes:

Year	Income
1	$80,000
2	92,000
3	106,000
4	112,000
5	114,000

Determine the store's after-tax yearly incomes if the tax schedule is

Taxable Income	Rate
Up to $50,000	10%
$50,000–$75,000	20%
$75,000–$100,000	30%
Beyond $100,000	40%

Solution

Follow the five-step procedure outlined in Subsection 13.5.1. Since the before-tax cashflow table is already given, step 1 is complete. The given tax schedule contains information about the incremental tax rates, so step 2 is also complete. Next, we determine the tax[19] for each year.

[19]Note that we are assuming that these rates will not change during the next five years.

Let us consider year 1 first. Referring to the tax schedule, the \$80,000 income falls in the third bracket. This income can be divided into three components[20] of \$50,000, \$25,000, and \$5,000. This will result in

$$\begin{aligned}\text{Year 1 tax} &= 10\% \text{ of } \$50{,}000 + 20\% \text{ of } \$25{,}000 + 30\% \text{ of } \$5{,}000 \\ &= \$5{,}000 + \$5{,}000 + \$1{,}500 \\ &= \$11{,}500\end{aligned}$$

One can find the taxes for the other years in a similar way, but such a process will prove to be lengthy.

An efficient way is to apply the incremental tax rate on additional incomes, while being careful about the applicable rate. For example, the \$92,000 income for year 2 can be looked upon as an incremental income of \$12,000 over that for year 1 (\$80,000). With the year 2 income of \$92,000 being less than \$100,000 (the upper cut-off for the third bracket), the tax rate remains at 30%. Thus, the incremental tax rate applicable to the additional \$12,000 income is 30%. This gives an additional tax of $0.3 \times \$12{,}000 = \$3{,}600$. Thus,

$$\begin{aligned}\text{Year 2 tax} &= \text{Year 1 tax} + \text{Additional tax for year 2} \\ &= \$11{,}500 + \$3{,}600 \\ &= \$15{,}100\end{aligned}$$

Other years' taxes can be determined the same way[21]. To keep track of the yearly taxes, add one more column to the given cashflow table:

Year	Before-Tax Income	Tax Payable[22]
1	\$80,000	\$11,500
2	92,000	15,100
3	106,000	19,900
4	112,000	22,300
5	114,000	23,100

The fourth step involves considering the effect of taxes on before-tax incomes. In this case, since the taxes are payable, they reduce the incomes. So subtract them from the corresponding incomes. Tabulate the results in another column.

[20] Based on the upper income limit of the tax brackets; see Example 13.1, if necessary.

[21] The incremental procedure may look to be clumsy, but with practice you will find it easier than the direct approach used for year 1.

[22] Note that for the incremental income of \$14,000 for year 3 over year 2 (\$106,000 − \$92,000), two different incremental rates are applicable: 30% on the first \$8,000 and 40% on the remaining \$6,000.

Year	Before-Tax Income	Tax Payable	After-Tax Income
1	$80,000	$11,500	$68,500
2	92,000	15,100	76,900
3	106,000	19,900	86,100
4	112,000	22,300	89,700
5	114,000	23,100	90,900

Note that though the before-tax income for the fifth year has increased by $34,000 over that for the first year, the after-tax income has increased only by $22,400 ($90,900 − $68,500). This illustrates how taxes can distort cashflows and therefore must be accounted for in any economic analysis.

The fifth step of the procedure is inapplicable, since analysis of the after-tax cashflows has not been asked for.

13.5.2 Depreciable Case

The procedure hereunder must be followed if depreciation data are given, even if the problem does not explicitly ask for a consideration of depreciation. When depreciation is to be accounted for, the before-tax cashflows should be modified *first* for depreciation. The resulting *after-depreciation incomes* are taxable, and hence the basis of the income tax obligation. The complete procedure involves the following six steps:

1. Gather the given before-tax data in a cashflow table.
2. Determine the depreciation charges and tabulate them.
3. Apply the depreciation charges to before-tax cashflows.
4. Determine taxes based on after-depreciation cashflows.
5. Apply the taxes to the before-tax cashflows.
6. Analyze the resulting after-tax cashflows.

Again, implement the steps by adding columns to the original cashflow table, as illustrated in Example 13.8.

EXAMPLE 13.8

An investment of $3,900 in a special tool is expected to result in savings of $1,000 during the first year, $1,500 during the second year, and $2,000 during the third year. At the end of the third year the tool will be salvaged for $300. Is this a good investment considering income tax obligations if present worth (PW) is the decision criterion? Assume straight-line depreciation, a 60% incremental tax rate, and an 8% MARR.

Solution
The savings represent a reduction in costs, which translates into increased income. The before-tax cashflows as given are

Year	Cashflow
0	−$3,900
1	1,000
2	1,500
3	2,000
	300 (salvage value)

This was step 1. In step 2, depreciation charges are determined. Following the straight-line method (refer to Chapter 12, if necessary), for the given data, $P = \$3{,}900$, $S = \$300$, and $N = 3$ years,

$$\begin{aligned}\text{Depreciation charge} &= (P - S)/N \\ &= (\$3{,}900 - \$300)/3 \\ &= \$1{,}200\end{aligned}$$

This yearly depreciation is charged to the before-tax incomes (step 3). Thus, the after-depreciation cashflows, which are taxable incomes[23], will be lower:

Year	Before-Tax Cashflow	Depreciation[24]	Taxable Income
0	−$3,900		
1	1,000	$1,200	−200
2	1,500	1,200	300
3	2,000	1,200	800
	300		

Next, as step 4, we determine the tax obligations for each year. The incremental tax rate is given to be 60%. The calculated taxes are tabulated in yet another column[25].

[23]We are assuming that the before-tax cashflows represent net savings (incomes) that have already taken the operating costs into account.

[24]For year 0, representing now, there is no depreciation charge or taxable income.

[25]In the tables in this example, the first row, corresponding to year 0, and the salvage value can initially be excluded, if they are confusing. The cashflow in this row represents the investment—cost incurred now—which is not affected by depreciation or income tax. Likewise, the salvage value remains unaffected by depreciation and tax considerations. However, these must be added later to complete the after-tax cashflow table, just prior to applying the decision criterion (PW in this case).

(1) Year	(2) Before-Tax Cashflow	(3) Depreciation	(4) Taxable Income (2) − (3)	(5) Tax @ 60%
0	−$3,900			
1	1,000	$1,200	−200	−120
2	1,500	1,200	300	180
3	2,000	1,200	800	480
	300			

As step 5, the effect of taxes on incomes (before-tax cashflows) is accounted for, generating after-tax cashflows in another column.

(1) Year	(2) Before-Tax Cashflow	(3) Depreciation	(4) Taxable Income (2) − (3)	(5) Tax @ 60%	(6) After-Tax Cashflow (2) − (5)
0	−$3,900				−$3,900
1	1,000	$1,200	−200	−120	1,120
2	1,500	1,200	300	180	1,320
3	2,000	1,200	800	480	1,520
	300				300

The after-tax cashflows in the last column have thus accounted for the tax benefits of asset depreciation. As the final (sixth) step, we analyze them for their PW to judge the project's investment worthiness.

$$\begin{aligned}\text{PW} &= -3{,}900 + 1{,}120(P/F, 8\%, 1) + 1{,}320(P/F, 8\%, 2) + 1{,}820(P/F, 8\%, 3)\\ &= -3{,}900 + 1{,}120 \times 0.9259 + 1{,}320 \times 0.8573 + 1{,}820 \times 0.7938\\ &= -3{,}900 + 1{,}037 + 1{,}132 + 1{,}445\\ &= -\$286\end{aligned}$$

Since the PW of the after-tax cashflows is negative, the investment in the special tool is not recommended.

After-tax analyses render economic decisions more realistic and reliable. Had we ignored the tax obligation in Example 13.8, we would have erred[26] by investing in the special tool.

[26]Had the tax obligations not been considered in Example 13.8, the PW would have been positive, as shown below, recommending investment in the tool.

$$\begin{aligned}\text{PW} &= -3{,}900 + 1{,}000(P/F, 8\%, 1) + 1{,}500(P/F, 8\%, 2) + 2{,}300(P/F, 8\%, 3)\\ &= -3{,}900 + 1{,}000 \times 0.9259 + 1{,}500 \times 0.8573 + 2{,}300 \times 0.7938\\ &= -3{,}900 + 926 + 1{,}286 + 1{,}826\\ &= \$138\end{aligned}$$

13.6 AFTER-TAX ROR ANALYSIS[27]

Though the after-tax analysis can be carried out under any of the decision criteria discussed in Part II, it is common with ROR.

The procedure for after-tax ROR analysis is very similar to that explained in Subsection 13.5.2. The underlying principle is a two-part process. First, modify the before-tax cashflows to account for depreciation and the associated income tax. This yields after-tax cashflows. Second, analyze the after-tax cashflows to judge the project's worthiness based on the rate of return. The ROR analysis follows the usual procedure discussed in Chapter 9, as illustrated in Example 13.9.

EXAMPLE 13.9

Woman Wardrobe, a garment manufacturer, plans to buy an old truck for business use for $13,000. The truck is expected to increase company earnings annually (through savings) by $4,000. The company will sell the truck for an estimated price of $7,000 at the end of three years. Assuming straight-line depreciation, should the company invest in the truck if the incremental income tax rate is 40% and the MARR is 12%?

Solution

Note that in the third year there will be two cash inflows, one the $4,000 annual benefit and the other the $7,000 salvage. The before-tax cashflow table for the project is

Year	Before-Tax Cashflow
0	−$13,000
1	4,000
2	4,000
3	4,000
	7,000 (salvage value)

Though the criterion for decision making is not explicit, there is a hint to use the ROR criterion, since the MARR is given. With a given MARR, we can avoid having to evaluate the ROR exactly. We can instead decide on the basis of the PW's sign corresponding to the MARR.

We first determine the after-tax cashflows by accounting for depreciation and the associated income tax. Let us follow the 6-step procedure explained earlier in Section 13.5.2 and applied in Example 13.8. We begin by determining the yearly straight-line depreciation charge as

$$\begin{aligned}\text{Depreciation charge} &= (P - S)/N \\ &= (\$13{,}000 - \$7{,}000)/3 \\ &= \$2{,}000\end{aligned}$$

[27]Though we discuss only the ROR analysis in this section, similar procedures are applicable to analyses under any of the other decision criteria covered in Part II.

The analyses for depreciation and taxes yield the following[28] table (see Example 13.8 for details on how to carry out the calculations).

(1) Year	(2) Before-Tax Cashflow	(3) Depreciation	(4) Taxable Income (2) − (3)	(5) Tax @ 40%	(6) After-Tax Cashflow (2) − (5)
0	−\$13,000				−\$13,000
1	4,000	\$2,000	\$2,000	\$800	3,200
2	4,000	2,000	2,000	800	3,200
3	4,000	2,000	2,000	800	3,200
	7,000				7,000

Now, determine[29] the PW of the after-tax cashflows corresponding to a 12% MARR, which is

$$\begin{aligned} \text{PW} &= -13{,}000 + 3{,}200(P/A, 12\%, 3) + 7{,}000(P/F, 12\%, 3) \\ &= -13{,}000 + 3{,}200 \times 2.402 + 7{,}000 \times 0.7118 \\ &= -13{,}000 + 7{,}686 + 4{,}983 \\ &= -\$331 \end{aligned}$$

Since the PW is negative, the after-tax ROR will be lower than the 12% MARR. Therefore, the company should not invest[30] in the old truck.

---

[28]Some analysts post the payable tax data (column 5 in the table) with a negative sign, considering them as costs. For year 1, their entry will be −\$800, instead of \$800. The signed tax data (column 5) is added algebraically to the before-tax data (column 2) to get the after-tax data (column 6). While this concept is mathematically more sound, we practice in this text what is easier to comprehend. We post the payable taxes with no sign (i.e., a positive sign) and subtract them from the before-tax data. That taxes are paid out and should therefore be subtracted from income is easier to understand.

There may arise situations where tax implications of an investment are beneficial, for example when the depreciation cost exceeds the before-tax income (Example 13.8 data for year 1). In such cases we sign the income tax as negative, meaning a refund rather than a payment. When subtracted from the before-tax cashflow, the negative tax renders the after-tax cashflow larger than the before-tax cashflow.

[29]It may be better to keep the salvage value separate from the last period cashflow. This avoids charging any tax on the salvage value. Moreover it can keep the evaluation simpler. Here, for example, the evaluation of PW is simplified by considering the \$3,200 after-tax data for period 3 with the annual cash inflows.

[30]Were the income-tax implications not considered, the PW would have been

$$\begin{aligned} \text{PW} &= -13{,}000 + 4{,}000(P/A, 12\%, 3) + 7{,}000(P/F, 12\%, 3) \\ &= -13{,}000 + 4{,}000 \times 2.402 + 7{,}000 \times 0.7118 \\ &= -13{,}000 + 9{,}608 + 4{,}983 \\ &= \$1{,}591 \end{aligned}$$

The positive PW would have led the engineer to invest in the truck—obviously an erroneous decision.

After-tax analysis may decline funding a project that would have been approved otherwise. This is so because taxation usually reduces the benefits, while the first cost remains unchanged.

13.7 COMBINED INCOME TAX

Some state governments[31] too collect income tax from individuals as well as from businesses. In such cases the implications of both state and federal income taxes should be considered in the analysis. The procedures discussed in the earlier sections, including the concept of incremental tax rate, remain applicable when taxes are due to both federal and state governments. The simplest approach is to create another column in the cashflow table and account for state tax as we did for the federal tax. However, the usual practice is to determine a *combined incremental tax rate* that takes both the taxes into account, and use this rate in the analysis.

The combined incremental tax rate is based on the common practice that the federal government allows state income tax as an itemized deduction. Under this practice,

$$\text{Combined incremental tax rate} = r_s + r_f(1 - r_s)$$

where r_s = state incremental tax rate
r_f = federal incremental tax rate

The basis of this relationship[32] can easily be explained. Since the state income tax is deducted from the taxable income for federal tax purposes, the factor $(1 - r_s)$ normalizes the federal tax obligation from r_f to $r_f(1 - r_s)$. The normalized rate is then added to the state rate r_s to yield the combined incremental tax rate, as illustrated in Example 13.10.

EXAMPLE 13.10

RST company operates in a state that charges no income tax. In investment analyses the company uses an incremental rate of 40% for federal income tax. It is considering moving its operations to a neighboring state that charges income tax. If the incremental rate for state income tax is 10%, what will the combined rate be to include both federal and state income tax obligations?

Solution

$$\text{Given } r_s = 10\% = 0.10$$
$$r_f = 40\% = 0.40$$

$$\begin{aligned}\text{Combined income tax rate} &= r_s + r_f(1 - r_s)\\ &= 0.10 + 0.40(1 - 0.10)\\ &= 0.10 + 0.36\\ &= 0.46\\ &= 46\%\end{aligned}$$

---

[31]In the United States, even some city governments levy income tax.

[32]For every dollar of taxable income, only $(1 - r_s)$ dollar is taxed at the federal rate r_f. Thus, the federal tax is $r_f(1 - r_s)$ while the state tax is r_s. This yields a total tax of $r_s + r_f(1 - r_s)$ for every taxable dollar.

Note that the combined rate of 46% is higher than the federal rate (40%), but less than the algebraic sum of the federal and state tax rates (40% + 10% = 50%).

If the state tax rate is negligible in comparison to the federal tax rate, one can ignore the effect of state income tax on economic analysis. This is obvious in the preceding relationship, where the combined tax rate $r_s + r_f(1 - r_s)$ reduces to r_f when r_s is zero. As an illustration, if the state rate is 3% and the federal rate is 50%, the combined rate works out to be 51.5%, which is close to the 50% federal rate. Ignoring the state tax implications in such a case may be acceptable.

EXAMPLE 13.11

The RST company of Example 13.10 is considering a project whose data are in Example 13.9. Is the project worthy of investment?

Solution

$$\text{Given } r_s = 10\% = 0.10$$
$$r_f = 40\% = 0.40$$

Since the state rate is not negligible in comparison to the federal rate, consider them both. From Example 13.10, the combined income tax rate is 46%.

Redo Example 13.9 with 46% tax rate. The results are summarized as

(1) Year	(2) Before-Tax Cashflow	(3) Depreciation	(4) Taxable Income (2) − (3)	(5) Tax @ 46%	(6) After-Tax Cashflow (2) − (5)
0	−$13,000				−$13,000
1	4,000	$2,000	2,000	$920	3,080
2	4,000	2,000	2,000	920	3,080
3	4,000	2,000	2,000	920	3,080
	7,000				7,000

Thus, the present worth of the after-tax cashflows, considering both the federal and state taxes, is

$$\begin{aligned} \text{PW} &= -13{,}000 + 3{,}080(P/A, 12\%, 3) + 7{,}000(P/F, 12\%, 3) \\ &= -13{,}000 + 3{,}080 \times 2.402 + 7{,}000 \times 0.7118 \\ &= -13{,}000 + 7{,}398 + 4{,}983 \\ &= -\$619 \end{aligned}$$

Since the PW is negative, the project is not worthy of investment.[33]

13.8 OTHER CONSIDERATIONS

As mentioned earlier, a major role of governments throughout the world is to actively participate[34] in the national economy. The income tax is one of the modes of direct participation. Another example is through taxation on capital gains and write-offs on capital losses.

13.8.1 Capital Gain and Loss

Companies may earn a profit or incur loss when a capital asset is disposed off. If a profit is realized, the resulting gain is taxable. A capital gain occurs when the asset is sold for a price above its book value. Thus,

$$\text{Capital gain} = \text{Selling price} - \text{Book value}$$

The capital gain is treated as an income in the year it is realized. It is accounted for by considering its impact on the cashflow, as illustrated in Example 13.12.

EXAMPLE 13.12

If in Example 13.8, the tool is sold for $1,500 at the end of its 3-year life, how much is the capital gain? What effect does this gain have on the net cashflows and what is the present worth?

Solution

Note that the capital consumed $(P - S)$ in the investment has been recovered over the years through the annual $1,200 depreciation. Thus, the book value equals the $300 salvage value. Since the tool is sold for $1,500—more than the book value—there is a capital gain.

$$\begin{aligned}\text{Capital gain} &= \$1{,}500 - \$300 \\ &= \$1{,}200\end{aligned}$$

---

[33]The decision remains unchanged in this case. But in cases where the PW is marginally positive, consideration of state income tax may render it negative, thus changing the decision.

[34]Though the participation is well-intentioned, it may hurt the economy, rather than help it, if the government policies are not implemented judiciously. The United States is one of the few countries that seems to understand the danger of playing a role bigger than necessary.

We follow the same procedure as in Example 13.8, but account for this capital gain by posting it in the year it is realized, that is, year 3. This has been done by adding columns (4) and (5).

(1) Year	(2) Before-Tax Cashflow	(3) Deprecia-tion	(4) After-Depreciation Cashflow (2) − (3)	(5) Capital Gain	(6) Taxable Income (4) + (5)	(7) Tax @ 60%	(8) After-Tax Cashflow (2) + (5) − (7)
0	−\$3,900						−\$3,900
1	1,000	\$1,200	−200		−200	−\$120	1,120
2	1,500	1,200	300		300	180	1,320
3	2,000	1,200	800	1,200	2,000	1,200	2,000
	300						300

The effect of capital gain is reflected in the increased taxable income for the third year from \$800 to \$2,000. This has increased the tax for the year from \$480 (Example 13.8) to \$1,200; the \$720 increase is 60% of the \$1,200 capital gain. The net effect, however, is the increased third-year after-tax cashflow, from \$1,520 to \$2,000. Thus, the cashflows in the last column account for depreciation and capital gain, as well as income taxes.

Once the cashflows have been modified to account for all the applicable tax obligations, the net after-tax cashflows are analyzed in the usual way. For this case,

$$\begin{aligned} PW &= -3{,}900 + 1{,}120(P/F, 8\%, 1) + 1{,}320(P/F, 8\%, 2) + (2000 + 300)(P/F, 8\%, 3) \\ &= -3{,}900 + 1{,}120 \times 0.9259 + 1{,}320 \times 0.8573 + 2{,}300 \times 0.7938 \\ &= -3{,}900 + 1{,}037 + 1{,}132 + 1{,}826 \\ &= \$95 \end{aligned}$$

Sometimes, a capital loss may occur when the asset is sold for less than its book value, that is,

$$\text{Capital loss} = \text{Book value} - \text{Selling price}$$

The capital loss is treated as a cost in the year it occurs, reducing the profit and hence the income tax. It is accounted for in the analysis by considering it accordingly, as illustrated in Example 13.13.

EXAMPLE 13.13

If in Example 13.8, the tool is sold for \$100 at the end of its 3-year life, what is the present worth of the project considering capital loss?

Solution
The tool's book value is $300. Since it is sold for $100, there occurs a capital loss.

$$\text{Capital loss} = \$300 - \$100$$
$$= \$200$$

We follow the procedure of Example 13.8 and account for this loss by posting it for the year it occurred, that is, for year 3. This has been done by adding columns (5) and (6).

(1) Year	(2) Before-Tax Cashflow	(3) Depreciation	(4) After-Depreciation Cashflow (2) − (3)	(5) Capital Loss	(6) Taxable Income (4) − (5)	(7) Tax @ 60%	(8) After-Tax Cashflow (2) − (5) − (7)
0	−$3,900						−$3,900
1	1,000	$1,200	−200		−200	−$120	1,120
2	1,500	1,200	300		300	180	1,320
3	2,000	1,200	800	200	600	360	1,440
	300						300

The capital loss has decreased the third-year taxable income by $200. This has the effect of reducing income tax for that year from $480 (Example 13.8) to $360. The $120 decrease represents 60% of the $200 capital loss. The ultimate effect is decreased third-year net cashflow in the last column. Once the cashflows have been modified to account for all the applicable tax implications, the net after-tax cashflows are analyzed the usual way. For this case,

$$\begin{aligned} \text{PW} &= -3{,}900 + 1{,}120(P/F, 8\%, 1) + 1{,}320(P/F, 8\%, 2) + (1{,}440 + 300)(P/F, 8\%, 3) \\ &= -3{,}900 + 1{,}120 \times 0.9259 + 1{,}320 \times 0.8573 + 1{,}740 \times 0.7938 \\ &= -3{,}900 + 1{,}037 + 1{,}132 + 1{,}381 \\ &= -\$350 \end{aligned}$$

Taxation rules governing capital gains and losses change continually, depending on the extent and frequency of government intervention in the economy. Engineers and technologists need to keep up-to-date[35] with these changes so as to incorporate them in economic analyses. The best information resources on the changes are usually the taxing authority and the company tax attorney or accountant.

[35]For example, as an incentive to small companies, the tax laws sometimes allow equipment within certain cost limits—currently $17,500 in the United States—to be written off in the year of the acquisition. In other words, the entire cost can be depreciated in the first year.

13.8.2 Investment Credit

When there is a slowdown in the economy, governments encourage investment by offering industry incentives of various types. One such incentive is an investment tax credit, which indirectly pays companies a certain percentage of the investment in equipment. Investment tax credits may vary from year to year; the relevant information can be obtained from the taxing authority or company attorney.

The investment tax credit must be claimed in the year the equipment is bought. It does not affect the initial cost of the equipment, and therefore depreciation charges remain unaffected. Again, the cashflows should be modified to include the effect of tax credit, as illustrated in Example 13.14.

EXAMPLE 13.14

If an investment tax credit of 10% is allowed in Example 13.9, how is the decision affected?

Solution

The overall procedure and the discussions of Example 13.9 remain applicable. The 10% investment tax credit (0.10 × \$13,000 = \$1,300) posted in the following *modified* cashflow table in the row[36] for year 1 under column 6 affects the after-tax cashflow (column 7). This cashflow is higher by \$1,300, from \$3,200 to \$4,500. Note that the investment tax credit does not affect depreciation.

(1) Year	(2) Before-Tax Cashflow	(3) Depreciation	(4) Taxable Income (2) − (3)	(5) Tax @ 40%	(6) Investment Credit	(7) After-Tax Cashflow (2) − (5) + (6)
0	−\$13,000					−\$13,000
1	4,000	\$2,000	2,000	\$800	\$1,300	4,500
2	4,000	2,000	2,000	800		3,200
3	4,000	2,000	2,000	800		3,200
	7,000					7,000

The present worth with the capital investment credit should be higher, since after-tax cash inflow has increased for year 1. As seen here, it is so (compared to Example 13.9). In fact, PW is now positive, rendering the project worthy of investment.

$$\begin{aligned} \text{PW} &= -13{,}000 + 1{,}300(P/F, 12\%, 1) + 3{,}200(P/A, 12\%, 3) + 7{,}000(P/F, 12\%, 3) \\ &= -13{,}000 + 1{,}300 \times 0.8929 + 3{,}200 \times 2.402 + 7{,}000 \times 0.7118 \\ &= -13{,}000 + 1{,}161 + 7{,}686 + 4{,}983 \\ &= \$830 \end{aligned}$$

[36]The investment tax credit is usually taken at the end of the year the asset is acquired.

As seen in this example, the investment tax credit can turn an otherwise unworthy project into a worthy one. Engineers should be on the lookout for any investment opportunity offered by the government and reassess the projects that were denied funding.

SUMMARY

Income taxes and other tax obligations may have significant implications for engineering investment decisions. To boost economic growth, governments throughout the world use income tax and other incentives—such as capital gains and losses, and investment tax credits—to encourage industry to invest in new equipment. These incentives may vary from year to year. Engineers should keep in touch with company management, which tracks such incentive programs. The company tax attorney and/or accountant are also useful resources on such matters.

Income taxes are easier to calculate based on the concept of incremental tax rate. The federal and state income tax rates can be considered jointly through a combined tax rate. The basic approach to accounting for taxes and tax incentives is to modify the original (before-tax) cashflows. Modifications are better carried out by adding columns to the cashflow table. The modified after-tax cashflows are then analyzed the usual way under any of the decision criteria discussed in Part II. Decisions can change when tax implications are considered. Most engineering economics analyses include depreciation, income taxes, and other factors that may affect the investment decision.

EXERCISES

Discussion Questions

13.1 Why should engineers be conversant with current laws on income tax and other government investment incentives?

13.2 What roles can company tax attorneys play in decision making by engineers and technologists pertaining to engineering projects?

13.3 Discuss in not more than 100 words how engineering economic analysis is carried out when income tax is to be accounted for.

13.4 Does the investment tax credit affect depreciation charges? Explain.

13.5 If the investment tax credit is spread over an asset's life, rather than taken in the year of investment, will it make the investment more attractive or less attractive? Explain your answer.

13.6 Explain the concept of incremental income tax rate.

13.7 Why is the combined income tax rate lower than the algebraic sum of the federal and state tax rates?

Multiple-Choice Questions (Circle the *best* answer.)

13.8 Taxable income is usually _____ the adjusted gross income.

a. greater than
b. the same as
c. less than
d. none of the above

13.9 In computing personal income tax, you can deduct the
a. state income tax from the federal taxable income.
b. federal income tax from the state taxable income.
c. either a or b, whichever reduces your total income tax
d. neither a nor b

13.10 While accounting for both depreciation and income tax, modify the cashflows
a. first for income tax and then for depreciation.
b. first for depreciation and then for income tax.
c. either a or b; it does not matter
d. either a or b, whichever reduces the tax obligation

13.11 Capital loss on equipment, if allowed, must be accounted for in the year
a. of the investment.
b. the asset undergoes its first major overhaul.
c. halfway through its useful life.
d. it is disposed of.

13.12 The best resource for information on the current income tax rules, while analyzing an engineering project, is
a. the county's property tax collector.
b. the boss who is a technical manager.
c. the company tax attorney.
d. last year's file on the subject.

13.13 The incremental income tax rate is the
a. increase this year in the tax rate over last year.
b. difference between federal and state income tax rates.
c. difference in individual and corporation tax rates.
d. none of the above

Numerical Problems

13.14 Universal Robotics' gross income for the current year is $265,000. It has invested in a special hand gripper costing $15,000, whose estimated useful life and salvage values are 3 years and $3,000. Its operating expense for the year has been $70,000. Estimate the payable income tax if the applicable tax rate schedule is

Taxable Income	Tax Rate
Below $100k	30%
$100k–$150k	40%
$150k–$200k	50%
Above $200k	60%

13.15 In Problem 13.14, if Universal Robotics can increase its income by $5,000 by investing in a special tool, what is the incremental tax rate?

13.16 A used truck is purchased for $12,000. It is estimated to have a useful life of five years, during which MACRS-based depreciation is charged. The truck is expected to annually save the company $2,000 in material-handling costs. Assuming zero salvage value and a tax rate of 50%, is it a good investment? Assume the MARR = 12%.

13.17 A company operating in Mississippi pays 5% state income tax. If its federal income tax rate is 50%, what is the combined income tax rate?

13.18 Sam's Salvage Store (3S) is considering buying a van for $35,000 so that it can offer free local delivery. The free delivery is likely to increase 3S's sale by $16,000 each year. The truck's annual operating cost will be $6,000. Assuming zero salvage value and a useful life of seven years, is this investment sound on the basis of present worth? The company uses SL depreciation and pays income tax at a combined incremental rate of 50%. Assume the MARR = 8%.

13.19 If the MARR is 10% in Problem 13.18, is the investment sound on the basis of after-tax rate of return?

13.20 What is the expected after-tax ROR in Example 13.19?

13.21 For an initial investment of $100k in R&D, Fancy Products expects its new solar-powered hair dryer to bring in annual revenue of $30k for the next 10 years. The production and marketing costs will be $5k per year. Assuming a 50% income tax rate and SL depreciation, determine the after-tax ROR if the salvage value is zero.

13.22 Sam's Snowmobiles is considering investing in an automatic tool changer that will reduce the setup time on one of its CNC mills. The investment cost and the resulting benefits are summarized in the following cashflow table. Note that the $350 cashflow for year 5 includes a salvage value of $200. If the company practices DDB depreciation and its income tax rate is 45%, should the project be funded? Assume the MARR = 10%.

Year	Before-Tax Cashflow
0	−$1,500
1	750
2	500
3	350
4	150
5	350

13.23 Precision Machinists believes in keeping its resources technologically updated. It is considering investing $250,000 in a new machining cell, which at the end of its five-year useful life is likely to fetch 20% of the initial cost. The benefits from the cell are estimated to be $50k in the first year, $70k in the second year, and $90k in each of the remaining three years. The cell is eligible for a 10% investment tax credit. Any capital gain is taxed, or loss tax-credited, at a rate of 25%. For a 10% MARR and 50% tax rate, is the investment desirable? Use MACRS for depreciation.

CHAPTER FOURTEEN

Replacement Analysis

IN THIS CHAPTER YOU WILL LEARN ABOUT

- The defender–challenger concept
- Analysis of the defender
- The defender's economic worth
- The remaining life of the defender
- Analysis of the challenger
- The challenger's economic life
- EUAC-based analysis
- Marginal-cost-based analysis
- MAPI method

In the economic analyses discussed so far, we implicitly assumed a "green field" scenario, which means that the plant is new and resources are procured for the first time. While this scenario prevails occasionally, engineers and technologists often make economic decisions to replace resources that already exist and are in use. Such a decision is called *replacement analysis*.

In replacement analyses, the reference for comparison is the existing resource—a machine or equipment. The basic question is, When should we replace the resource[1]? In fact, the more focused question is, Should we replace the equipment now (during the next fiscal year) or sometime later? Replacement analysis offers an answer to such a question. While this question can be raised anytime, it is usually entertained once a year at the time of capital budgeting. Thus, the precise question is, *Should we budget now to replace the resource during the next fiscal year?*

---

[1]The term *resource* is used in a generic way to mean anything that is used in business such as a machine, equipment, a building, or a tool.

14.1 THE DEFENDER–CHALLENGER CONCEPT

The existing resource is analyzed in relation to what the market has to offer as a replacement. If the market has nothing to offer, the replacement question does not arise. The existing resource is continued as long as its operational cost is acceptable, or else it is retired.

Replacement analysis can be conceptualized better by considering the existing resource as a *defender*—as if it were trying to defend its continued use. The one being considered to replace the defender is appropriately called the *challenger*. The replacement analysis can thus be thought of as a "match" between the defender and the challenger, with the engineering economist as the referee.

The challenger is selected from among the alternatives by any of the methods discussed in Part II, for example present worth (Chapter 6) or rate of return (Chapter 9). Usually, the challenger is economically the best the market has to offer at the time of replacement. Engineering economists sometimes postpone the replacement decision to wait for a better challenger likely to be available in the market in the near future.

A complete replacement analysis involves three tasks:

1. Selection of the defender and its analysis
2. Selection of the challenger and its analysis
3. Defender–challenger comparison

We discuss these tasks, illustrating them through examples where appropriate, in the next five sections (14.2 through 14.6).

14.2 THE DEFENDER

Companies usually look into replacement needs once a year as part of the budget process. Department managers request funds for replacing the troublesome resources. Although they might like to replace all the resources impeding productivity, the company's limited capital budget does not allow it. Budget limitations and the severity of productivity bottlenecks created by the failing resources largely dictate the selection of the defender(s). These bottlenecks are measured by the frequency of past breakdowns, repair costs, average repair time, mean time between failures, effect on product quality, and so on, which eventually impact the cost of production and hence the profit.

Once a defender has been selected for replacement, the next relevant question is, What is its economic worth?

14.2.1 Economic Worth

Several types of costs or values are associated with a resource.

Initial Cost The initial cost is the sum spent when the defender was acquired. It usually includes shipping, installation, and other *one-time costs* incurred at the time of procurement. It is also called *original cost* or *first cost*.

Trade-in Value The trade-in value is the price offered for the defender when the challenger is being negotiated for acquisition. This value is usually "inflated," that is, higher than the real value. You may be aware of the high price car dealers offer for an old car as a trade-in.

Book Value Since the defender has been in use for some time, depreciation may have been charged against it. The book value of an asset is the difference between the first cost and the cumulative depreciation. For example, if the first cost of a 3-year-old defender is $75,000 and the annual depreciation charge is $15,000, then the book value at the end of third year of its life is $75,000 − (3 × $15,000) = $30,000.

Market Value The defender's market value is what the defender will fetch if sold in the open market. It is based on free sale, with no strings attached. If the dealer is willing to pay you $1,000 for a trade-in that is worth no more than $400 (market value), the attached string is represented by the profit he is trying to make on the new car you are buying. The price of the new car has certainly been marked up, at least by $600—the loss the dealer is ready to suffer on the trade-in.

Present Cost The present cost of the defender is the current price of the same type and model if bought new. Due to inflation, it is usually higher than the defender's first cost. At times, it may be the same or even lower, due to technological advances that reduce the price by more than the inflationary increase. For products undergoing rapid technological changes, such as personal computers, prices of new models may be lower, even with enhanced capabilities.

Which of these costs or values is the *real* economic worth of the defender, especially for the purpose of replacement analysis? The answer is, *market value*, in most cases[2].

The market value of an asset may at times be lower than its book value, the difference being the capital loss. Thus,

$$\text{Capital loss} = \text{Book value} - \text{Market value}$$

Often a capital gain may be realized, since the market value is more than the book value, where

$$\text{Capital gain} = \text{Market value} - \text{Book value}$$

Capital losses and other such losses are collectively called *sunk cost*. Analysts may be tempted to recover the sunk cost by including it in the replacement analysis, but that is erroneous. Economic analysis should never include the sunk cost (past losses); only

[2]The trade-in may at times be attractive. Consider that the market value of a defender is $3,000 and that the challenger costs $10,000 if bought in the open market. A distributor offers a trade-in of $5,000, with the challenger at $11,500. Trading in is thus preferable, since the challenger's additional cost of $1,500 over the open market price is offset by the additional $2,000 on the trade-in. The *real* economic worth of the defender, however, remains the same—its market value of $3,000.

the present and estimated future costs are relevant. Accordingly, *all sunk costs are ignored in economic analyses*. Though no attempt should be made to recover the sunk cost, the capital loss may be partially recovered through reduced taxes (see Chapter 13).

EXAMPLE 14.1

A replacement analysis of Bahadur, a pick-and-place robot purchased three years ago for $20,000, has been initiated by the jobshop. At the time of purchase Bahadur's useful life was estimated to be five years, with an expected salvage value of $5,000. During the past three years, SL depreciation has been charged.

Due to recent innovations in microprocessor technology, a new robot is available for $15,000. The robot manufacturer is offering $7,000 for the defender as a trade-in. The jobshop has researched to conclude that Bahadur can be sold in the open market for $5,000. Doing so will enable the jobshop to buy the new versatile robot that will not only do Bahadur's job but also facilitate integration among the other computer-controlled machines of the plant. Determine Bahadur's

a. book value,
b. economic worth, and
c. sunk cost.

Solution

Bahadur has already been in use for three years of its five-year useful life. Let us assume that its analysis is being conducted during the end of the third year for possible replacement in the fourth year.

a. The book value is the difference between the first cost and the cumulative depreciation over three years[3].

$$\begin{aligned}\text{Annual straight-line depreciation} &= (P - S)/N \\ &= (\$20{,}000 - \$5{,}000)/5 \\ &= \$3{,}000\end{aligned}$$

$$\begin{aligned}\text{Cumulative depreciation over three years} &= \$3{,}000 \times 3 \\ &= \$9{,}000\end{aligned}$$

Thus,

$$\begin{aligned}\text{Book value} &= \text{First cost} - \text{Cumulative depreciation} \\ &= \$20{,}000 - \$9{,}000 \\ &= \$11{,}000\end{aligned}$$

b. The economic worth of the defender is its market value, which is $5,000 (given).

[3]It is being assumed that the third-year depreciation will be charged by the year end.

c. If the jobshop replaces Bahadur, it will suffer a capital loss, since the market value is lower than the book value, given by

$$\begin{aligned} \text{Capital loss} &= \text{Book value} - \text{Market value} \\ &= \$11{,}000 - \$5{,}000 \\ &= \$6{,}000 \end{aligned}$$

This $6,000 loss in capital at the time of replacement is the sunk cost.

14.2.2 Remaining Life

The defender being considered for replacement can in most cases be continued in use, at least theoretically[4], provided we do not mind its operational cost of repairs, breakdowns, productivity loss, and so on. In reality, however, the operational cost becomes prohibitive.

How long a defender can be continued is decided upon usually on the basis of the cost of its use. This cost comprises two basic elements: capital cost and maintenance cost.

1. *Capital Cost.* Capital cost is the cost of current investment in the existing resource[5], which is based on its economic worth and the prevailing interest rate or MARR. By not selling off the defender, a capital equal to its market value remains "tied up." This realizable capital has the potential of use elsewhere in the company.
2. *Maintenance Cost.* The existing machine might break down, resulting in costs for repairs and lost production. Even when it is operational, its age may contribute to higher scrap, rework, or customer returns. It also costs money for its usual supplies. All such costs, along with those on insurance, space, utilities, and others, are grouped together as *maintenance* or *operational cost*.

The cost to keep the existing resource in use is thus the total of the capital cost and maintenance cost:

$$\text{Defender's cost} = \text{Capital cost} + \text{Maintenance cost}$$

Noting that the time period in replacement analysis is usually a year, these costs are considered annual unless mentioned otherwise. Example 14.2 illustrates the discussions of this subsection.

---

[4]The defender may become completely useless, for example what it was doing is no longer required, or it may become redundant due to technological breakthroughs. In such cases, the resource is simply *retired*, rather than replaced. The analysis for retirement is conceptually similar to that for replacement, and hence most discussions in this chapter are valid for both. Section 14.7 briefly discusses retirement analysis.

[5]The terms *resource, machine*, and *equipment* are used synonymously.

EXAMPLE 14.2

Make an economic comparison between the following two options pertaining to Bahadur, the pick-and-place robot of Example 14.1:

a. Sell it off.

b. Keep it operational for one more year, during which its maintenance cost is estimated to be $200. The salvage value at the end of the next year is estimated to be $4,000. Assume the MARR = 10%.

Solution

a. By selling the robot, the company can recover the "tied up" capital[6], which is Bahadur's market value of $5,000. This capital can be used elsewhere, earning at least the 10% MARR.

b. The second option of keeping the robot operational for one more year will cost money, which is the total of capital and maintenance costs for the year. The maintenance cost is given to be $200. We need to determine the capital cost for the year.

By not selling off the robot, the jobshop keeps $5,000 of its capital "tied up," which would have earned interest at the (given) 10% MARR. Thus, an interest income of $500 (10% of $5,000) is forgone for the year. Moreover, the robot's salvage value of $4,000 at the end of the next year is lower than this year's salvage value of $5,000. So by not selling off this year, a loss of $1,000 in the value of the robot (capital) is also incurred. Hence, the cost of capital in keeping the robot in use for one more year is

$$\begin{aligned}\text{Loss in capital} + \text{Loss in interest income} &= \$1{,}000 + \$500 \\ &= \$1{,}500\end{aligned}$$

Thus, the total cost of keeping the robot for one more year is

$$\begin{aligned}\text{Capital cost} + \text{Maintenance cost} &= \$1{,}500 + \$200 \\ &= \$1{,}700\end{aligned}$$

A comparison of the two options shows that option (a) generates a capital of $5,000 now (this year) with the loss of Bahadur as a resource, while option (b) keeps Bahadur in use for a year at a cost of $1,700, and releases a capital of $4,000 at the end of the year from salvaging.

The $1,700 cost in option (b) can also be explained through the cashflow diagram in Fig. 14.1. The upper diagram represents option (a). The selling-off of the robot generates a cash inflow of $5,000 now (year 0), shown by its upward vector. Option (b) is represented by the lower diagram, where not selling it this year is equivalent to an investment of $5,000. This is represented by the downward vector at year 0. The $200 downward vector at year 1 represents the maintenance cost for the year (following the period-end convention of cashflow posting), while the $4,000 upward vector denotes the cash inflow from salvaging the robot next year.

---

[6]The term *capital* to denote the market value is appropriate, since, once recovered by selling off the robot, this $5,000 becomes capital for the company. Note that the same sum may be described by different terms—market value, salvage value, or capital—depending on the context.

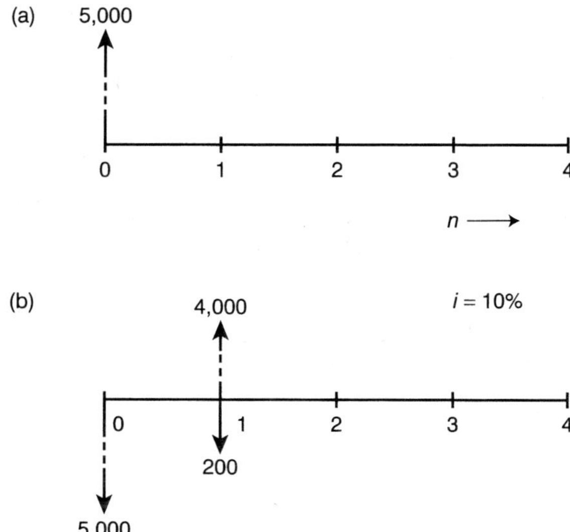

Figure 14.1 Cost of Keeping the Defender for a Year

Referring to the lower diagram for option (b), the \$5,000 investment at year 0 is equivalent to its future value F at year 1 as

$$\begin{aligned} F &= P(F/P, i, n) \\ &= \$5{,}000(F/P, 10\%, 1) \\ &= \$5{,}000 \times 1.1 \\ &= \$5{,}500 \end{aligned}$$

Thus, instead of the \$5,000 vector at year 0, we can consider a downward vector of \$5,500 (not shown) at year 1. The net effect of this transferred \$5,500 vector and the other two vectors already existing at year 1 is (following the sign convention learned in Chapter 2)

$$-\$5{,}500 - \$200 + \$4{,}000 = -\$1{,}700$$

Being negative, this \$1,700 represents a cost at year 1. Thus, the cost of keeping the robot in use for a year[7] is \$1,700.

We can determine the present worth of the \$1,700 cost by transferring it to year 0 as

$$\begin{aligned} P &= \$1{,}700(P/F, 10\%, 1) \\ &= \$1{,}700 \times 0.9091 \\ &= \$1{,}545 \end{aligned}$$

[7]Based on the year-end cashflow convention, the \$1,700 cost at year 1 represents the cost from now (year 0) until the end of the year (year 1).

This $1,545 is the present value of the cost of keeping the robot for one more year. In contrast, option (a) results in a present benefit of $5,000, but with the loss of the resource for use.

The discussions in Example 14.2 have illustrated that while analyzing the defender for its cost implications (as well as other considerations of the replacement analysis) the time value of money must be accounted for. Rather than the asset's present or future worth of the costs, its annual cost is usually the basis of decision in replacement analysis. The *equivalent uniform annual cost* (EUAC)[8] serves this purpose. Example 14.3 illustrates the evaluation and role of EUAC in analyzing a defender.

EXAMPLE 14.3

Determine the annual cost (EUAC) of operating the pick-and-place robot of Example 14.1 for each of the next five years. The year-end salvage values and the maintenance costs are

Year	Year-End Salvage Value	Maintenance Cost
1	$4,000	$200
2	3,500	300
3	3,000	400
4	2,000	500
5	1,000	600

Solution

The given data may be difficult to comprehend. What they mean, for example for year 1, is that if we keep the robot for one more year, then its maintenance cost for the year will be $200 and at the end of the year its salvage value will be $4,000.

On the other hand, the data for year 3 say that if we keep the robot for three years, then its maintenance costs will be $200 for year 1, $300 for year 2, and $400 for year 3, and at the end of the third year its salvage value will be $3,000. That means, as far as the data for year 3 are concerned, the $4,000 and $3,500 data in column 2 for years 1 and 2 are irrelevant.

Thus, the appropriate cashflow table, after posting the market value as year-0 cash outflow, for analyzing the robot to keep it for three years is

---

[8]See Chapter 8 if you find EUAC unfamiliar.

Year	Cashflow
0	−$5,000
1	−200
2	−300
3	−400
	3,000 (salvage value)

Similarly, the cashflow table appropriate for the five-year analysis is

Year	Cashflow
0	−$5,000
1	−200
2	−300
3	−400
4	−500
5	−600
	1,000 (salvage value)

Thus, the data given in this example is in a concise format, wherefrom the cashflow table appropriate to the particular analysis should be derived.

We begin by creating a modified table based on the above discussions. The table will also contain results of the analysis for each year under the heading EUAC.

Remaining Life n, Years	Year-End Value	Maintenance Cost	EUAC
0	$5,000		
1	4,000	$200	
2	3,500	300	
3	3,000	400	
4	2,000	500	
5	1,000	600	

Note that we have changed the title of column 1 to reflect more accurately the meaning of year by calling it the *remaining life* of the robot. *By remaining life we mean the length of time the defender is to be kept.* The second column, *Year-End Value*, denotes the salvage value at the end of the year, while the maintenance costs appear in column 3. Also note the introduction of a new row for year 0, containing the robot's current market value of $5,000. Note that this sum, and all the others under column 2, represent the investment, for the corresponding remaining life, in the defender if it is kept in use (not sold off).

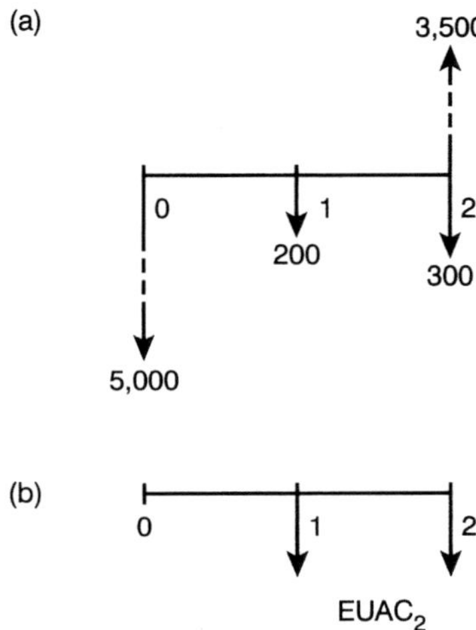

Figure 14.2 Diagram for Example 14.3

We now analyze the data, following the procedure explained in Example 14.2, to determine the EUAC for each of the five options—keeping the robot for one year, keeping the robot for two years, . . . , and keeping the robot for five years. The calculations to determine[9] the EUAC can be approached by one of the following two ways.

1. Sketch and analyze the appropriate cashflow diagram. We illustrate this approach for the option of keeping the robot for two years, for which the appropriate cashflow diagram is Fig. 14.2. In (a), the downward vector at year 0 represents the current investment. The maintenance costs are shown at the appropriate time periods, while the $3,500 salvage value is sketched as an upward vector at year 2, when the robot will be sold off.

Our task is to determine an equivalent uniform annual cost $EUAC_2$ for the diagram in (a). In other words, what value[10] of $EUAC_2$ in diagram (b) will render the two diagrams (a) and (b) equivalent?
Thus,

$$\begin{aligned} EUAC_2 &= 5{,}000(A/P, 10\%, 2) + 200(P/F, 10\%, 1)(A/P, 10\%, 2) \\ &\qquad - (3{,}500 - 300)(A/F, 10\%, 2) \\ &= 5{,}000 \times 0.5762 + 200 \times 0.9091 \times 0.5762 - 3{,}200 \times 0.4762 \\ &= 2{,}881 + 105 - 1{,}524 \\ &= \$1{,}462 \end{aligned}$$

[9]Refer to Chapter 8 if you can't follow the discussions.

[10]Note that the EUAC relates to two years only, not five years, since the plan is to keep the robot for two years only. For a four-year analysis, that is, the option of keeping the robot for four years, EUAC will be spread over four years.

In a similar way the EUAC for the other four options[11] can be calculated[12] by referring to their cashflow diagrams, and the results can be summarized in the table as

Remaining Life n, Years	Year-End Value	Maintenance Cost	EUAC
0	\$5,000		
1	4,000	\$200	\$1,700
2	3,500	300	1,462
3	3,000	400	1,398
4	2,000	500	1,485
5	1,000	600	1,536

What do the EUAC values mean in this table? They are simply the total annual cost for each of the five options (keeping for one year, keeping for two years, . . . , keeping for five years).

For example, if the robot is kept for four years, then the total cost of doing so will be \$1,485 per year during the next four years. That means, for this option, the entries \$1,700, \$1,462, and \$1,398 in the table under EUAC are irrelevant, and so is the \$1,536 entry for year 5.

2. Equation-based calculations (cashflow diagrams not essential). The equation-based approach attempts to do away with the need for sketching the cashflow diagrams. Here, the calculations for EUAC are done in two parts: one for the capital cost or recovery, and the other for the maintenance cost. The two parts are then added together to evaluate the EUAC for the year under consideration, as in part (b) of Example 14.2.

The capital cost is evaluated by determining an equivalent annual cost of the capital P tied up in the asset and the salvage cash inflow S. It is the annual recovery of the "used-up" capital represented[13] by $P - S$.

Refer to the cashflow diagram in Fig. 14.3 showing an investment P (now) and a salvage value S at year n. Thus the capital used up over n years is $P - S$. The annual cost of this amount of used capital in terms of EUAC can be obtained by subtracting the annual worth of S from that of P, as

$$\text{EUAC}_n = P(A/P, i, n) - S(A/F, i, n)$$

[11]A separate diagram is required for each option, with different timeline lengths. For example, for the option of keeping the robot for four years, there will be four periods in the diagram. While calculating the EUAC, exploit the arithmetic-series pattern in the maintenance costs. For example, for year 4, the most efficient solution is

$$\begin{aligned}\text{EUAC}_4 &= 5{,}000(A/P, 10\%, 4) + 200 + 100(A/G, 10\%, 4) - 2{,}000(A/F, 10\%, 4)\\ &= 5{,}000 \times 0.3155 + 200 + 100 \times 1.381 - 2{,}000 \times 0.2155\\ &= 1{,}578 + 200 + 138 - 431\\ &= 1{,}485\end{aligned}$$

[12]The value for year 1 already worked out in Example 14.2.

[13]Note that P and S occur at two different periods, necessitating a consideration of their time values.

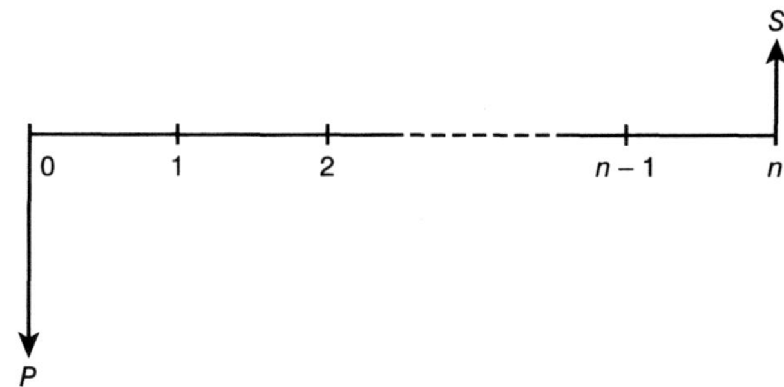

Figure 14.3 Effect of P and S on the EUAC

The use of this relationship requires evaluating two functional notations. For efficient solution, this can be reduced to one by modifying it (see Chapter 3) as follows, since $(A/F, i, n) = (A/P, i, n,) - i$.

$$\begin{aligned} \text{EUAC}_n &= P(A/P, i, n) - S(A/F, i, n) \\ &= P(A/P, i, n) - S[A/P, i, n) - i] \\ &= (P - S)(A/P, i, n) - iS \qquad (14.1) \end{aligned}$$

Equation (14.1) can be used to determine the EUAC—the annual cost[14] of using up the capital $(P - S)$—corresponding to any period by substituting for n.

To facilitate the posting of the two parts of EUAC, the previous table is modified by having three subcolumns under EUAC:

			EUAC		
Remaining Life n, Years	Year-End Value, S	Maintenance Cost	Capital Recovery	Maintenance	Total
0	P = $5,000				
1	4,000	$200			
2	3,500	300			
3	3,000	400			
4	2,000	500			
5	1,000	600			

In the first row, column 2, note that the robot's market value of $5,000 has been denoted by P. The salvage value S corresponding to the remaining lives (years) can be denoted by $S_1, S_2, \ldots$ to distinguish one from the other. These notations avoid any confusion in using Equation (14.1). So from the table $S_3 = \$3{,}000$ while $S_5 = \$1{,}000$.

[14]When equipment is so old that its salvage value depends mostly on the scrap value of the material it is made of, the salvage value does not change much with its age. In other words, S does not change, and the investment P (the market value) in the defender almost equals S; thus, $(P - S)$ tends to zero. In such cases, from Equation (14.1), the EUAC of capital recovery is simply iS, since the first term is almost zero.

Let us now illustrate the calculations for year 2 as an example, based on the equation approach. For year 2, the salvage value $S_2 = \$3{,}500$. The EUAC of capital recovery from Equation (14.1) is

$$\begin{aligned}\text{EUAC}_2 &= (P - S_2)(A/P, i, 2) + iS_2\\ &= (P - S_2)(A/P, 10\%, 2) + 0.1\,S_2\\ &= (5{,}000 - 3{,}500) \times 0.5762 + 0.1 \times 3{,}500\\ &= 864 + 350\\ &= \$1{,}214\end{aligned}$$

In a similar way, the EUAC of capital recovery for the other years can be calculated.

$$\begin{aligned}\text{EUAC}_1 &= (P - S_1)(A/P, 10\%, 1) + 0.1S_1\\ &= (5{,}000 - 4{,}000) \times 1.1 + 0.1 \times 4{,}000\\ &= 1{,}100 + 400\\ &= \$1{,}500\end{aligned}$$

$$\begin{aligned}\text{EUAC}_3 &= (P - S_3)(A/P, 10\%, 3) + 0.1S_3\\ &= (5{,}000 - 3{,}000) \times 0.4021 + 0.1 \times 3{,}000\\ &= 804 + 300\\ &= \$1{,}104\end{aligned}$$

$$\begin{aligned}\text{EUAC}_4 &= (P - S_4)(A/P, 10\%, 4) + 0.1S_4\\ &= (5{,}000 - 2{,}000) \times 0.3155 + 0.1 \times 2{,}000\\ &= 947 + 200\\ &= \$1{,}147\end{aligned}$$

$$\begin{aligned}\text{EUAC}_5 &= (P - S_5)(A/P, 10\%, 5) + 0.1S_5\\ &= (5{,}000 - 1{,}000) \times 0.2638 + 0.1 \times 1{,}000\\ &= 1{,}055 + 100\\ &= \$1{,}155\end{aligned}$$

On posting these capital recovery parts of the EUAC, the table looks like this:

			EUAC		
Remaining Life *n*, Years	Year-End Value, *S*	Maintenance Cost	Capital Recovery	Maintenance	Total
0	*P* = \$5,000				
1	4,000	\$200	\$1,500		
2	3,500	300	1,214		
3	3,000	400	1,104		
4	2,000	500	1,147		
5	1,000	600	1,155		

Next, the EUACs of the maintenance are calculated. Illustrated here is the calculation for year 2. Looking at the pattern[15] in the maintenance costs, they follow an arithmetic series, for which we have a functional notation (Chapter 4). Thus[16],

$$\begin{aligned}\text{EUAC}_{2\text{ maintenance}} &= 200 + 100(A/G, 10\%, 2)\\ &= 200 + 100 \times 0.476\\ &= 200 + 48\\ &= \$248\end{aligned}$$

In a similar way, the maintenance EUACs for the other years are calculated and posted in the table.

Finally, by adding the EUAC of the capital recovery to that of the maintenance, the total EUAC for the year is obtained, as shown in the last column.

Remaining Life n, Years	Year-End Value, S	Maintenance Cost	EUAC Capital Recovery	EUAC Maintenance	EUAC Total
0	P = $5,000				
1	4,000	$200	$1,500	$200	$1,700
2	3,500	300	1,214	248	1,462
3	3,000	400	1,104	294	1,398
4	2,000	500	1,147	338	1,485
5	1,000	600	1,155	381	1,536

[15]If the maintenance costs do not follow any pattern—either as given or as "tailored"—then EUAC is determined by the straightforward method, i.e., by considering the time value of each cost individually.

[16]We could have done

$$\begin{aligned}\text{EUAC}_{2\text{ maintenance}} &= [200(P/F, 10\%, 1) + 300(P/F, 10\%, 2)](A/P, 10\%, 2)\\ &= (200 \times 0.9091 + 300 \times 0.8264) \times 0.5762\\ &= (181.82 + 247.92) \times 0.5762\\ &= 248\end{aligned}$$

or

$$\begin{aligned}\text{EUAC}_{2\text{ maintenance}} &= 200(P/F, 10\%, 1)(A/P, 10\%, 2) + 300(A/F,) 10\%, 2)\\ &= (200 \times 0.9091 \times 0.5762) + 300 \times 0.4762\\ &= 104.76 + 142.86\\ &= 248\end{aligned}$$

or

$$\begin{aligned}\text{EUAC}_{2\text{ maintenance}} &= [(200(F/P, 10\%, 1) + 300](A/F, 10\%, 2)\\ &= (200 \times 1.1 + 300) \times 0.4762\\ &= 248\end{aligned}$$

As expected, the EUAC results based on the equation are the same as the previous ones based on cashflow diagrams. You can use either of the two approaches you feel comfortable with to evaluate EUAC in replacement analysis. Also note that the EUAC in this chapter differs from that in Chapter 8. This EUAC is basically the negatively signed AW of the defender, while the EUAC in Chapter 8 is AW_{costs}.

14.2.3 Economic Life

The various costs of keeping a defender in use can be expressed as an equivalent uniform annual cost (EUAC), as explained in the preceding subsection. The variation in EUAC with the remaining life of the defender shows the cost implication of keeping it in use. Obviously, the defender should be kept in use for the duration yielding minimum EUAC. In some cases, this minimum occurs at year 0 (now) since EUAC increases with n, as depicted by the variation (a) in Fig. 14.4. In such a case the equipment is replaced now.

In other cases, EUAC reaches a minimum at some future time, as depicted by the variation (b) in Fig. 14.4, which is when the defender should be replaced. The remaining life for which EUAC is minimum is called the *economic life* of the defender. In Example 14.3, browsing through the total EUAC in the last column of the final table, Bahadur's economic life is 3 years, when its EUAC is minimum at $1,398. Rather than browse through, one can plot the EUAC values to generate a curve like (b) in Fig. 14.4 and read the economic life off the x-axis. This is illustrated in Fig. 14.5 for the results of Example 14.3, wherefrom Bahadur's economic life is 2.9 years.

Whether the EUAC is minimum now or at some time later depends on the rates at which the salvage value decreases and maintenance cost increases as the asset ages.

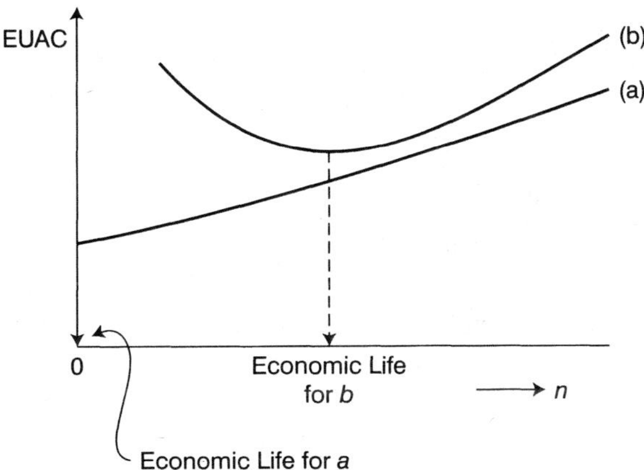

Figure 14.4 EUAC as a Function of Remaining Life

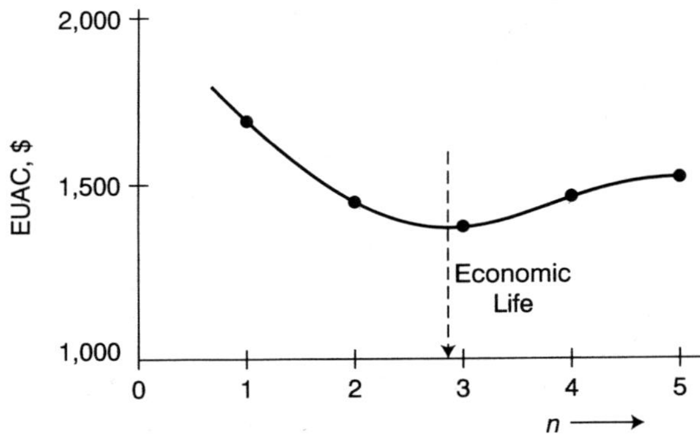

Figure 14.5 Economic Life Corresponds to Minimum EUAC

14.3 THE CHALLENGER

Replacement analysis begins with separate analyses of the defender and the challenger, and ends with their comparison. In the previous section we discussed the analysis of the defender. In this section, we do that for the challenger.

14.3.1 Selection

The selection of the challenger involves deciding the best resource available in the market as a possible replacement of the defender. In the simplest scenario, the manufacturer of the existing equipment offers a functionally equivalent or better replacement. If the equipment is specialized, then there may not be a choice in the market, and what the manufacturer offers becomes the challenger.

In cases where the defending equipment is of the general-purpose type, several alternatives may be available in the market, some of which may be from other manufacturers. In such a case, all the possible[17] alternatives are analyzed by one or more methods discussed in Part II of the text to select the best one. This best becomes the challenger and is analyzed further.

14.3.2 Analysis

The analysis of the challenger is similar to that of the defender discussed in Section 14.2.2. Its objective is to evaluate the challenger's economic life, which may not be the same as its useful life, as illustrated in Example 14.4.

---

[17]One of the options in economic analysis, as pointed out earlier in other chapters, is *do nothing*, i.e., maintain the status quo. In replacement analysis, this option does not exist, since the defender has already signaled the need of replacement through unacceptable breakdowns and the associated costs. In fact, the analysis of the defender is triggered by the absence of the do-nothing option.

EXAMPLE 14.4

The manufacturer of Bahadur, the pick-and-place robot, now markets an updated version, Virbahadur. Virbahadur costs \$15,000, which is \$5,000 less than what Bahadur cost three years ago. It comes with free maintenance for the first two years. For the following salvage values and maintenance costs, determine Virbahadur's economic life, assuming i to be 10%.

Year	Year-End Salvage Value	Maintenance Cost
1	\$13,000	\$0
2	10,000	0
3	6,000	200
4	4,500	300
5	3,000	800
6	1,500	200
7	0	300

The high maintenance cost of \$800 at year 5 is due to an overhaul, which keeps the subsequent years' costs low.

Solution

The analysis of the challenger is similar to that of the defender, so the procedure discussed in Example 14.3 is applicable. Let us follow the equation-based approach. However, keep the appropriate cashflow diagrams in mind while determining the maintenance EUACs.

The economic life corresponds to the minimum EUAC. Each year's EUAC is obtained by determining the EUAC of the capital recovery using Equation (14.1) and adding it to the corresponding maintenance EUAC.

EUACs of Capital Recovery

$$\begin{aligned}\text{EUAC}_{1\text{ capital recovery}} &= (P - S_1)(A/P, 10\%, 1) + 0.1S_1 \\ &= (15{,}000 - 13{,}000) \times 1.1 + 0.1 \times 13{,}000 \\ &= 2{,}200 + 1{,}300 \\ &= \$3{,}500\end{aligned}$$

$$\begin{aligned}\text{EUAC}_{2\text{ capital recovery}} &= (P - S_2)(A/P, 10\%, 2) + 0.1S_2 \\ &= (15{,}000 - 10{,}000) \times 0.5762 + 0.1 \times 10{,}000 \\ &= 2{,}881 + 1{,}000 \\ &= \$3{,}881\end{aligned}$$

$$\begin{aligned}\text{EUAC}_{3\text{ capital recovery}} &= (P - S_3)(A/P, 10\%, 3) + 0.1S_3 \\ &= (15{,}000 - 6{,}000) \times 0.4021 + 0.1 \times 6{,}000 \\ &= 3{,}619 + 600 \\ &= \$4{,}219\end{aligned}$$

$$\begin{aligned}\text{EUAC}_{4\text{ capital recovery}} &= (P - S_4)(A/P, 10\%, 4) + 0.1S_4 \\ &= (15{,}000 - 4{,}500) \times 0.3155 + 0.1 \times 4{,}500 \\ &= 3{,}313 + 450 \\ &= \$3{,}763\end{aligned}$$

$$\begin{aligned}\text{EUAC}_{5\text{ capital recovery}} &= (P - S_5)(A/P, 10\%, 5) + 0.1S_5 \\ &= (15{,}000 - 3{,}000) \times 0.2638 + 0.1 \times 3{,}000 \\ &= 3{,}166 + 300 \\ &= \$3{,}466\end{aligned}$$

$$\begin{aligned}\text{EUAC}_{6\text{ capital recovery}} &= (P - S_6)(A/P, 10\%, 6) + 0.1S_6 \\ &= (15{,}000 - 1{,}500) \times 0.2296 + 0.1 \times 1{,}500 \\ &= 3{,}100 + 150 \\ &= \$3{,}250\end{aligned}$$

$$\begin{aligned}\text{EUAC}_{7\text{ capital recovery}} &= (P - S_7)(A/P, 10\%, 7) + 0.1S_7 \\ &= (15{,}000 - 0) \times 0.2054 + 0.1 \times 0 \\ &= 3{,}081 + 0 \\ &= \$3{,}081\end{aligned}$$

EUACs of Maintenance[18]

$$\text{EUAC}_{1\text{ maintenance}} = 0 \quad \text{(given)}$$

$$\text{EUAC}_{2\text{ maintenance}} = 0 \quad \text{(given)}$$

$$\begin{aligned}\text{EUAC}_{3\text{ maintenance}} &= 200(A/F, 10\%, 3) \\ &= 200 \times 0.3021 \\ &= \$60\end{aligned}$$

$$\begin{aligned}\text{EUAC}_{4\text{ maintenance}} &= [200(F/P, 10\%, 1) + 300](A/F, 10\%, 4) \\ &= (200 \times 1.1 + 300) \times 0.2155 \\ &= 520 \times 0.2155 \\ &= \$112\end{aligned}$$

$$\begin{aligned}\text{EUAC}_{5\text{ maintenance}} &= [200(F/P, 10\%, 2) + 300(F/P, 10\%, 1) \\ &\qquad + 800](A/F, 10\%, 5) \\ &= (200 \times 1.21 + 300 \times 1.1 + 800) \times 0.1638\end{aligned}$$

[18]Be careful while evaluating maintenance EUACs for the fourth through seventh year. What is within the brackets, [ ], yields the F-value, which is then converted into annual value.

$$= 1{,}372 \times 0.1638$$
$$= \$225$$

$$\begin{aligned} \text{EUAC}_{6\text{ maintenance}} &= [200(F/P, 10\%, 3) + 300(F/P, 10\%, 2) \\ &\quad + 800(F/P, 10\%, 1) + 200](A/F, 10\%, 6) \\ &= (200 \times 1.331 + 300 \times 1.21 + 800 \times 1.1 + 200) \times 0.1296 \\ &= 1{,}709 \times 0.1296 \\ &= \$221 \end{aligned}$$

$$\begin{aligned} \text{EUAC}_{7\text{ maintenance}} &= [200(F/P, 10\%, 4) + 300(F/P, 10\%, 3) + 800(F/P, 10\%, 2) \\ &\quad + 200(F/P, 10\%, 1) + 300](A/F, 10\%, 7) \\ &= (200 \times 1.464 + 300 \times 1.331 + 800 \times 1.21 + 200 \times 1.1 \\ &\quad + 300) \times 0.1054 \\ &= 2{,}180 \times 0.1054 \\ &= \$230 \end{aligned}$$

Carefully post the EUACs in their columns and add them for each year to determine the total EUAC, as follows.

			EUAC		
Remaining Life n, Years	Year-End Value, S	Maintenance Cost	Capital Recovery	Maintenance	Total
0	P = \$15,000				
1	13,000	\$0	\$3,500	\$0	\$3,500
2	10,000	0	3,881	0	3,881
3	6,000	200	4,219	60	4,279
4	4,500	300	3,763	112	3,875
5	3,000	800	3,466	225	3,691
6	1,500	200	3,250	221	3,471
7	0	300	3,081	230	3,311

The economic life is obtained by browsing through the data in the last column and locating the minimum EUAC. This is evidently 7 years, corresponding to the EUAC of \$3,311.

14.4 EUAC-BASED ANALYSIS

As mentioned earlier, replacement analysis involves three tasks:

1. Defender analysis
2. Challenger analysis, and
3. Defender–challenger comparison

We discussed the first two tasks in the previous two sections. We now embark on the third task, namely comparison of the defender with the challenger to decide whether to replace or not. There are two approaches to carrying out this task. In this section we discuss the first approach based on EUAC. The second approach, based on marginal cost, is presented in the next section.

The defender–challenger comparison can proceed only after both have been analyzed individually for their economic lives, as discussed in Sections 14.2 and 14.3. Two different situations can arise, as discussed in the next two subsections. In the first one, the defender's and challenger's economic lives are equal, whereas in the other, which is relatively more difficult to analyze, their economic lives are unequal.

14.4.1 Equal Lives

When the defender's remaining economic life is equal to the challenger's economic life, the comparison is simpler. In fact the problem reduces to the analysis of two alternatives. Any of the analyses discussed in the text, including the after-tax ROR, can be carried out. Examples 14.5 and 14.6 illustrate two of the possible replacement scenarios.

EXAMPLE 14.5

A five-year-old lathe has been analyzed as a defender and found to have an economic life of five years. Its current market value is $5,000, and the maintenance cost is $500 per year. Its salvage value at the end of its economic life is estimated to be $500.

The challenger is a modern CNC lathe whose analysis indicated its economic life to be five years. The CNC lathe is capable of operating as a turning cell due to its automatic tool changer, enhanced integratability, and automatic palletizing. Its first cost is $55,000, and the annual maintenance cost is $400. It is expected to save $14,000 per year through higher productivity. Its estimated salvage value at the end of five years is $5,000.

Should the defender be replaced if the MARR = 12%?

Solution

Note that the defender's economic life of five years is the same as that of the challenger. With EUAC as the basis of comparison, our task is to determine the EUAC of the defender and compare it with that of the challenger. The one with the lower EUAC should win. For both, i = MARR = 12%, and n = 5 years.

Defender (lathe)

$$P = \text{current market value} = \$5{,}000$$
$$S = \$500$$
$$\text{AM} = \text{annual maintenance} = \$500$$

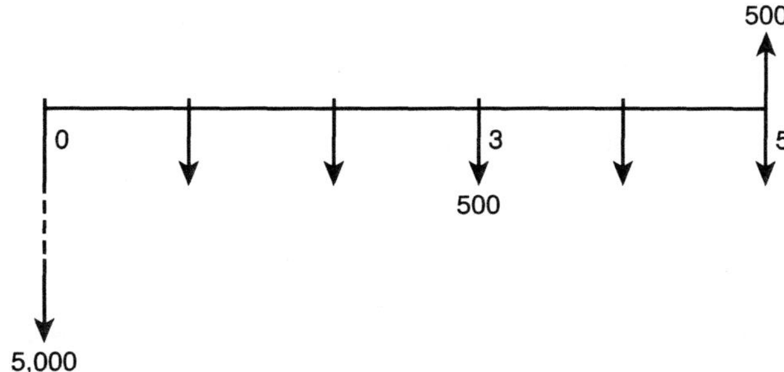

Figure 14.6 Diagram for Example 14.5

From the defender's cashflow diagram in Fig. 14.6[19],

$$\begin{aligned}\text{EUAC}_{\text{defender}} &= 500 + 5{,}000(A/P, 12\%, 5) - 500(A/F, 12\%, 5)\\ &= 500 + 5{,}000 \times 0.2774 - 500 \times 0.1574\\ &= 500 + 1{,}387 - 79\\ &= \$1{,}808\end{aligned}$$

Alternatively, we can use Equation (14.1), from which

$$\begin{aligned}\text{EUAC}_{\text{defender}} &= \text{Capital recovery cost} + \text{Maintenance cost}\\ &= [(P - S)(A/P, 12\%, 5) + iS] + \text{AM}\\ &= (5000 - 500) \times 0.2774 + 0.12 \times 500 + 500\\ &= 4{,}500 \times 0.2774 + 60 + 500\\ &= \$1{,}808\end{aligned}$$

Challenger (CNC lathe)

$$\begin{aligned}P &= \$55{,}000\\ S &= \$5{,}000\\ \text{AM} &= \text{annual maintenance} = \$400\\ \text{AB} &= \text{annual benefit} = \$14{,}000\end{aligned}$$

Again, we can follow either the first-principles approach based on a cashflow diagram, or use Equation (14.1). From the equation, we get

$$\begin{aligned}\text{EUAC}_{\text{challenger}} &= \text{Capital recovery cost} + \text{Maintenance} - \text{Benefit}\\ &= [(P - S)(A/P, 12\%, 5) + iS] + \text{AM} - \text{AB}\\ &= (55{,}000 - 5{,}000) \times 0.2774 + 0.12 \times 5{,}000 + 400 - 14{,}000\end{aligned}$$

[19]Since the two cashflows at period 5 cancel each other, you may be tempted to evaluate

$$\text{EUAC}_{\text{defender}} = 500 + 5{,}000(A/P, 12\%, 4)$$

But this is incorrect, since this $\text{EUAC}_{\text{defender}}$ covers only four years. What we need to determine is the value of $\text{EUAC}_{\text{defender}}$ spread over the five-year economic life.

$$= 50{,}000 \times 0.2774 + 600 - 13{,}600$$
$$= \$870$$

Since the challenger's EUAC is lower, replace the defender with the CNC lathe.

EXAMPLE 14.6

The replacement analysis of a welding robot purchased three years ago at an initial cost of $20,000 has determined its economic life to be five years and its current market value to be $5,000. The robot began straight-line depreciation over its then-estimated useful life of five years, with zero salvage value. Its future maintenance cost is predicted to be $200 per year.

It (the defender) is being considered for replacement by a more accurate deluxe model (the challenger), whose analysis yields an economic life of five years. The challenger is priced at $25,000 and is estimated to cost $300 annually in maintenance. It will also be straight-line depreciated over its useful life, with zero salvage value. An analysis has determined the challenger's economic life to be its useful life. In relation to the defender the deluxe model is estimated to generate an annual income of $10,000 through savings. With the tax obligation at a 50% rate, should the deluxe model replace the defender if $i = 15\%$?

Solution

Note that the defender's five-year economic life equals that of the challenger. The analysis period should therefore be five years. Also note that the defender has already been depreciated for three years.

Since no comparison criterion is specified, we can choose any of those discussed in the text. Let us opt for EUAC. Our task is to determine the defender's EUACs and compare them with those of the challenger, taking depreciation and income tax into account. The one with the lower EUAC should be selected. For both, $i = 15\%$, and $n = 5$ years.

Since the redundant data in the problem statement can be confusing, it is preferable to summarize the relevant data.

Defender (existing welding robot)

Initial cost = $20,000
Then-estimated life = 5 years
Then-estimated salvage = 0
Annual maintenance = $200
Current market value = $5,000

Annual depreciation[20] = (20,000 − 0)/5
= $4,000

---

[20]Determined when the defender was bought and estimated to have a useful life of five years; three of these years have already passed.

Since depreciation has already been charged for three years,

$$\begin{aligned}\text{Book value} &= \text{First cost} - \text{Cumulative depreciation}\\ &= \$20{,}000 - 3 \times \$4{,}000\\ &= \$8{,}000\end{aligned}$$

Since the defender's market value of $5,000 is lower than its $8,000 book value, a capital loss occurs in replacing the defender now.

$$\begin{aligned}\text{Capital loss} &= \text{Book value} - \text{Current market value}\\ &= \$8{,}000 - \$5{,}000\\ &= \$3{,}000\end{aligned}$$

We can summarize the results in a table. An explanation of the entries is provided following the table. Note that the columns have been selected on the basis of what we learned in Chapters 12 and 13. The $3,000 capital loss can be used to reduce income tax now (year 0 in the table).

(1) Year	(2) Before-Tax Cashflow	(3) Depreciation	(4) Capital Loss	(5) Taxable Income (2) − (3)	(6) Tax @ 50%	(7) After-Tax Cashflow (2) − (6)
0	−$5,000		$3,000		$1,500	−$6,500
1	−200	$4,000		−4,200	−2,100	1,900
2	−200	4,000		−4,200	−2,100	1,900
3	−200	0		−200	−100	−100
4	−200	0		−200	−100	−100
5	−200	0		−200	−100	−100

Explanatory Notes

1. In the first column, year 0 means now. The past three years of the defender are of no relevance. The $5,000 entry in the first row, second column represents the defender's current market value.
2. In the fourth column, for year 0, there is an entry[21] of $3,000. Since replacing the defender now would result in a capital loss of $3,000, saving the company $1,500 in income tax, by not doing so an opportunity to reduce tax is forfeited. Thus, the defender represents an investment[22] of $5,000 + $1,500 = $6,500.
3. In the third column, the $4,000 SL depreciation for the remaining two years have been charged.

[21]The posting of a $3,000 capital loss in row 0 means that the associated tax saving is being realized at the time the investment is made (year 0). If the saving is delayed due to accounting or the tax cycle, then this entry should be made in the following year (year 1) and calculations modified accordingly.

[22]The capital loss would reduce income tax for the year. The forfeiture of this opportunity if the defender is not replaced is equivalent to an increase in taxable income by that amount. The equivalency is based on the fact that by not realizing the capital loss now, the company is not reducing its taxable income by $3,000, and thus its tax by $1,500. In other words, not replacing the defender results in $1,500 more payable tax.

4. For year 1, the taxable income is equal to the before-tax cashflow of −$200 (maintenance) minus the depreciation charge of $4,000, yielding −$4,200. The income tax on this is −$2,100. The negative sign indicates a tax refund rather than a payment. Entries in other rows have similar sense.
5. The data in the last column are the after-tax cashflows. For each row, the entry equals before-tax cashflow minus tax. For example, for year 1, it is −$200 − (−$2,100) = $1,900; its positive sign indicates a cash inflow (due to the $2,100 tax refund).

We next determine the EUAC of the after-tax cashflows, as below. Note that the negative signs of these cashflows are being treated as positive since EUAC represents cost. If you have difficulty comprehending the following equation[23], help yourself by sketching an appropriate cashflow diagram.

$$\begin{aligned}\text{EUAC}_{\text{defender}} &= [6{,}500 - 2{,}000(P/A, 15\%, 2)](A/P, 15\%, 5) + 100\\ &= (6{,}500 - 2{,}000 \times 1.626) \times 0.2983 + 100\\ &= (6{,}500 - 3{,}252) \times 0.2983 + 100\\ &= 969 + 100\\ &= \$1{,}069\end{aligned}$$

Challenger (deluxe model)

$$\begin{aligned}\text{Initial cost} &= \$25{,}000\\ \text{Useful life} &= 5 \text{ years}\\ \text{Estimated salvage} &= 0\\ \text{Annual maintenance} &= \$300\\ \text{Annual savings (benefit)} &= \$10{,}000\\ \text{Annual depreciation} &= (25{,}000 - 0)/5\\ &= \$5{,}000\\ \text{Net annual benefit} &= \text{Savings} - \text{Maintenance cost}\\ &= \$10{,}000 - \$300\\ &= \$9{,}700\end{aligned}$$

A table similar to that for the defender is prepared for the challenger. The explanation given earlier on the entries in the defender's table should be helpful in comprehending the entries in this table too.

---

[23]To tailor the cashflow pattern, consider the $1,900 cashflow to be divided into two components: $2,000 and −$100. This makes the $100 cash outflow form a uniform series. In the absence of tailoring, evaluation would have been longer:

$$\begin{aligned}\text{EUAC}_{\text{defender}} &= [6{,}500 - 1{,}900(P/A, 15\%, 2)](A/P, 15\%, 5)\\ &\quad + 100(F/A, 15\%, 3)(A/F, 15\%, 5)\\ &= (6{,}500 - 1{,}900 \times 1.626) \times 0.2983 + 100 \times 3.472 \times 0.1483\\ &= 1{,}017 + 51\\ &= \$1{,}068\end{aligned}$$

Year	Before-Tax Cashflow	Depreciation	Taxable Income	Tax @ 50%	After-Tax Cashflow
0	−$25,000				−$25,000
1	9,700	$5,000	$4,700	$2,350	7,350
2	9,700	5,000	4,700	2,350	7,350
3	9,700	5,000	4,700	2,350	7,350
4	9,700	5,000	4,700	2,350	7,350
5	9,700	5,000	4,700	2,350	7,350

The EUAC of the after-tax cashflows for the challenger is

$$\begin{aligned} \text{EUAC}_{\text{challenger}} &= 25{,}000\ (A/P,\ 15\%,\ 5) - 7{,}350 \\ &= 25{,}000 \times 0.2983 - 7{,}350 \\ &= \$108 \end{aligned}$$

With its lower EUAC, the challenger is attractive, and hence the deluxe model should replace the existing welding robot.

If the decision based on the replacement analysis is to keep a defender whose economic life is greater than one year, the defender should be reanalyzed closer to the end of its economic life. However, if the salvage and maintenance cost data are less reliable or have changed since the last analysis, or a stronger challenger has become available in the market, the defender should be reanalyzed earlier, for example the following year during capital budgeting.

14.4.2 Unequal Lives

When the defender's remaining economic life is different from the challenger's economic life, the comparison becomes slightly more complex. We have discussed the analysis of two alternatives of unequal lives in earlier chapters in Part II. In replacement analysis, the defender and the challenger are the two alternatives. We determine a common analysis period based on the least common multiple (LCM) of their lives and follow the procedures outlined in Part II for two alternatives.

14.5 MARGINAL-COST-BASED ANALYSIS

The EUAC-based approach discussed in the previous section is not appropriate in all cases. We now discuss another approach—based on marginal cost. We begin with the question, When do we use one approach instead of the other?

As mentioned earlier, an asset involves two costs: one relating to the recovery of the capital invested in it and the other to its operation and maintenance. The capital recovery cost decreases with the age of the asset. The operating cost, on the other hand,

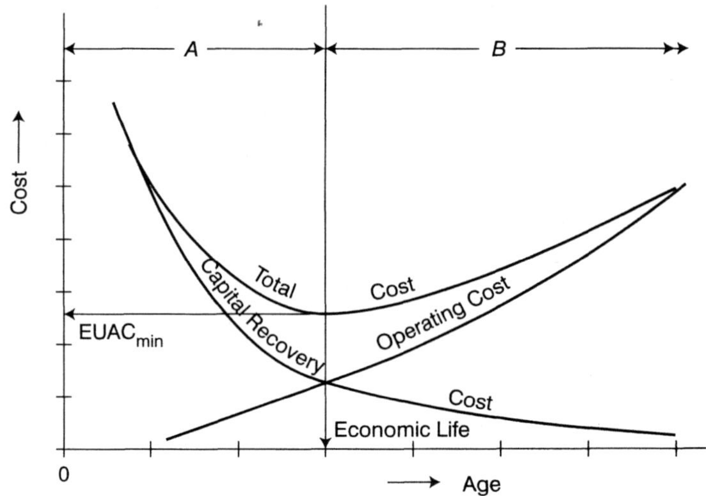

Figure 14.7 Total Cost of an Asset as It Ages

increases with age as maintenance becomes more expensive. In general, the total of the two costs initially decreases with age and then increases, as shown in Fig. 14.7. The total-cost curve usually displays a minimum, the coordinates of which are $EUAC_{min}$ and economic life. Example 14.3 and Fig. 14.5 illustrated this earlier.

The total-cost curve in Fig. 14.7 displays two distinct zones. In zone *A*, when the equipment is relatively new, the total cost decreases, primarily due to faster recovery of the capital arising from a rapid decrease in the salvage value. In zone *B*, when the equipment is relatively old, the total cost increases due to a sharper increase in the operating cost. In this zone, the salvage value remains almost constant[24], and thus the capital recovery cost is insignificant. Note that the operating cost is usually nonlinear, increasing exponentially.

The EUAC-based replacement analysis discussed in Section 14.4 is applicable when a replacement decision is being made within zone *A* and the early part of zone *B*. If the replacement analysis encompasses the latter part of zone *B*, in which the total cost continues to increase with the asset age, with no minimum, the marginal-cost-based approach is applicable.

The marginal cost of an asset is the year-by-year total cost of keeping it operational. It differs from the EUAC. It relates only to the following year, while the EUAC can relate to any time duration. The marginal cost too is the total cost (comprising capital recovery cost and operating cost), but only for the next year. Thus, the time period to which marginal cost refers is always one year.

[24]When the asset is old (latter part of zone *B*) its salvage value depends mostly on the value of material it is made of. The salvage value therefore remains almost constant, resulting in a capital recovery cost that is insignificant in comparison to the operating cost.

In many replacement analyses, the marginal cost of the defender makes better sense, since it looks at the next year's cost only. Example 14.7 illustrates how marginal costs are determined, while Example 14.8 illustrates marginal-cost-based replacement analysis.

EXAMPLE 14.7

A vision-based robotic system is being considered for acquisition at a cost of $30,000. Its annual maintenance cost for the first five years is estimated to be $1,000. Thereafter it will increase each year by $1,000. The cost of insurance and utilities for the first year is expected to be $2,000, increasing thereafter by $750 each year. For a 15% MARR, determine the asset's marginal cost over its 10-year useful life if its estimated market values are

Year	Market Value
0	$30,000 (first cost)
1	24,000
2	19,000
3	15,000
4	12,000
5	10,000
6	8,500
7	7,500
8	6,750
9	6,250
10	6,000

Solution

Let us first gather the various costs of operating the asset and post them in a table.

Year	Market Value	Maintenance Cost	Utility Costs
0	$30,000		
1	24,000	$1,000	$2,000
2	19,000	1,000	2,750
3	15,000	1,000	3,500
4	12,000	1,000	4,250
5	10,000	1,000	5,000
6	8,500	2,000	5,750
7	7,500	3,000	6,500
8	6,750	4,000	7,250
9	6,250	5,000	8,000
10	6,000	6,000	8,750

Next, we determine the marginal cost for each year as

$$\begin{aligned}\text{Marginal cost} &= \text{Total cost for the year}\\ &= \text{Capital recovery cost} + \text{Operating cost}\end{aligned}$$

For Year 1

$$\begin{aligned}\text{Capital recovery cost} &= \text{Loss in market value} + \text{Interest lost}\\ &= (30{,}000 - 24{,}000) + 15\% \text{ of } \$30{,}000\\ &= 6{,}000 + 4{,}500\\ &= \$10{,}500\end{aligned}$$

$$\begin{aligned}\text{Operating cost} &= \text{Maintenance and utility costs}\\ &= \$1{,}000 + \$2{,}000\\ &= \$3{,}000\end{aligned}$$

$$\begin{aligned}\text{Marginal cost} &= \$10{,}500 + \$3{,}000\\ &= \$13{,}500\end{aligned}$$

For Year 2

$$\begin{aligned}\text{Capital recovery cost} &= \text{Loss in market value} + \text{Interest lost}\\ &= (24{,}000 - 19{,}000) + 15\% \text{ of } \$24{,}000\\ &= 5{,}000 + 3{,}600\\ &= \$8{,}600\end{aligned}$$

$$\begin{aligned}\text{Operating cost} &= \text{Maintence and utility costs}\\ &= \$1{,}000 + \$2{,}750\\ &= \$3{,}750\end{aligned}$$

$$\begin{aligned}\text{Marginal cost} &= \$8{,}600 + \$3{,}750\\ &= \$12{,}350\end{aligned}$$

For Year 3

$$\begin{aligned}\text{Capital recovery cost} &= \text{Loss in market value} + \text{Interest lost}\\ &= (19{,}000 - 15{,}000) + 15\% \text{ of } \$19{,}000\\ &= 4{,}000 + 2{,}850\\ &= \$6{,}850\end{aligned}$$

$$\begin{aligned}\text{Operating cost} &= \text{Maintenance and utility costs}\\ &= \$1{,}000 + \$3{,}500\\ &= \$4{,}500\end{aligned}$$

$$\begin{aligned}\text{Marginal cost} &= \$6{,}850 + \$4{,}500\\ &= \$11{,}350\end{aligned}$$

The marginal cost for the other years can be determined the same way, and the results summarized as follows.

Year	Loss in Market Value	Interest Foregone	Operating Cost	Marginal Cost
1	$6,000	$4,500	$1,000 + $2,000	$13,500
2	5,000	3,600	1,000 + 2,750	12,350
3	4,000	2,850	1,000 + 3,500	11,350
4	3,000	2,250	1,000 + 4,250	10,500
5	2,000	1,800	1,000 + 5,000	9,800
6	1,500	1,500	2,000 + 5,750	10,750
7	1,000	1,275	3,000 + 6,500	11,775
8	750	1,125	4,000 + 7,250	13,125
9	500	1,013	5,000 + 8,000	14,513
10	250	938	6,000 + 8,750	15,938

In this example, the marginal cost decreases initially, reaching a minimum of $9,800 at year 5, and then increases. Therefore, the EUAC-based approach of Section 14.4 is appropriate for replacement analysis of this vision-based robotic system. However, if a defender's marginal cost increases each year, with no minimum, then the rule of replacement decision is, *Keep the defender if its marginal cost is lower than the challenger's EUAC; if not, replace it with the challenger.* Example 14.8 illustrates the procedure.

EXAMPLE 14.8

The following data pertain to a defender:

Purchase price (three years ago)	$50,000
Current market value	$10,000
Market value to decrease over next 6 years by	$750/year
Current operating cost	$7,500
Operating cost to increase by	$1,000/year

Considering the system of Example 14.7 to be its challenger, and assuming a 15% MARR, should the defender be replaced?

Solution

We first analyze the defender by determining its marginal costs following the procedure of Example 14.7. The results are summarized in the following table. With their marginal costs known, the defender and the challenger are compared later.

Year	Loss in Market Value	Interest Forgone	Operating Cost	Marginal Cost
1	$750	$0.15 \times 10{,}000 = \$1{,}500$	$7,500	$9,750
2	750	$0.15 \times 9{,}250 = 1{,}388$	8,500	10,638
3	750	$0.15 \times 8{,}500 = 1{,}275$	9,500	11,525
4	750	$0.15 \times 7{,}750 = 1{,}163$	10,500	12,413
5	750	$0.15 \times 7{,}000 = 1{,}050$	11,500	13,300
6	750	$0.15 \times 6{,}250 = 938$	12,500	14,188

Since the defender's marginal cost increases during the six-year analysis period, we compare its marginal cost, year by year, with the corresponding EUAC of the challenger to decide when to replace it.

However, what we have in the last column of the final table in Example 14.7 are the marginal costs, not the EUACs, of the challenger. From these marginal cost data, we determine the EUAC corresponding to the year and then compare it with the defender's marginal cost. The defender is replaced when its marginal cost exceeds the challenger's EUAC.

Year 1

The analysis period for determining the challenger's EUAC is one year. From Example 14.7, its marginal cost for year 1 is $13,500. Since there is no other cashflow, the EUAC equals the marginal cost. With the defender's $9,750 marginal cost for year 1 (see the table) being less than the challenger's $13,500 EUAC, the defender is not replaced.

Year 2

The analysis period for determining the challenger's EUAC is two years. The marginal costs of the challenger (Example 14.7) for years 1 and 2 are $13,500 and $12,350 respectively. These costs at periods 1 and 2 are converted (sketch a cashflow diagram, if helpful) to EUAC by first determining their P value and then converting the P value into an A value.

$$\begin{aligned}\text{EUAC}_2 &= [13{,}500(P/F, 15\%, 1) + 12{,}350(P/F, 15\%, 2)](A/P, 15\%, 2)\\ &= (13{,}500 \times 0.8696 + 12{,}350 \times 0.7561) \times 0.6151\\ &= (11{,}740 + 9{,}338) \times 0.6151\\ &= \$12{,}965\end{aligned}$$

Since the defender's $10,638 marginal cost for year 2 is less than the EUAC_2 of the challenger, the defender is not replaced.

Year 3

The marginal costs of the challenger (Example 14.7) for years 1, 2, and 3 are $13,500, $12,350, and $11,350 respectively. These costs at periods 1, 2, and 3 are converted to their EUAC by first determining their P value and then converting the P value into an A value.

$$\begin{aligned}\text{EUAC}_3 &= [13{,}500(P/F, 15\%, 1) + 12{,}350(P/F, 15\%, 2) \\ &\qquad + 11{,}350(P/F, 15\%, 3)](A/P, 15\%, 3) \\ &= (13{,}500 \times 0.8696 + 12{,}350 \times 0.7561 + 11{,}350 \times 0.6575) \times 0.4380 \\ &= (11{,}738 + 9{,}338 + 7{,}463) \times 0.4380 \\ &= \$12{,}500\end{aligned}$$

Since the defender's marginal cost for year 3 ($11,525 from the table) is less than the EUAC_3 of the challenger ($12,500), the defender is not replaced.

The calculations for the other years are carried out the same way. A table is used to summarize the results and make the replacement decision.

Year (n)	Defender's Marginal Cost	Challenger's EUAC	Decision to Replace
1	$9,750	$13,500	No
2	10,638	12,965	No
3	11,525	12,500	No
4	12,413	12,100	Yes

Since the defender's marginal cost at year 4 ($12,413) exceeds the EUAC_4 of the challenger ($12,100), the defender is replaced at year 4. Thus, the defender is kept operational for four years, at the end of which it is replaced by its challenger (of Example 14.7).

14.6 THE MAPI METHOD

To facilitate economic analyses of engineering projects, some companies develop their own manuals or monographs. For example, AT&T's manual[25] has been in use for a long time. The primary purposes of developing manuals are to streamline, simplify, and guide the decision-making process throughout the company on the methods, calculations, and other associated matters such as sources of data and their reliability. One major topic within such manuals invariably is replacement analysis.

Replacement analysis is often cumbersome, as was seen in the previous sections. Companies try to make such analyses easier for their decision makers by developing how-to procedures. An attractive feature of such step-by-step procedures[26] is the simplicity derived from a streamlined analysis with minimum mathematical efforts. This is achieved by using tables and charts rather than equations. In the case of replacement

[25] AT&T, *Engineering Economy*, third edition, McGraw-Hill, 1977.

[26] If your company practices any special procedure or uses a manual for replacement analysis, or for that matter any economic decision making, that procedure is usually preferred to other methods. However, such procedures must not violate the basic principles of engineering economics discussed throughout the text.

analysis, a popular method has been the one developed by George Terborgh for the Machinery and Allied Products Institute (MAPI), a trade organization. In this section, we discuss the MAPI method and illustrate its use through an example.

The MAPI method is fully described in the MAPI publication *Business Investment Management*. It relies on the use of worksheets or forms for organizing the given data. The forms are filled in, leading to the determination of after-tax ROR based on the incremental approach discussed in Chapter 9. The filling in of the worksheets requires referring to one of the two charts provided. Thus, the use of the MAPI method requires knowing which chart[27] to refer to and how to fill in the forms for the given data. It is this simplicity, primarily due to the absence of equations, that has popularized the MAPI method with practicing engineers. Note, however, that the charts and forms are based on the fundamental principles discussed throughout the text.

14.6.1 The Concept of Capital Consumption

According to the MAPI method, investment in and use of any resource is basically *capital consumption*. The cost incurred in procuring equipment and keeping it operational can be conceptualized as the cost of buying a service. A resource's service life starts with its installation and is 100% at the beginning. As it is used, the percentage of remaining service life decreases, becoming zero by the end of useful life. This concept is illustrated in Fig. 14.8. The horizontal axis represents the percentage of service life expired, which is zero in the beginning (when installed) and 100% at the end of useful life. The vertical axis represents the projected before-tax earnings as a percentage of the initial investment, which is 100% in the beginning and zero at the end. The straight-line variation assumes that the earnings decrease linearly as the service life expired[28] increases.

The utility of the equipment, or the service it provides, is derived from the consumption of the capital invested. The capital consumption for each period decreases in such a way that the cumulative capital consumption by the end of the useful life equals the initial investment.

In the MAPI method we determine annual capital consumption, which is the difference between unrecovered investment at the beginning of the year and the equipment value at the end of the year. The determination is based on the following assumed allocations of the before-tax earnings:

1. Pay income tax at a 50% rate on the difference between the earnings and the capital cost. The capital cost is the sum of an asset's depreciation and the interest paid on the debt portion of the investment.

---

[27]The MAPI method makes several assumptions on which the charts and forms are based. The user must be aware of these assumptions. One should develop another set of charts and forms if the assumptions differ from those in the MAPI method. The discussions in Section 14.6 and the associated problems at the end of this chapter are limited to those assumptions pertaining to MAPI.

[28]Service life expired is the opposite of remaining life. The former increases as the resource ages, while the latter decreases. In the beginning, service life expired is 0%, while the remaining life is 100%. By the end of the useful life, they respectively become 100% and 0%.

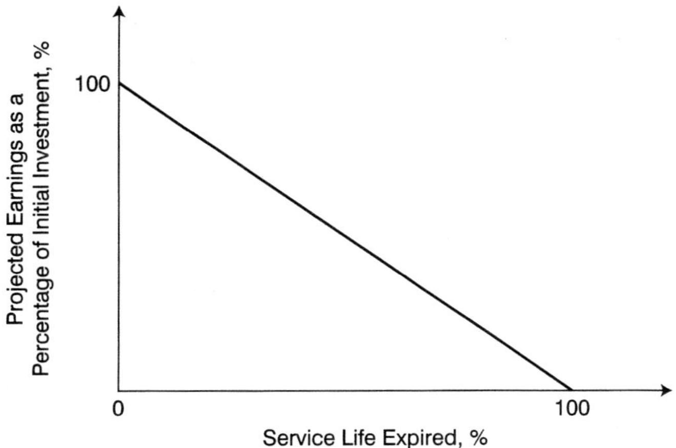

Figure 14.8 Concept of Capital Consumption

2. Pay 3% interest on one fourth of the unrecovered investment, which is assumed borrowed.
3. Provide a 10% after-tax return on the three fourths of the unrecovered investment that is assumed equity capital (raised within the company).

Based on these[29] allocations, the MAPI method relates before-tax earnings with the initial investment. For a given investment and other relevant data, a set of before-tax earnings for each year of use yield capital consumptions in such a way that the cumulative capital consumption equals the investment by the end of the service life.

For the purpose of illustration, let us consider an investment of $1,000, five-year useful life, straight-line depreciation, and zero salvage value. The computations for the yearly capital consumptions are summarized in Table 14.1, with the following explanation about the entries.

Explanation

1. The before-tax earnings (row 1), for example $475 for year 1, were calculated based on MAPI assumptions. They render the sum[30] of capital consumptions for the five years of service (row 9) equal to the $1,000 initial investment (note that the salvage value is zero). The data in row 1 are based on the linear relationship in Fig. 14.8, as illustrated in Fig. 14.9. For example, the before-tax earning from Fig. 14.9 for year 3 is 60% of that for year 1, that is, $0.60 \times \$475 = \$285$.
2. In row 2, the straight-line depreciation for zero salvage value and a five-year useful life is obtained from $(P - S)/N = (\$1{,}000 - 0)/5 = \$200$.
3. Row 3 data are obtained by charging 3% interest on 25% of the unrecovered investment. For example, for year 1, the unrecovered investment in row 7 is $1,000. The 3% interest on 25% of this sum $= 0.03 \times (0.25 \times \$1{,}000) = 0.03 \times \$250 \approx \$8$.

[29]For allocations other than these, another set of charts and forms can, and should, be developed and used.

[30]$259 + 231 + 202 + 171 + 137 = 1{,}000$

Table 14.1 Computation of Yearly Capital Consumptions (Initial investment = $1,000, 5-year service life, straight-line depreciation, no salvage value)

	Year 1	2	3	4	5
1. Before-tax earnings	$475	$380	$285	$190	$95
2. Straight-line depreciation	200	200	200	200	200
3. Interest payment	8	6	4	2	1
4. Taxable earnings	267	174	81	−12	−106
5. Income tax @50%	134	87	41	−6	−53
6. After-tax earnings	341	293	244	196	148
7. Unrecovered investment	1,000	741	510	308	137
8. Equity capital return @10%	75	56	38	23	10
9. Capital consumption	259	231	202	171	137

The computation of interest for the other years is slightly complex, since it requires the value of unrecovered investment for the year. Let us consider year 3. The unrecovered investment for this year is equal to the initial investment of $1,000 minus the sum of capital consumptions for years 1 and 2. From row 9, these consumptions are $259 and $231. Thus the unrecovered investment at the end of year 2 is $1,000 − ($259 + $231) = $510. This $510 becomes the unrecovered investment for year 3, 25% of which equals $127.50. Thus, the interest for year 3 = 3% of $127.50 ≈ $4.

4. The data in the fourth row (taxable earnings) are simply before-tax earnings minus the sum of interest and the depreciation charge. For example, for the second year, taxable earnings are $380 − ($200 + $6) = $174. Note that the taxable earnings for years 4 and 5 are negative.
5. Row 5 contains income tax obligations, which are 50% of the taxable earnings. For example, for the third year, tax owed is 50% of $81 ≈ $41.
6. The data in the sixth row are the differences between before-tax earnings and the corresponding income tax (row 1 minus row 5). For year 2, for example, the difference is $380 − $87 = $293. For year 4, it is $190 − (−$6) = $196.
7. The unrecovered investment for the first year is simply the initial investment. For other years, it is the value of the investment remaining after cumulative capital consumption up to the previous year. For example, for year 2, it is $1,000 minus the capital consumption for year 1 which (in row 9) is $259. Thus, the entry in row 7 under year 2 = $1,000 − $259 = $741. For year 4, the unrecovered investment[31] likewise is $1,000 minus the cumulative capital consumption up to year 3 = $1,000 − ($259 + $231 + $202) = $308.

---

[31]The unrecovered investment at the beginning of year 4 can be evaluated alternatively, in fact easily, as being

Unrecovered investment at the beginning of year 3 − Capital consumption for year 3
= $510 − $202
= $308

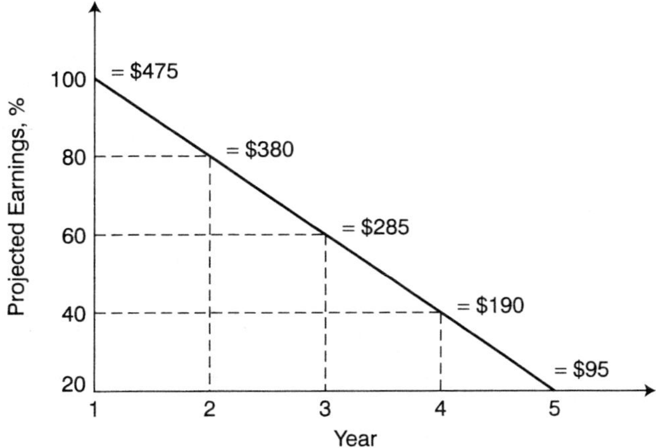

Figure 14.9 Consumption Rate of $1,000 of Capital

8. The data in the eighth row represent a 10% return on the equity capital, which in the MAPI method is assumed to be 75% of the unrecovered investment. For the third year, for example, the equity capital, being 75% of the unrecovered investment of \$510, is \$382.50. Thus, the return for the third year is 10% of \$382.50 ≈ \$38.
9. The data in the last row are the amounts of capital consumed each year. For any year, these are equal to after-tax earnings (row 6) minus the interest paid (row 3) minus the return on equity capital (row 8). Thus, for year 4, it is \$196 − \$2 − \$23 = \$171. Note that the total of capital consumptions (sum of the data in row 9) equals the initial investment of \$1,000, a basic premise of the MAPI method. Such an equality check ascertains the accuracy of the computations.

14.6.2 Retention Value

During the course of determining the year-by-year capital consumption, as illustrated earlier, equipment's retention values get computed. The retention value is simply the value of the asset at the year end. Thus,

$$\text{Retention value at year end} = \text{Retention value at the beginning of the year} \\ - \text{Capital consumption for the year}$$

Let us reconsider the illustration of the previous subsection, in which the initial cost is \$1,000. For the first year, capital consumption is \$259; hence, the retention value[32] at the end of the first year = \$1,000 − \$259 = \$741. The retention value is sometimes expressed as a percentage of the initial cost. For year 1, it is \$741/\$1,000 = 0.741 = 74.1% For year 2, the capital consumption is \$231. Thus,

[32]Same as unrecovered investment for the following year (row 7 under year 2) in Table 14.1. Thus, the retention value at the end of any year is the unrecovered investment for the following year.

Retention value at the end of year 2
= Retention value at the *beginning of year* 2
 − Capital consumption for year 2
= Retention value at the *end of year* 1
 − Capital consumption for year 2
= \$741 − \$231
= \$510
= 51% of the \$1,000 initial cost

The retention values for the other years[33] are determined the same way, yielding[34]

Year	Retention Value, %
1	74.1
2	51.0
3	30.8
4	13.7
5	0

The retention values are plotted in a chart, as in Fig. 14.10, where the y-axis denotes the percentage and the x-axis the service life. Such a chart does away with the need to compute retention values for a given problem; the value is simply read off the chart. The curves in Fig. 14.10 correspond to a set of salvage ratios. The salvage ratio is the ratio of an asset's salvage value to its initial cost, and is expressed as a percentage. For the illustrative example under discussion, the salvage value is zero, yielding a salvage ratio of zero. For a service life of 5 years, the retention value can be read off the zero salvage-ratio curve as 74%, approximately the same as computed earlier. The instruction for using the chart is given in the figure caption. Note that this chart is valid for a *one-year comparison period* under straight-line depreciation.

For comparison periods longer than one year, another chart is used. Such a chart is shown in Fig. 14.11, where the x-axis represents the comparison period as a percentage of service life. Again, the chart includes instructions for use. For a 5-year life, the comparison period for the first year is obviously 20% of the service life. For the other years, it is 40% for year 2, 60% for year 3, 80% for year 4, and 100% for year 5. The retention-value percentages computed in the illustrative example earlier are easily read off the zero-salvage-ratio curve of Fig. 14.11, corresponding to the comparison period percentages on the x-axis. For example, for year 1 (20% on the x-axis the retention value on the zero-salvage-ratio curve is 74.1%, while for year 3 (60% on the x-axis) it is 30.8%. Note that while Fig. 14.11 can be used for one-year comparison periods, Fig. 14.10 gives higher resolution in reading off the retention values, and hence is more accurate for such periods.

---

[33]Note that the percentage values relate to the data in row 7 of Table 14.1, being equal to the data for the year divided by \$1,000—the investment.

[34]The retention value decreases to zero at the end of useful life, as can be noted in the table.

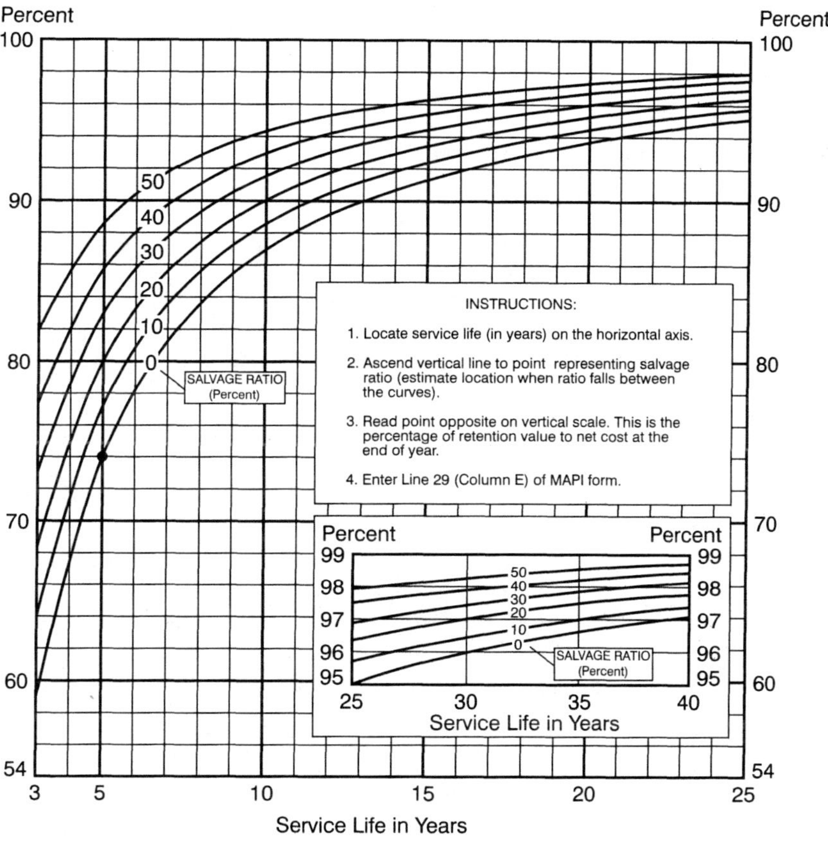

Figure 14.10 Retention Values—Service Life (1-year comparison)

14.6.2 Summary Forms

Along with the charts of retention values, a worksheet called the *MAPI Summary Form*[35] is used in replacement analysis by the MAPI method. Such a form is shown in Fig. 14.12. The summary form contains forty items grouped in two parts. Part I, comprising items 1 through 25, enables the computation of challenger's operating advantage, while Part II, comprising items 26 through 40, details the investments and returns. Each part has three sections: A, B, and C. The before-tax and after-tax rates of return get calculated at items 34 and 40. A step-by-step procedure of filling in the summary form leads to the evaluation of a project's after-tax ROR, the objective of the MAPI method.

---

[35]George Terborgh, *Business Investment Management*, 1967. Reprint by permission of the Manufacturers Alliance/MAPI (formerly the Machinery and Allied Products Institute), 1525 Wilson Boulevard, Arlington, Virginia 22209.

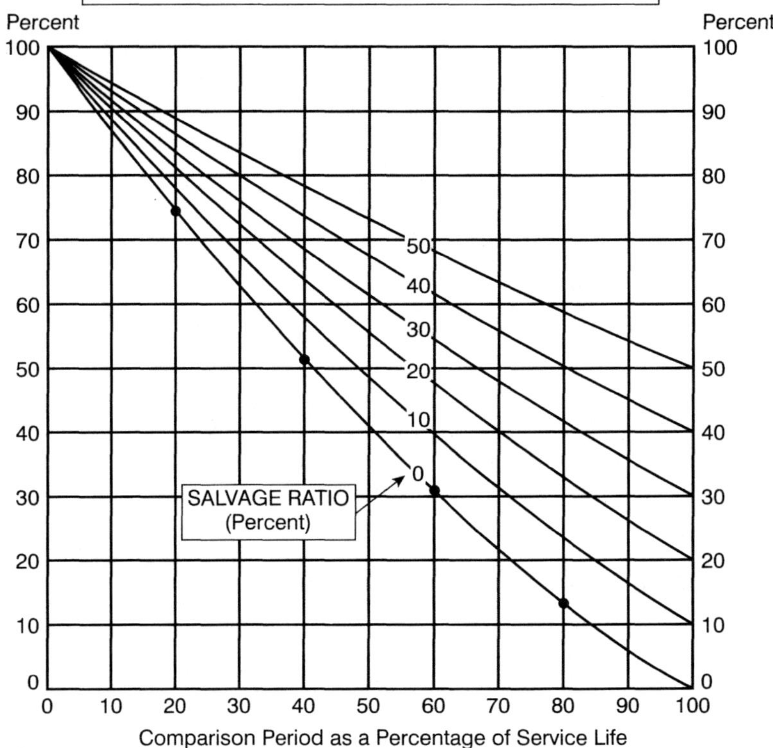

Figure 14.11 Retention Values—Comparison Periods (%)

PROJECT NO.______________ SHEET 1

MAPI SUMMARY FORM
(AVERAGING SHORTCUT)

PROJECT______________

ALTERNATIVE______________

COMPARISON PERIOD (YEARS) (P)______________

ASSUMED OPERATING RATE OF PROJECT (HOURS PER YEAR) ______________

I. OPERATING ADVANTAGE
(NEXT-YEAR FOR A 1-YEAR COMPARISON PERIOD,* ANNUAL AVERAGES FOR LONGER PERIODS)

A. EFFECT OF PROJECT ON REVENUE

		INCREASE	DECREASE	
1	FROM CHANGE IN QUALITY OF PRODUCTS	$	$	1
2	FROM CHANGE IN VOLUME OF OUTPUT			2
3	TOTAL	$ X	$ Y	3

B. EFFECT ON OPERATING COSTS

		INCREASE	DECREASE	
4	DIRECT LABOR	$	$	4
5	INDIRECT LABOR			5
6	FRINGE BENEFITS			6
7	MAINTENANCE			7
8	TOOLING			8
9	MATERIALS AND SUPPLIES			9
10	INSPECTION			10
11	ASSEMBLY			11
12	SCRAP AND REWORK			12
13	DOWN TIME			13
14	POWER			14
15	FLOOR SPACE			15
16	PROPERTY TAXES AND INSURANCE			16
17	SUBCONTRACTING			17
18	INVENTORY			18
19	SAFETY			19
20	FLEXIBILITY			20
21	OTHER			21
22	TOTAL	$ Y	$ X	22

C. COMBINED EFFECT

23	NET INCREASE IN REVENUE (3X – 3Y)	$	23
24	NET DECREASE IN OPERATING COSTS (22X – 22Y)	$	24
25	ANNUAL OPERATING ADVANTAGE (23 + 24)	$	25

* Next year means the first year of project operation. For projects with a significant break-in period, use performance after break-in.

Figure 14.12 MAPI Summary Form (continued)

SHEET 2

II. INVESTMENT AND RETURN

A. INITIAL INVESTMENT

26 INSTALLED COST OF PROJECT $______
MINUS INITIAL TAX BENEFIT OF $______ (Net Cost) $______ 26
27 INVESTMENT IN ALTERNATIVE
CAPITAL ADDITIONS MINUS INITIAL TAX BENEFIT $______
PLUS: DISPOSAL VALUE OF ASSETS RETIRED
BY PROJECT* $______ $______ 27
28 INITIAL NET INVESTMENT (26 – 27) $______ 28

B. TERMINAL INVESTMENT

29 RETENTION VALUE OF PROJECT AT END OF COMPARISON PERIOD
(ESTIMATE FOR ASSETS, IF ANY, THAT CANNOT BE DEPRECIATED OR EXPENSED; FOR OTHERS, ESTIMATE OR USE MAPI CHARTS.)

Item or Group	Installed Cost, Minus Initial Tax Benefit (Net Cost) A	Service Life (Years) B	Disposal Value, End of Life (Percent of Net Cost) C	MAPI Chart Number D	Chart Percentage E	Retention Value $\left(\frac{A \times E}{100}\right)$ F
	$					$

ESTIMATED FROM CHARTS (TOTAL OF COLUMN F) $______
PLUS: OTHERWISE ESTIMATED $______ $______ 29
30 DISPOSAL VALUE OF ALTERNATIVE AT END OF PERIOD* $______ 30
31 TERMINAL NET INVESTMENT (29 – 30) $______ 31

C. RETURN

32 AVERAGE NET CAPITAL CONSUMPTION $\left(\frac{28 - 31}{P}\right)$ $______ 32

33 AVERAGE NET INVESTMENT $\left(\frac{28 + 31}{2}\right)$ $______ 33

34 BEFORE-TAX RETURN $\left(\frac{25 - 32}{33} \times 100\right)$ %______ 34

35 INCREASE IN DEPRECIATION AND INTEREST DEDUCTIONS $______ 35
36 TAXABLE OPERATING ADVANTAGE (25 – 35) $______ 36
37 INCREASE IN INCOME TAX (36 × TAX RATE) $______ 37
38 AFTER-TAX OPERATING ADVANTAGE (25 – 37) $______ 38
39 AVAILABLE FOR RETURN ON INVESTMENT (38 – 32) $______ 39

40 AFTER-TAX RETURN $\left(\frac{39}{33} \times 100\right)$ %______ 40

* After terminal tax adjustments.

Figure 14.12 MAPI Summary Form (continued)

An explanation of the summary form (Fig. 14.12) may be helpful. The analysis begins by filling in the relevant information, such as project title and comparison period[36], at the top of the form.

Under Part I: OPERATING ADVANTAGE, we enter in section A the financial impact of the project on revenues; for example item 1 allows for such an impact due to change in product quality, which may increase or decrease the revenue. The effects on operating costs are considered in section B (items 4 through 22). In the non-MAPI methods, several of these costs might get left out unless the analysis is comprehensive and the analyst is alert. The existence of a list, as here, prompts the analyst to consider all the possible costs and revenues. Section C considers the combined effect of all the revenues and costs, summarizing the investment's annual operating advantage at item 25.

Under Part II: INVESTMENT AND RETURN, the initial net investment is worked out in section A, the terminal investment in B. The retention value percentage from one of the charts (Fig. 14.10 or Fig. 14.11) is entered in section B as item 29E. In section C, the before-tax return is worked out at item 34, the after-tax return at item 40. Note that the variable P in item 32 is the comparison period used, whose value is one in the case of one-year comparative analysis.

Using the charts and the summary form, a step-by-step procedure of analyzing projects by the MAPI method is illustrated in Example 14.9.

EXAMPLE 14.9

A pick-and-place robot whose economic life is one year is being considered for replacement by its deluxe model. The operating advantages of the deluxe model have been posted in Part I of the MAPI summary form as in Fig. 14.13. The deluxe model costs $40,000 and qualifies for a 25% tax credit.
Assume straight-line depreciation and

10 years as the estimated service life,
10% of the net cost as the salvage value,
$1,000 as the next-year increase in depreciation and interest deductions,
$5,000 as the defender's disposal value, and
$4,000 as the challenger's salvage value.

Based on the MAPI method, should the defender be replaced if the MARR is 10% and the tax rate is 50%?

Solution

MAPI analysis involves a careful filling in of the Summary Form with the given data, as presented in Fig. 14.14, along with the associated simple computations. The procedure for Part I comprises:

---

[36]The comparison period is usually a year, for which Fig. 14.10 is applicable. For longer comparison periods, Fig. 14.11 is used.

PROJECT NO.________ SHEET 1

MAPI SUMMARY FORM
(AVERAGING SHORTCUT)

PROJECT________

ALTERNATIVE________

COMPARISON PERIOD (YEARS) (P) 1 year

ASSUMED OPERATING RATE OF PROJECT (HOURS PER YEAR) ________

I. OPERATING ADVANTAGE
(NEXT-YEAR FOR A 1-YEAR COMPARISON PERIOD,* ANNUAL AVERAGES FOR LONGER PERIODS)

A. EFFECT OF PROJECT ON REVENUE

		INCREASE	DECREASE	
1	FROM CHANGE IN QUALITY OF PRODUCTS	$ 6000	$	1
2	FROM CHANGE IN VOLUME OF OUTPUT	1500		2
3	TOTAL	$ X	$ Y	3

B. EFFECT ON OPERATING COSTS

		INCREASE	DECREASE	
4	DIRECT LABOR	$	$ 200	4
5	INDIRECT LABOR	300		5
6	FRINGE BENEFITS	50		6
7	MAINTENANCE		500	7
8	TOOLING		200	8
9	MATERIALS AND SUPPLIES		1400	9
10	INSPECTION		100	10
11	ASSEMBLY			11
12	SCRAP AND REWORK		500	12
13	DOWN TIME			13
14	POWER		50	14
15	FLOOR SPACE			15
16	PROPERTY TAXES AND INSURANCE	200		16
17	SUBCONTRACTING			17
18	INVENTORY			18
19	SAFETY		100	19
20	FLEXIBILITY		100	20
21	OTHER			21
22	TOTAL	$ Y	$ X	22

C. COMBINED EFFECT

23	NET INCREASE IN REVENUE (3X – 3Y)	$	23
24	NET DECREASE IN OPERATING COSTS (22X – 22Y)	$	24
25	ANNUAL OPERATING ADVANTAGE (23 + 24)	$	25

* Next year means the first year of project operation. For projects with a significant break-in period, use performance after break-in.

Figure 14.13 Data for Example 14.9

PROJECT NO.________ SHEET 1

MAPI SUMMARY FORM
(AVERAGING SHORTCUT)

PROJECT *Pick-and-Place Robot-Deluxe Model*

ALTERNATIVE *Continue with the Current Model*

COMPARISON PERIOD (YEARS) (P) *1 year*

ASSUMED OPERATING RATE OF PROJECT (HOURS PER YEAR) ________

I. OPERATING ADVANTAGE

(NEXT-YEAR FOR A 1-YEAR COMPARISON PERIOD,* ANNUAL AVERAGES FOR LONGER PERIODS)

A. EFFECT OF PROJECT ON REVENUE

		INCREASE	DECREASE	
1	FROM CHANGE IN QUALITY OF PRODUCTS	$ 6,000	$	1
2	FROM CHANGE IN VOLUME OF OUTPUT	1,500		2
3	TOTAL	$ 7,500 X	$ Y	3

B. EFFECT ON OPERATING COSTS

		INCREASE	DECREASE	
4	DIRECT LABOR	$	$ 200	4
5	INDIRECT LABOR	300		5
6	FRINGE BENEFITS	50		6
7	MAINTENANCE		500	7
8	TOOLING		200	8
9	MATERIALS AND SUPPLIES		1,400	9
10	INSPECTION		100	10
11	ASSEMBLY			11
12	SCRAP AND REWORK		500	12
13	DOWN TIME			13
14	POWER		50	14
15	FLOOR SPACE			15
16	PROPERTY TAXES AND INSURANCE	200		16
17	SUBCONTRACTING			17
18	INVENTORY			18
19	SAFETY		100	19
20	FLEXIBILITY		100	20
21	OTHER			21
22	TOTAL	$ 550 Y	$ 3,150 X	22

C. COMBINED EFFECT

23	NET INCREASE IN REVENUE (3X – 3Y)	$ 7,500	23
24	NET DECREASE IN OPERATING COSTS (22X – 22Y)	$ 2,600	24
25	ANNUAL OPERATING ADVANTAGE (23 + 24)	$ 10,100	25

* Next year means the first year of project operation. For projects with a significant break-in period, use performance after break-in.

Figure 14.14 Solution of Example 14.9 (continued)

II. INVESTMENT AND RETURN

A. INITIAL INVESTMENT

26 INSTALLED COST OF PROJECT $ 40,000
MINUS INITIAL TAX BENEFIT OF 25% credit $ 10,000 (Net Cost) $ 30,000 26

27 INVESTMENT IN ALTERNATIVE
CAPITAL ADDITIONS MINUS INITIAL TAX BENEFIT $ 0
PLUS: DISPOSAL VALUE OF ASSETS RETIRED
BY PROJECT* $ 5,000 $ 5,000 27

28 INITIAL NET INVESTMENT (26 – 27) $ 25,000 28

B. TERMINAL INVESTMENT

29 RETENTION VALUE OF PROJECT AT END OF COMPARISON PERIOD
(ESTIMATE FOR ASSETS, IF ANY, THAT CANNOT BE DEPRECIATED OR EXPENSED; FOR OTHERS, ESTIMATE OR USE MAPI CHARTS.)

Item or Group	Installed Cost, Minus Initial Tax Benefit (Net Cost) A	Service Life (Years) B	Disposal Value, End of Life (Percent of Net Cost) C	MAPI Chart Number D	Chart Percentage E	Retention Value $\left(\frac{A \times E}{100}\right)$ F
Deluxe model robot	$ 30,000	10	10%	3A (Fig. 14.10)	88.8	$ 26,640

ESTIMATED FROM CHARTS (TOTAL OF COLUMN F) $ 26,640
PLUS: OTHERWISE ESTIMATED $ – $ 26,640 29

30 DISPOSAL VALUE OF ALTERNATIVE AT END OF PERIOD* $ 4,000 30

31 TERMINAL NET INVESTMENT (29 – 30) $ 22,640 31

C. RETURN

32 AVERAGE NET CAPITAL CONSUMPTION $\left(\frac{28 - 31}{P}\right)$ $ 2,360 32

33 AVERAGE NET INVESTMENT $\left(\frac{28 + 31}{2}\right)$ $ 23,820 33

34 BEFORE-TAX RETURN $\left(\frac{25 - 32}{33} \times 100\right)$ % 32.5 34

35 INCREASE IN DEPRECIATION AND INTEREST DEDUCTIONS $ 1,000 35

36 TAXABLE OPERATING ADVANTAGE (25 – 35) $ 9,100 36

37 INCREASE IN INCOME TAX (36 × TAX RATE) 50% $ 4,550 37

38 AFTER-TAX OPERATING ADVANTAGE (25 – 37) $ 5,550 38

39 AVAILABLE FOR RETURN ON INVESTMENT (38 – 32) $ 3,190 39

40 AFTER-TAX RETURN $\left(\frac{39}{33} \times 100\right)$ % 13.4% 40

* After terminal tax adjustments.

Figure 14.14 Solution of Example 14.9 (continued)

Fill in the top portion of the Summary Form.
Do the total of items 1 and 2 at item 3.
Do the total of items 4 through 21 at item 22.
Compute the values at items 23, 24, and then 25.

For the given data, the annual operating advantage, as computed at item 25, is \$10,100.

Next, Part II of the form is filled in with the appropriate data. The \$40,000 installed cost and the 25% tax credit yield \$30,000 as the net cost in Section A. With \$5,000 as the defender's disposal value, the initial net investment is found at item 28 to be \$25,000. In Section B, the 88.8% data at item 29E has been obtained from Fig. 14.10 corresponding to a 10-year service life and 10% salvage ratio (\$4,000/\$40,000 = 0.1 = 10%). The terminal net investment as computed at item 31 is \$22,640. The value of P at item 32 is 1, since it is a 1-year comparison. Finally, the after-tax return as computed at item 40 is 13.4%. Since it is higher than the 10% MARR, the deluxe model should replace the defender.

This example has illustrated the simplicity and comprehensiveness of the MAPI method for replacement analysis.

14.7 RETIREMENT

At times, engineers and technologists make the decision to retire a resource. Retirement does not involve replacement. It is simply a decision of whether to continue operating the resource or not, as illustrated in Example 14.10.

EXAMPLE 14.10

A private company is considering whether to retire an irrigation canal it built 50 years ago at a cost of \$2 million. The original estimated life of the canal was 75 years. The construction of an interstate highway has reduced the water discharge in the canal. Revenue collection from irrigation fees is estimated to be \$120,000 per year. The cost to maintain the canal is \$80,000 per year, which is likely to increase each year by \$5,000. Assuming zero salvage value for the canal, should it be retired (abandoned!) if the applicable annual compound interest rate is 6%?

Solution

The initial construction cost of \$2 million, being sunk cost, is irrelevant. Since a decision criterion is not specified, we can choose to use any of the six discussed in the text. Let us use the PW criterion, under which we abandon the canal if its PW is negative.

Noting that the canal's remaining life is 25 years, $n = 25$. As given, $i = 6\%$. For the given cashflow data (sketch a diagram if necessary),

$$\begin{aligned} \text{PW} &= \text{PW}_{\text{benefits}} - \text{PW}_{\text{costs}} \\ &= 120{,}000(P/A, 6\%, 25) - [80{,}000(P/A, 6\%, 25) + 5{,}000(P/G, 6\%, 25)] \end{aligned}$$

$$= (120{,}000 - 80{,}000)(P/A, 6\%, 25) - 5{,}000(P/G, 6\%, 25)$$
$$= 40{,}000 \times 12.783 - 5{,}000 \times 115.9732$$
$$= 511{,}320 - 579{,}866$$
$$= -\$68{,}546$$

Since its present worth is negative, the canal should be retired.

SUMMARY

Replacement analysis is a specialized engineering decision-making process in which existing equipment, called the defender, is analyzed for possible replacement by its challenger. It is based on the basic principles of engineering economics discussed in the earlier chapters. The analyst should ensure that sunk costs are not included in the analysis. The defender is analyzed for its economic worth—usually its market value—and remaining economic life. Likewise, the challenger is analyzed for its economic life and costs and benefits. An asset's economic life corresponds to the time period at which its equivalent uniform annual cost (EUAC) is minimum. In the EUAC-based replacement analysis, the defender and the challenger are compared for their EUACs and the one with lower EUAC is preferred. Problems involving unequal economic lives of the defender and challenger are relatively more complex. For assets whose total cost increases each year, with no minimum, replacement analysis is based on marginal cost, which is the year-by-year total cost of keeping the asset operational. In marginal-cost-based analyses too, the defender and the challenger are compared, and the defender is replaced if its marginal cost exceeds the challenger's EUAC.

Companies sometimes use their own "home-grown" procedure to simplify replacement analysis. Company-specific procedures, such as the MAPI method, are based on specific assumptions and the associated charts and forms, rather than on equations. Based on the concept of capital consumption, the MAPI method is relatively simple and yet comprehensive. Retirement analysis, in which the asset is retired or abandoned, rather than replaced, is a special type of replacement analysis, and is usually simpler.

EXERCISES

Discussion Questions

14.1 Why is market value the best indicator of a defender's economic worth? Illustrate through an example from everyday life, such as selling an old car.

14.2 How useful is the concept of defender and challenger in a replacement analysis? Does it make the analysis any simpler?

14.3 Why has replacement analysis been given prominence in the text by devoting a complete chapter on it?

14.4 List the major assumptions of the MAPI method.

14.5 Compare the MAPI method with the EUAC-based and marginal-cost-based replacement methods.

14.6 Under what circumstance will you opt for marginal-cost-based replacement analysis?

14.7 Why are sunk costs considered irrelevant in replacement analysis?

Multiple-Choice Questions (Circle the *best* answer.)

14.8 A defender's economic worth is usually its
   a. book value.
   b. sunk cost.
   c. market value.
   d. trade-in value.

14.9 In a replacement analysis, the existing equipment is the
   a. challenger.
   b. defender.
   c. either a or b, depending on whether its remaining life is longer than the useful life of the new one.
   d. either a or b, depending on whether its cost is higher than that of the new one.

14.10 Trade-in value is usually ________ the market value.
   a. equal to
   b. smaller than
   c. greater than
   d. half of

14.11 In a replacement analysis, the defender is considered an investment worth its
   a. original cost minus the cumulative depreciation.
   b. market value.
   c. trade-in value.
   d. book value.

14.12 In general, there are two choices in a replacement analysis. One is to replace the defender now, and the other is to keep it
   a. for its remaining useful life.
   b. for another year.
   c. until a better challenger is available.
   d. none of the above

14.13 Economic life corresponds to
   a. minimum EUAC.
   b. maximum EUAC.
   c. minimum PW.
   d. maximum PW.

14.14 Depreciation affects the
   a. first cost.
   b. salvage value.
   c. market value.
   d. book value.

Numerical Problems

14.15 You are considering the replacement of a 20-year-old stamping machine, for which you visit the local dealer, where a salesperson says:

The new machine you are interested in has been very popular. Unfortunately, its list price has gone up to $29,500 only last month. But I think I would be able to convince the manager, who graduated from the same college as you, to sell it to you at the old price of $28,000, thus saving you $1,500. For your old machine our trade-in offer is $8,000, which is $2,000 better than its market value you have checked to be $6,000. Thus, the new machine will cost you only $20,000. If you don't buy from us, you will be paying $23,500 ($29,500 − $6,000). Thus, buying from us saves you $3,500—a real good bargain.

From the viewpoint of replacement analysis,

a. What is the defender's economic worth?

b. What is the challenger's first cost?

14.16 Best Bread purchased a personal computer system at $20,000 three years ago. At that time the system was estimated to have a useful life of 5 years and a salvage value of $10,000. The company uses MACRS for depreciation. It invested another $5,000 this year in software upgrades. The current operating cost is $3,000 per year. The anticipated salvage value at the end of the computer's useful life is now reduced to $4,000. The system is being analyzed for possible replacement, which has determined that it can be sold for $9,000. What value is relevant in the analysis? For an income tax rate of 50%, what is the net cashflow from the disposal of the computer system?

14.17 Since Mira cannot afford to replace her old car, worth $500, now, she decides to keep it for one more year. Based on the car's repair history, she estimates its repair cost for the next year to be $650. She will not be able to sell the car next year for more than $200. How much will it cost her to keep the car operational for one more year if $i = 10\%$ per year?

14.18 A forklift truck is being analyzed as a challenger. It costs $30,000 to buy. Its salvage value and maintenance costs are as follows:

Year	Salvage Value	Maintenance Cost
1	$25,000	$250
2	20,000	750
3	15,000	1,500
4	10,000	2,500
5	5,000	4,000

Assuming a MARR of 8%, what is the truck's economic life?

14.19 An old lathe with market value of $5,000 is undergoing replacement analysis. Determine its economic life if $i = 8\%$ per year and its estimated salvage values and maintenance costs are as follows.

Remaining Life n, Years	Salvage Value, S at the End of Year n	Maintenance Cost
0	$5,000	
1	4,000	$100
2	3,500	200
3	3,000	350
4	2,500	550
5	2,000	800
6	1,500	1,100
7	1,500	1,450
8	1,500	1,900

14.20 A consulting firm maintains a small aircraft for its use. The aircraft was bought 15 years ago for $300,000. The accounting department feels that it is costing too much to keep this aircraft operational. The firm is considering replacing it with an identical aircraft. The maintenance and cost estimating departments have gathered the following data.

Year	Maintenance Cost	Salvage Value
1	$5,000	$30,000
2	7,000	20,000
3	9,000	12,000
4	11,000	5,000
5	13,000	0

Besides the maintenance there are other costs as well, which together total $15,000 per year. Assuming that the aircraft has already been fully depreciated and its market value is $60,000, what is its economic life if MARR is 20%?

14.21 A defender with a market value of $1,000 is being analyzed for replacement by a challenger whose first cost is $1,500. Both have an economic life of 3 years and no salvage value. During the three years the defender's operating costs are estimated to be $350, $400, and $500. These data for the challenger are $100, $200, and $300 respectively. For a MARR of 15%, should the defender be replaced?

14.22 In Problem 14.21, assume that a similar challenger is available from the distributor who offers a $1,250 trade-in for the defender. The challenger, however, costs $1,800. Should the distributor's offer be accepted?

14.23 A CNC lathe purchased five years ago for $120,000 is being considered for replacement by a turning cell. The market value of the CNC lathe is $25,000, but its value will decrease during the next five years as per the following table, which also shows the operating costs. If the MARR is 15%, when should the CNC be replaced?

Year	Salvage Value	Operating Cost
1	$23,000	$3,500
2	20,000	4,000
3	16,000	4,750
4	9,000	5,750
5	0	7,000

14.24 The replacement analysis of a concrete mixer yielded the following results:

Year (n)	Defender's EUAC if Kept n Years	Challenger's EUAC if Kept n Years
1	$3,675	$6,500
2	4,500	5,875
3	5,230	5,125
4	5,985	4,580
5	6,450	6,150

What are the economic lives of the defender and the challenger? When should the defender be replaced?

14.25 A stamping machine was installed five years ago at a cost of $47,500. Its annual operating cost was then estimated to be $2,500, increasing each year by $1,000. Its market value was estimated to decrease each year by 10%. The current projection is that the machine can be used for the next five years. However, the supplier of the machine offers an improved version for $60,000. This cost includes free maintenance for the next five years; its EUAC will be minimum at the fifth year. Ignoring other operating costs, and assuming a MARR of 12%, when should the defender be replaced?

14.26 A company is considering replacing its fork lift truck by an automated guided vehicle (AGV) for its material handling needs. The truck's economic life is one year, and it can be sold this year for $10,000.

The AGV costs $120,000 and qualifies for a 30% investment tax credit under the government's Manufacturing Modernization Mandate. It is estimated to be in service for 10 years, and its end-of-life disposal value is estimated to be $10,000. The increase in next year's depreciation and interest charges is likely to be $5,000. The AGV will be depreciated by the straight-line method. It offers the following annual operating advantages and disadvantages:

Direct labor down by	$25,000
Maintenance down by	$300
Benefit from improved flexibility	$250
Benefit from reduced downtime	$150

What is the after-tax ROR from investment in the AGV as per the MAPI method, if the company's tax obligations are 50%?

14.27 A U.S. company built a chemical plant in Ethiopia 25 years ago at a cost of $4 million. The original estimated life of the plant was 75 years. The net profit from the operations is estimated to be $250,000 per year. The annual operating cost is $125,000, which is likely to increase annually by $10,000. If the plant has zero salvage value, should it be abandoned? Assume an annually compounded interest rate of 10% per year.

CHAPTER FIFTEEN

Other Considerations

IN THIS CHAPTER YOU WILL LEARN ABOUT

- How to account for inflation
- Analysis of government projects
- Rationing of capital
- Selection of a MARR
- Uncertainties in the data
- The computer as a tool
- An automation project case study

Throughout this text the basic principles of engineering economics have been emphasized. They have been illustrated through their applications in decision making pertaining to investment in engineering resources. What you have learned so far is sufficient to analyze a large variety of industrial projects. Some projects, however, may require considerations beyond what we have covered so far. Some of these considerations are of an advanced nature; their comprehensive discussions are beyond the scope of this text. However, the important adjunct considerations are briefly presented in this last chapter to apprise you of their ramifications for economic analyses. Engineering economists and managers need to be aware of their possible effects on the decision, especially if the best alternative is only marginally better than the second best.

15.1 INFLATION

Inflation is a fact of life. It denotes a general increase in the prices of goods and services. Governments throughout the world try to keep inflation under control. Well-managed economies under stable governments, for example that of the United States, usually succeed in doing so. Inflation in the United States in the 1990s has been in the 2 to 3% range. In many countries, inflation is in double digits (10% and above). At times, it can get out of control, as in Europe during the second World War, in the United States during the 1970s oil embargo, and in Russia in the 1990s following the demise of the Soviet Union. The term *hyperinflation* is used to describe extremely high inflation, usually out of control.

Prices may at times decrease rather than increase, resulting in *deflation,* as happened following the second World War. Deflation results when the supply of goods or services is greater than the demand. A deflationary economy may occur when the industrial infrastructure undergoes severe perturbation, usually due to political instability or collapse. In stable times, inflation is the norm.

15.1.1 Inflation Rate

Inflation is expressed as the *rate of increase* in prices. The rate is measured by considering the prices of a "basket" of consumer goods and services. The use of the term *basket* is metaphorical, encompassing those items that fulfill the basic needs of an average consumer. If the cost of the items in this basket was $350 a year ago, but is $370 now, then the prices have increased by a factor of $370/350 = 1.057$, that is at an annual *inflation rate* of 5.7%.

Another yardstick for measuring inflation is the *consumer price index* (CPI). The prices prevailing in an arbitrary (reference) year are represented by an index of 100. The prices in subsequent years are related to this index. In the United States, an index of 100 was allocated most recently to the general prices of 1983. Since then the CPI has increased as per the following table. Based on the previous year's CPI, the annual increase for any year can be evaluated, as expressed in the last column. For example, for 1995 the increase has been $(CPI_{1995} - CPI_{1994})/CPI_{1994} = (152.4 - 148.2)/148.2 = 0.0283 = 2.83\%$. The percentage in the last column is thus a measure of the annual inflation rate.

Year	CPI	Annual Increase, %
1983	100	
⋮	⋮	
1990	130.7	
1991	136.2	4.21
1992	140.3	3.01
1993	144.5	2.99
1994	148.2	2.56
1995	152.4	2.83
1996	156.9	2.95
1997	160.3	2.16
1998	163.5	2.00
1999	167.8	2.00 (estimate, not actual)

When the value of the CPI gets large over time, the reference (or base) year is reset, that is, advanced to another recent year, to which a new price index of 100 is allocated.

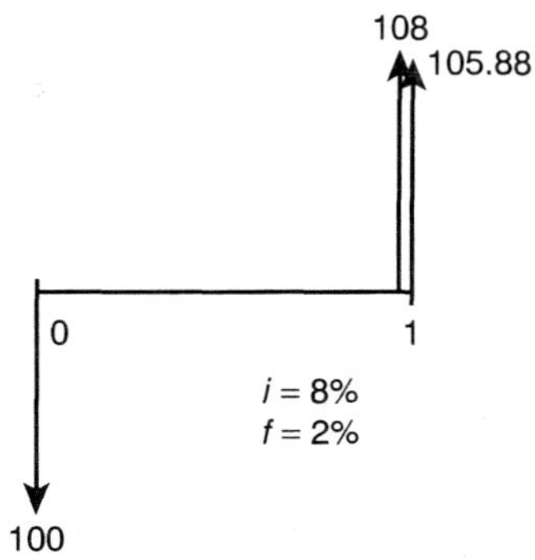

Figure 15.1 Inflation-Modified Cashflow

This advancing of the base year keeps the CPI values in three digits (before the decimal). The CPI-based measurement of the inflation rate is practiced by other countries as well, but the base year may vary.

The inflation rate in a country is the result of a complex interaction among several variables, such as money supply, government policies, currency exchange rates, quality of infrastructure, and the level of political stability.

15.1.2 Analysis

Inflation diminishes the purchasing power of future sums. Consider, for example, an investment last year of \$100. At an 8% interest rate, this has become \$108 today—\$8 more. During the year, however, prices have risen too. That means that, due to inflation, what can be purchased this year with a sum is less than what could have been bought last year with that sum. If the inflation rate f is assumed to be 2%, for example, then the inflation factor is $102/100 = 1.02$. Thus, with reference to last year's purchasing power, the *worth* of today's $\$108 = \$108/1.02 = \$105.88$. We denote \$108 as *actual* dollars and \$105.88 as *inflation-modified* dollars. Referring to the cashflow diagram in Fig. 15.1, the vector \$108 *has not* accounted for inflation, while the \$105.88 vector has. When the future dollars (cashflows) of a project are inflation-modified, they are called *current* dollars to signify that they represent the same purchasing power as of now. From an inflation viewpoint, they are the *real* dollars.

To account for inflation in engineering economic analyses, the cashflows are first modified. The inflation-modified cashflows are then analyzed the usual way, by any of the methods discussed in the text. Examples 15.1 through 15.3 illustrate the procedure.

EXAMPLE 15.1

For a 10% MARR, determine the present worth of the following cashflows by considering (a) no inflation, and (b) 4% annual inflation.

Year	Cashflow
0	−$500
1	300
2	300
3	300
4	300

Solution

a. The cashflows remain as given. The $500 investment yields a $300 annual benefit over four years ($n = 4$). Thus,

$$\begin{aligned} PW &= PW_{\text{benefits}} - PW_{\text{costs}} \\ &= 300(P/A, 10\%, 4) - 500 \\ &= 300 \times 3.170 - 500 \\ &= 951 - 500 \\ &= \$451 \end{aligned}$$

b. To account for inflation, the cashflows are modified. Since the rate is 4%, this is done by dividing the future cashflows by 1.04 for each year into the future. Note that a 4% inflation rate means an inflation factor of one plus the fractional value of the inflation rate, that is, $1 + 0.04 = 1.04$. For an inflation rate f expressed as a percentage, the factor is $(1 + f/100)$; where f is a fraction, it is $(1 + f)$. Thus, the inflation-modified value of the

$$\begin{aligned} \text{First-year \$300 cashflow} &= 300/1.04 \\ &= 288 \end{aligned}$$

$$\begin{aligned} \text{Second-year \$300 cashflow} &= 300/(1.04 \times 1.04) \\ &= 277 \end{aligned}$$

Proceeding the same way, the inflation-modified value of the

$$\begin{aligned} \text{Third-year \$300 cashflow} &= 300/(1.04 \times 1.04 \times 1.04) \\ &= 266 \end{aligned}$$

$$\begin{aligned} \text{Fourth-year \$300 cashflow} &= 300/(1.04 \times 1.04 \times 1.04 \times 1.04) \\ &= 256 \end{aligned}$$

A more efficient way to modify the cashflows is to divide the cashflow-of-now by the inflation factor $(1 + f)^n$, where n is the year. For example, the third year's modified cashflow $= 300/(1 + 0.04)^3 = 300/1.04^3 = 266$.

Another way is to use the previous year's modified value and divide it by $(1 + f)$. For example, the third year's modified cashflow = second year's modified cashflow/1.04 = 277/1.04 = 266. The inflation-modified cashflow table becomes

Year	Cashflow
0	−$500
1	288
2	277
3	266
4	256

Note that the consideration of inflation has disturbed the uniform-series pattern in the given cash inflows (benefits). Therefore, the functional notation $(P/A, i, n)$ is no longer applicable. We have to handle each benefit separately using the functional notation $(P/F, i, n)$. Thus, the present worth of the cashflows, taking inflation into account, is

$$\begin{aligned} \text{PW} &= \text{PW}_{\text{benefits}} - \text{PW}_{\text{costs}} \\ &= [288(P/F, 10\%, 1) + 277(P/F, 10\%, 2) + 266(P/F, 10\%, 3) \\ &\quad + 256(P/F, 10\%, 4)] - 500 \\ &= (288 \times 0.9091 + 277 \times 0.8264 + 266 \times 0.7513 + 256 \times 0.6830) - 500 \\ &= (262 + 229 + 200 + 175) - 500 \\ &= 866 - 500 \\ &= \$366 \end{aligned}$$

Note that the inflation-modified PW ($366) is lower than that in part (a) where inflation was not considered ($451). By ignoring the inflation, the analyst might make an erroneous decision about a project, especially if the selected alternative is marginally better than the second best.

EXAMPLE 15.2

A hardness testing machine is being considered for acquisition at a cost of $10,000. Its useful life and salvage value are estimated to be 5 years and $3,000. The annual saving from in-house testing using this machine is likely to be $3,000. Prepare[1] the cashflow table if annual inflation during the machine's useful life is projected to be 4%.

Solution

To prepare the cashflow table, we follow the procedure learned in Chapter 2. The skeleton of the table is as follows, where the first column is for the time period and the second column for the cashflows.

---

[1]As a rule of thumb, if inflation is given in a problem, then it must be considered in the analysis even if not explicitly asked for.

Year	Cashflow
0	
1	
2	
3	
4	
5	

With the given cost and benefits posted, the cashflow table with no consideration of inflation becomes

Year	Cashflow	
0	−$10,000	
1	3,000	
2	3,000	
3	3,000	
4	3,000	
5	3,000	
	3,000	(salvage)

Since inflation's effect on the cashflows is to be accounted for, we add two more columns for this purpose and also combine the two cashflows for period 5. Under the third column we post the inflation factors, while the last column contains the inflation-modified cashflows.

Year	Cashflow	Inflation Factor	Modified Cashflow
0	−$10,000		
1	3,000		
2	3,000		
3	3,000		
4	3,000		
5	6,000		

Since annual inflation is 4%, the inflation factor[2] is $(1 + 0.04)^n$, where n is the year.

[2]The inflation factor equals $(1 + f)^n$, where f is the inflation rate and n the time period; n is usually expressed in years.

On posting the given data, the cashflow table looks like this:

Year	Cashflow	Inflation Factor	Modified Cashflow
0	−\$10,000		−\$10,000
1	3,000	1.04^1	
2	3,000	1.04^2	
3	3,000	1.04^3	
4	3,000	1.04^4	
5	6,000	1.04^5	

Note that the \$10,000 cashflow is of now, and hence not modified. Only the future cashflows are modified, by dividing them by their inflation factors. For example, for year 3, inflation-modified cashflow = third-year cashflow/$(1 + f)^3$ = 3,000/1.04^3 = 2,667. The complete results are in the following table.

Year	Cashflow	Inflation Factor	Modified Cashflow
0	−\$10,000		−\$10,000
1	3,000	1.04^1	2,885
2	3,000	1.04^2	2,774
3	3,000	1.04^3	2,667
4	3,000	1.04^4	2,564
5	6,000	1.04^5	5,466

EXAMPLE 15.3

A small hydraulic machine is being planned for acquisition at a cost of \$5,000. During its useful life of seven years, it is expected to generate an annual profit of \$3,400. Its annual maintenance cost is projected to be \$300. The cost of space to house the machine and of the utilities is \$900 per year. Determine the payback period for the machine if the annual inflation rate is expected to be 5%.

Solution

This is Example 5.1, except that inflation is to be considered here. To evaluate the payback period, we need to determine the duration in which the \$5,000 investment will be recovered.

The recurring expenditure comprises the space and utility cost of \$900 and the maintenance cost of \$300, both annual. Thus, the total operating cost is \$900 + \$300 = \$1,200 per year. The net benefit from the investment in the machine is obtained by subtracting this total from the profit. This yields a net benefit of \$3,400 − \$1,200 = \$2,200 per year, which needs to be modified for inflation.

The modification for inflation disturbs the uniform-series pattern of benefits, as seen in Example 15.1. Consequently, the payback-period formula applicable for uniform cash inflows cannot be used. Instead, the approach explained in Example 5.2 for non-uniform benefits, where the *cumulative* value of the net benefit is tracked, is applicable. At the payback period the cumulative net benefit equals the \$5,000 investment. The calculations can be streamlined by tabulating the data and the results.

Year	Net Benefit	Inflation Factor	Modified Net Benefit	Cumulative (Modified) Net Benefit
1	2,200	1.05^1	2,095	2,095
2	2,200	1.05^2	1,995	4,090
3	2,200	1.05^3	1,900	5,990
4	2,200	1.05^4		
5	2,200	1.05^5		
6	2,200	1.05^6		
7	2,200	1.05^7		

The inflation factors in the third column should be obvious. The modified net benefit is obtained by dividing the net benefit by the corresponding inflation factor. For example, for year 2, it is $2{,}200/1.05^2 = 1{,}995$.

It is not essential to compute all the results for the last two columns of the table. To save time and effort, one should keep track of the cumulative value (last column) as calculations proceed. Once this value is equal to or greater than the investment (\$5,000), the calculation ceases because the payback period is then known.

The table shows that by the end of the third year, with a cumulative net benefit of \$5,990, the \$5,000 investment is more than recovered. Hence the payback period is less than three years. The year-2 value of \$4,090 suggests that it is greater than two years. Therefore, interpolate these two data, getting a payback period of 2.48 years.

$$
\begin{aligned}
\text{Payback period} &= 2 + (5{,}000 - 4{,}090)/(5{,}990 - 4{,}090) \\
&= 2 + 910/1{,}900 \\
&= 2 + 0.48 \\
&= 2.48 \text{ years}
\end{aligned}
$$

Should one interpolate to get the exact period for the payback? That depends on the accounting practice. If the bookkeeping follows the year-end convention—usually so—then the payback period in this example is 3 years. However, if the revenues are realized year-round, and so accounted, then the exact answer of 2.48 years makes better sense.

In Example 5.1, where inflation was ignored, the payback period was determined to be 2.27 years. This example's payback period of 2.48 years is longer, since inflation delays the recovery of investment by diminishing the value or purchasing power of future benefits.

The inflation rate f decreases the value of future sums by the factor $(1 + f)^n$. The interest rate i, on the other hand, increases a present sum by $(1 + i)^n$. Since the effect of f on future cashflows is also exponential, the compound interest tables can be used to read off the values of inflation factors. The functional factor $(P/F, i, n)$ can be used to account for the effect of inflation on future cashflows by considering it as the inflation factor $(P/F, f, n)$. For example, under a 6% inflation rate, a \$350 future sum five years hence will in current dollars be $350(P/F, 6\%, 5) = 350 \times 0.7473 = \$261.56$.

In Chapters 1–14, we have been using interest rates with no consideration of inflation. Such a rate is called *market interest rate*, since the market does not account for inflation while quoting the interest rates prevalent at the time. Inflation modifies the market interest rate as per

$$i_f = (i - f)/(1 + f) \qquad (15.1)$$

where

i_f = "real" interest rate that compensates for inflation
i = market interest rate, and
f = inflation rate.

Consider that a local bank offers its savers annually compounded 8% interest rate per year, that is, $i = 8\%$. If inflation (f) is 2%, then from Equation (15.1), substituting 0.08 for i and 0.02 for f, we have

$$\begin{aligned} i_f &= (0.08 - 0.02)/(1 + 0.02) \\ &= 0.0588 \\ &= 5.88\% \end{aligned}$$

In other words, savers *really* earn only 5.88%, not 8%. The inflation has "eaten" into the interest. Smart savers are aware of this fact and while investing consider the *real* interest rate i_f that hedges against inflation.

If f is low compared to i, then from Equation (15.1) the real interest rate i_f approximately equals the market interest rate i. This is the situation we have been assuming in Chapters 1 through 14.

In Example 15.3, we saw that inflation extends the payback period of an investment. When ROR analysis is carried out to account for inflation, there is a similar effect: the after-tax ROR is reduced. Thus, from the investor's viewpoint, inflation is bad, since investment takes longer to recover or the rate of return is reduced. From the borrower's viewpoint, however, inflation is good. The payments made in the future will be of reduced purchasing power. This is one reason why people borrow to finance their homes and other assets, especially in an inflationary economy.

If the project's analysis period is short, say 3 to 5 years, and the inflation rate is low, then the effect of inflation on the cashflows can be ignored. This is what has been assumed in Chapters 1 through 14. Moreover, since the operational costs and benefits of all the alternatives of a project are likely to be affected to the same degree, inflation's impact on the cashflows may cancel out. Thus, in most cases, a decision without considering inflation is likely to be the same as that with inflation considered. However, if the best alternative is only marginally better than the second best, and their cashflow patterns are different, a reanalysis accounting for inflation should be carried out to confirm the validity of the decision.

15.2 PUBLIC PROJECTS

Public projects are funded not for profit, but for the public good. They are usually government projects or those of nonprofit organizations. A variety of organizations such as school districts, charity hospitals, and the International Red Cross deal with public projects. The basic concept of costs and benefits applies to public projects as much as to private projects. But the profit-oriented criteria such as ROR and payback period become irrelevant in public projects. Public projects are analyzed on the basis of the benefit–cost ratio (or difference), as discussed in Chapter 10. For a project to be acceptable, its benefit–cost ratio should be greater than one. In the case of multialternative projects, the analyst must use the incremental method, accepting the higher-cost alternative only if its incremental benefit–cost ratio is greater than one.

Three major questions arise while analyzing government projects for funding considerations:

1. Who should pay for the project?
2. Who should derive the benefits?
3. What interest rate should be used in the analysis?

Based on common sense, only those who will benefit from the project should pay. But how do you establish and implement that? If the school district is planning to build a new classroom, should the in-district senior citizens with no school-age children pay toward the construction? As another example, should the city restrict its park only to those who pay city taxes? How do you ensure that only those who paid for the facility use it? How do you police it? Should the city issue identity cards to its residents and require the card for admission into the park? The first two questions are really complex, with no easy answers. Since they involve social and political considerations, the analysis of public projects may be more than economic.

The third question is equally perplexing. If the money to be spent on a government project has been raised through taxes, is the interest rate zero, since the money is not borrowed? Or should the rate be what taxpayers would have earned had they not been taxed? According to the U.S. federal government's Office of Management and Budget, it is economically unsound to take money from taxpayers, who would have earned 12%, and invest it in government projects that yield 4%!

Another difficulty experienced in analyzing government projects is the treatment of the operating costs. Should the operating costs be paid out of the future revenues, or should they be budgeted in the beginning, and paid out later, as part of the investment cost? The benefit–cost ratio of a project differs depending on how the operating costs are treated. This has been illustrated in Chapter 10 through Example 10.2 and its appertaining discussions.

15.3 CAPITAL RATIONING

Most often than not, companies have limited capital and several competing projects, and thus fail to fund them all. In such situations capital is rationed, that is, allocated to the most attractive projects. The rationing procedure involves evaluating the projects' RORs and ranking them in descending order. The allocation of the capital begins from the top of the list, going down until all the available capital has been assigned. Although Example 15.4 illustrates the procedure for the ROR criterion, capital can be rationed under any of the other five criteria.

EXAMPLE 15.4

ABC Incorporated is considering the following projects for possible investment next year. Each project has a five-year useful life and no salvage value. For a 15% MARR and a capital outlay of $25,000, which projects should be funded under ROR?

Project	1	2	3	4	5	6	7	8
First Cost	$5000	4500	2500	3000	5000	5500	6000	7000
Annual Benefit	$1319	1673	836	771	1492	1759	1963	2341

Solution
The solution procedure comprises the following steps. If the RORs are given, skip step 1.

Step 1: ***Evaluate the ROR of each project.***
All the projects have a five-year useful life, that is, $n = 5$. Calculate the ROR of each project as in Chapter 9. For example, for project 3, we have

$$2{,}500 = 836(P/A, i, 5)$$

Thus,

$$(P/A, i, 5) = 2{,}500/836 = 2.9904$$

From the interest tables, for this value of the factor, $i \approx 20\%$. Thus, the rate of return for the third project is 20%. Once all the RORs have been evaluated, they are tabulated as follows:

Project	First Cost	Annual Benefit	Computed ROR, %
1	$5,000	$1,319	10
2	4,500	1,673	25
3	2,500	836	20
4	3,000	771	9
5	5,000	1,492	15
6	5,500	1,759	18
7	6,000	1,963	19
8	7,000	2,341	20

Step 2: ***Discard the unattractive projects.***
Check to see which of the projects have a ROR less than the desired MARR, and discard them. In this case, for a 15% MARR, projects 1 and 4 are discarded.

Step 3: ***Rank the remaining projects.***
Arrange the remaining projects in descending order of ROR.

Project	First Cost	Annual Benefit	Computed ROR, %
2	$4,500	$1,673	25
3	2,500	836	20
8	7,000	2,341	20
7	6,000	1,963	19
6	5,500	1,759	18
5	5,000	1,492	15

Step 4: ***Allocate the available capital.***
Begin to allocate the available capital. Since project 2 tops the list, due to its highest ROR, its required capital of $4,500 is allocated first. The next two projects, 3 and 8, have the same ROR, so they are allocated together, but 3 is preferred due to its lower capital need. The allocations can be tabulated in another column.

Project	First Cost	Annual Benefit	Computed ROR, %	Allocation
2	$4,500	$1,673	25	$4,500
3	2,500	836	20	2,500
8	7,000	2,341	20	7,000
7	6,000	1,963	19	
6	5,500	1,759	18	
5	5,000	1,492	15	

The allocation process is continued until all the available capital has been used up. To keep track of this constraint another column is created to record the cumulative allocation.

Project	First Cost	Annual Benefit	Computed ROR, %	Allocation	Cumulative Allocation
2	$4,500	$1,673	25	$4,500	$4,500
3	2,500	836	20	2,500	7,000
8	7,000	2,341	20	7,000	14,000
7	6,000	1,963	19		
6	5,500	1,759	18		
5	5,000	1,492	15		

As seen in the last column, we have thus far allocated $14,000 out of the available $25,000. So the allocation continues. With the next allocation, the cumulative allocation increases to $20,000.

Step 5: Is there any remaining capital?
After allocating to as many projects as possible, you may be left with some capital. What should be done with the remaining capital? There are two choices, depending on how much is left.

a. If the remaining capital is significantly less than what is needed to fund the next best project, it is usually returned to the company capital pool for use elsewhere.
b. If the remaining capital is closer to what is needed to fund the next best project, then try to acquire the required additional capital and fund the next project.

In this particular case, after allocating the best four projects at a total cost of $20,000, we are left with $5,000 ($25,000 − $20,000). The next best project, namely 6, requires $5,500 of capital. If we had another $500 we could fund project 6 too. This fact should be discussed with the capital budgeting authority through the department manager. Negotiate with the budget officer, providing additional information on the project if needed and highlighting its merits. The budgeting authority might release the additional $500 provided project 6 is among the other worthy ones within the company.

The best of the projects that could not be funded represents an investment opportunity lost due to a paucity of capital. In the example, given $25,000 capital and assuming no success with acquiring the additional $500, project 6 was the best unfunded project. The rate of return of the best unfunded project is called *opportunity cost.* In Example 15.4, therefore, the opportunity cost is 18%. Note that the unit of opportunity cost is percent, not dollars. If the additional $500 could be acquired and project 6 funded, the opportunity cost would be 15%, the ROR of project 5.

15.4 MARR SELECTION

The MARR is the minimum rate of return a project must be expected to earn before it can be approved for funding. It is fixed by upper management on the basis of prevailing interest rates in the financial marketplace and other business considerations. Engineers or engineering technologists, especially in medium and larger companies, may not be involved in fixing the MARR. However, if they are working in small companies or in their own firms, they may face the task of deciding the value of the MARR to be used in economic analyses. Appertaining to this complex task we discuss some basic considerations in this section.

The capital for investment may be available internally, from the profits generated within the company. However, while deciding to use its own profit as capital, the company must consider the investment opportunities outside. The MARR to be used in the own-profit-as-capital case must be greater than, or at least equal to, the interest the capital can earn in the financial marketplace. For example, if the capital can be loaned to a financial institution or another company at 10%, then the MARR must be 10% or more. How much more should depend on the relative risks in loaning the capital or investing it within the company on engineering projects.

Alternatively, the capital for investment may be borrowed from financial institutions. In such cases, the MARR should obviously be more than the interest rate for the borrowed capital. Again, how much more should depend on the risk in the engineering investment. Often, the required capital may have to be raised from more than one financial institution at different interest rates. In such a case, a weighted average interest rate is determined to account for the variations in interest rates of borrowed capital; the MARR should be greater than this average interest rate.

After considering the opportunity cost discussed in Section 15.3, the opportunity of loaning the capital to the financial marketplace, the cost or interest rate of borrowed capital, and any government investment incentive available, the value of the MARR is decided upon. This value also takes into account the risk and uncertainty involved with the engineering investment, as well as the overall prevailing business climate, both within the company and without. As and when the financial market conditions change, the MARR is reviewed and readjusted if necessary. The decision about what MARR to use is thus complex, dynamic, and challenging.

15.5 DATA UNCERTAINTY

An important consideration in economic analyses is to ensure that the data being used are reliable. Throughout the text we have assumed the data to be deterministic. If we state that the salvage value of machine A is \$5,000, we are implying a 100% certainty[3] about this value. How could we be so sure? We know that the salvage value relates to the future; we can be certain about its value only after it would be salvaged. But we

---

[3]To avoid giving this impression, we have preferred to use in the text statements such as, The salvage value of machine A is *expected to be* \$5,000.

need this data for analysis now—at the present time—when the invest[illegible]
sion is being made. The only data whose values we can be certain about are [illegible] the past or pertaining to the present. For example, we can be sure of the cost [illegible] machine under consideration for investment, since the cost is based on a binding q[illegible]tation. This cost as a cashflow is 100% certain because it is paid following the pur[illegible]chase.

Future data are indeterministic; they can only be *estimated.* One cannot be 100% sure of their values. The further the data are into the future, more the uncertainty in their values. For example, the useful life of a machine can only be estimated. The reliability of estimation is higher if the useful life, for example, is 3 years than if it is 12 years.

In engineering economics, as in other disciplines, data values should be estimated carefully, using statistical techniques wherever necessary. The basic approach is to analyze the past data, if available, and use the results to predict the future value. This process is called forecasting. It is also called *time-series analysis* because the past data form a series that occurred at different times.

A comprehensive economic analysis that accounts for uncertainty in the data and allows for statistical analyses in their estimation is beyond the scope of this text. However, some elementary considerations are presented in the next two subsections.

15.5.1 Optimistic–Pessimistic Approach

The statistical nature of the data can be accounted for by considering their extreme possible values along with the most likely value. Let us say that you are not sure of the useful life of a hydraulic machine being analyzed. You have determined its most likely value to be 6 years. Others in the company think it is likely to be 7 years, while some suggest it to be 5 years. Those who think it will last for 7 years are optimists by nature, while those who suggest 5 years are pessimists. Thus, there are three different values for the same data: 6 years as the most likely value, 7 years as the optimistic value, and 5 years as the pessimistic value. The question is, Which value is relevant and should be considered? The analyst may prefer to use the most likely value in making the decision, as was done in Chapters 1 through 14. However, the optimists and pessimists consider the most likely value erroneous. Should the analyst ignore their values? These issues require statistical considerations in the analysis.

In the most elementary approach, the statistics are kept simple. The data values are incorporated in the analysis by using the mean value of the variable(s). The mean value is obtained by weighting the data, which may give equal or unequal importance to the most likely and pessimistic–optimistic values. With equal importance to all the values, we obtain the arithmetic mean. One form of the weighted mean is based on the assumption that the data satisfy a β-distribution. For such a case,

$$\text{Mean} = \frac{\text{Optimistic value} + 4(\text{Most likely value}) + \text{Pessimistic value}}{6}$$

Example 15.5 illustrates the application of the most likely and optimist–pessimist data values in decision making. Note how the solution has become lengthier even with simplistic statistics.

FIFTEEN

'LE 15.5

raulic machine is being analyzed for acquisition. The engineering econo- only of its cost, which is $5,000. For the values of the other three data, he ne research. Based on input from an insurance company about the ma- life and salvage value, and from within the company about net annual s determined the following most likely, optimistic and pessimistic values.

	Optimistic	Most Likely	Pessimistic
Net annual benefit	$3,000	$2,200	$1,000
Useful life, years	7	6	5
Salvage value	$500	$300	0

Should the machine be funded for investment if the payback period is not to exceed 2 years and the MARR is 10%? Assume that the data follow a β-distribution.

Solution

Under the β-distribution, the most likely value is given fourfold greater weighting than the optimistic or pessimistic value. In other words, the most likely value is counted four times, giving a total of six data values. The mean value of the data is therefore obtained by adding these six data values and dividing the total by 6:

$$\text{Mean} = \frac{\text{Optimistic value} + 4(\text{Most likely value}) + \text{Pessimistic value}}{6}$$

For the net annual benefit, the mean value is

$$\begin{aligned}\text{Benefit}_{\text{mean}} &= \frac{\$3{,}000 + 4 \times \$2{,}200 + \$1{,}000}{6}\\ &= \$2{,}133\end{aligned}$$

The mean value of useful life is

$$\begin{aligned}\text{Life}_{\text{mean}} &= \frac{5 + 4 \times 6 + 7}{6}\\ &= 6 \text{ years}\end{aligned}$$

The mean salvage value is

$$\begin{aligned}\text{Salvage}_{\text{mean}} &= \frac{\$500 + 4 \times \$300 + 0}{6}\\ &= \$283\end{aligned}$$

The problem is now solved in the usual way, as in other chapters, using the mean values of the parameters. With $A = \$2{,}133$,

$$\begin{aligned}\text{Payback period} &= \text{Investment/Annual benefit}\\ &= \$5{,}000/\$2{,}133\\ &= 2.34 \text{ years}\end{aligned}$$

Since the payback period is greater than the desirable two years, the machine should not be funded for investment. (Note that the salvage value was redundant data here.)

15.5.2 Statistical Analysis

In more precise statistical analyses, the past data are fully analyzed for their statistical characteristics. *Statistics* is concerned with the collection, analysis, and interpretation of quantitative data. Data may be of the attribute type or variable type. *Attribute data* can have only two values: conforming or nonconforming, pass or fail, go or no-go gauge. *Variable data,* on the other hand, can have any value in steps of the measuring system's least count. Most engineering economics data are of the variable type. The useful life of a machine, expressed as 5 years for example, is variable data. The value of the MARR is more like attribute data, a number above which the project is funded and below which it is not.

After collecting the data, they are described through tables, charts, or graphs, and analyzed. The graphical analysis summarizes the data in the form of a *frequency distribution,* which is modeled mathematically. The model is used to estimate the value of the parameter and to make predictions. For example, the useful life of a machine may be modeled by a *normal distribution*[4], also called the *bell curve*. Such a distribution applies to a variety of industrial as well as natural processes. The height of students in a class follows a normal distribution—so does the salvage value of a machine.

How do we estimate a machine's salvage value statistically? It is done by analyzing the salvage values of a large number of such machines used in the past. The past data are collected and their frequencies plotted as illustrated in Fig. 15.2. Note that the salvage values vary between \$200 and \$600 but seem to cluster around \$400. This \$400 is the mean salvage value.[5] In statistical analyses we incorporate the data variation as well. We therefore quantify both the mean and the dispersion based on the approximating normal distribution, or any other distribution that fits the data.

---

[4]The data that follow this distribution display two characteristics: *central tendency* and *dispersion*. The central tendency is measured in terms of mean, median, or mode, while the dispersion is measured in terms of range or standard deviation.

[5]We have used such a mean value throughout the text. In other words, we excluded any consideration of statistical variation, which may be acceptable in many industrial projects.

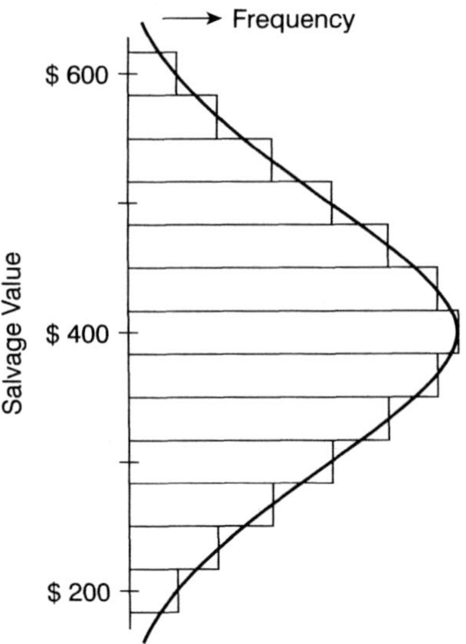

Figure 15.2 Concept of Normal Distribution

The normal distribution is mathematically defined as

$$y(x) = \frac{1}{\sigma\sqrt{2\pi}} e^{-(x-\mu)/(2\sigma)}$$

where $y(x)$ = ordinate of the curve corresponding to x,
μ = mean of all the x-values, and
σ = standard deviation.

The standard deviation is a measure of dispersion in the values of the variable x. Thus, by using μ and σ we are able to describe the salvage value in greater detail. We are not saying any more that the salvage value is $400, as was done earlier in Chapters 1 through 14. We are instead saying that the salvage value is likely to be $400, but this could be lower or higher as dictated by the value of σ.

15.6 THE COMPUTER AS A TOOL

Computers play an important role in engineering economics, since they can perform calculations almost instantly. They are especially useful in ROR analyses, where the calculations are repetitive and tedious. Recall how lengthy the trial and error approach to evaluating ROR becomes. However, computers must be programmed to carry out the calculations. The sequence of operations the hardware carries out under the control of a program is called software.

As a tool, the computer has had a phenomenal impact on engineering economics analyses. A plethora of software running on PCs helps engineers make faster and better economic decisions. The ROR method, the most complex of the six, has benefitted enormously from software-based analysis. While users may develop their own software by programming the relevant equations in a PC or programmable calculator, most prefer the commercial software products that have been developed professionally and are user friendly and frequently upgraded. Besides offering a precise, accurate, and complete solution, the software offers what-if capabilities that lead to the optimum decision. Users should, however, be aware of the GIGO (garbage-in, garbage-out) nature of the computer environment. A few step-by-step sample calculations should be done on calculators to confirm the computer results.

While sophisticated software products with advanced graphics capabilities are available for analyzing large engineering projects, spreadsheet analysis is usually sufficient for most needs. The spreadsheet is a collection of rows and columns in which equations can be programmed to act on the data contained in certain cells of the spreadsheet. The results are displayed in other cells. The spreadsheet-based analysis is relatively more user friendly.

Any computer can be used for economic analysis. The personal computer (PC), however, is the most popular platform for engineering economists. A variety of PC-based software products are available in the market. Buy the one that suits your needs. Talk to people who have used the software you intend to buy. Computer magazines and professional magazines occasionally publish a comparative evaluation of the various software packages available in the market.

15.7 AUTOMATION PROJECT—A CASE STUDY

Throughout the text we have emphasized the fundamental principles of engineering economics and their applications to decision making. To achieve this, the problems were kept manageable. Most of the pertinent data were provided in the problem statement. In the real world of industry, however, such data may not be readily available; they may have to be collected, rendering the decision-making task lengthy and difficult. Most projects may require efforts by several persons working as a team. As an engineer or engineering technologist, you may be a member of such a team. Other members may come from backgrounds different from yours, for example business, law, or computers.

In this section, we illustrate the economic analysis of a real-world automation project.[6] Since it involves the implementation of a robotic system, it has also been called a robotic system, and a new system or process. For simplicity of analysis, we apply the payback-period criterion. As discussed earlier, the analysis will require two items of data, namely the cost of investment and the expected future benefits. These data may not be

[6]Adapted from *Molding Systems*, Society of Manufacturing Engineers, September 1998, pp. 33–34.

easily available and may have to be estimated. The engineer's task is to cost-justify the system based on the payback criterion.

The first question the engineer faces is, What payback period is acceptable? His manager may offer a figure based on the company's investment policies. The next question is, Could a longer payback period be justified, since the robotic system also offers a nontangible benefit of enhancing the company image as a high-tech firm?

The analysis must consider the likely benefits from the system in relation to the current operation. It has to account for the current costs and determine the likely costs with the robotic system. The difference between these costs represents the net benefits from the investment.

The pertinent data are gathered from company books and various departments, the robot vendor and manufacturer, and other sources—both within and without. For example, the robot will cost $35,000 including shipping to the company plant. There are several other costs before the system becomes operational:

Installation Cost The installation cost includes the costs of moving the robot from where it will be delivered to its location in the plant, connection for electricity, any foundational need recommended by the manufacturer, and so on. These costs are gathered or calculated, adding up to $2,500.

Training Cost The operator and other personnel associated with the robotic system will be trained. Determine the total hours of training required for each one and their hourly rates. In this case, an operator, his assistant, and the plant safety officer are to be trained. Note that the hourly rate for an employee is higher than the wage rate because of fringe benefits, such as health care, vacation, retirement, social security taxes, and others. The operator is trained at a nearby community college under an arrangement with the robot vendor. The hourly rate for the operator is $20, though his wage rate is only $15. He takes 25 hours of training at the college, costing the company $500 in lost production. His travel, boarding, and lodging expenses are $450. On his return from the college training, he will train his assistant and the plant safety officer once the robot is operational. The cost of the assistant's and safety officer's training is $550. Thus, the total cost of training is $500 + $450 + $550 = $1,500.

End-of-Arm Tooling A closer look at the variety of parts the robot will handle indicated a need to purchase some off-the-shelf toolings at a cost of $1,850. Other required toolings can be built in-house, for which the estimated cost is $1,650. Thus, the total cost for the end-of-arm tooling is $3,500.

Maintenance Equipment The robotic system will require a $250 vibrometer for predictive maintenance.

Safety Guard According to the plant safety officer, a guard is essential to comply with the Occupational Safety and Health Administration's (OSHA) requirements. The guard, for which the material and labor costs are $500 and $1,000, will be fabricated in-house. Thus, the total cost for guarding is $1,500.

These costs, being essential for rendering the robotic system operational, are part of the investment. They can be summarized in a table, as follows. A summation of the costs shows that the total investment[7] in the system is $44,250.

Robot	$35,000
Installation	2,500
Training	1,500
End-of-arm tooling	3,500
Maintenance equipment	250
Guard	1,500
Total:	$44,250

In a similar way, we need to collect data on the benefits expected from the robotic system. The benefits are the savings from the system, which can be evaluated by comparing the cost of the current process with that of the new system. All types of cost items likely to be affected by the new system should be considered by listing them and collecting the associated data. Such a task is usually referred to as *cost accounting.* For the present case study, we carry out this task in a table containing four columns with proper headings. The first column is for the cost item; columns two and three are for cost data on the current and new process (after automation). The last column will contain their difference as a benefit.

Cost Item	Current Process	After Automation	Benefit

The number and type of cost items gathered will vary from project to project. For the present case, the following items are applicable.

Value of Part The part produced by the current process is valued at $0.50, which will remain unaffected by the automation. This value represents the revenue generated by the part. In many projects the value of the part may increase if the new system uses less material or extends the part's life.

Effect on Output The effect of automation on the output, if any, is determined next. To do this, compare the production rate under the current process with the new one. The current process produces 15,000 parts per week. The plant operates three 8-hour shifts each day, Monday through Friday, or 120 hours per week. Thus, the total time the process is operational is

$$120 \text{ hours/week} \times 60 \text{ minutes/hour} \times 60 \text{ seconds/minute} = 432{,}000 \text{ seconds/week}$$

$$\begin{aligned}\text{Average cycle time} &= \text{Total time/Production rate}\\ &= (432{,}000 \text{ seconds/week})/(15{,}000\text{/week})\\ &= 28.8 \text{ seconds}\end{aligned}$$

[7]We have called such data first cost or initial cost and invariably supplied them in the problem statements. Note how extensive the task can be in the real world, as illustrated in this case study.

It is estimated that the use of the robotic system will reduce[8] the cycle time by 2 seconds to 26.8 seconds. Thus, production will increase to (432,000 seconds/week)/(26.8 seconds) ≈ 16,120 per week.

Thus, the number of parts produced per year due to automation will increase from 780,000 (15,000/week × 52 weeks/year) to 838,240 (16,120/week × 52 weeks/per year). The increased production[9] of 838,240 − 780,000 = 58,240 parts will generate an additional revenue of $29,120, since each part sells for $0.50.

Let us post these data in the table before continuing with further data collection.

Item	Current Process	After Automation	Benefit
Value of part	$0.50	$0.50	
Annual production	780,000	838,240	$29,120

Labor Cost Currently one operator is used for each shift. Thus, each day three operators are required, whose annual cost is $121,680. It is estimated that the robotic system will decrease this cost by 25%. Thus, the saving will be 0.25 × $121,680 = $30,420. The labor cost after automation will be 0.75 × $121,680 = $91,260.

The labor cost per part is obtained by dividing the annual labor cost by the annual production. For the given data, it is

$$\text{Current process} = \$121{,}680/780{,}000 = \$0.16$$

$$\text{Automated process} = \$91{,}260/838{,}240 = \$0.11$$

These data can also be posted by updating the table as

Item	Current Process	After Automation	Benefit
Value of part	$0.50	$0.50	
Annual production	780,000	838,240	$29,120
Labor cost	$121,680	$91,260	$30,420
Labor per part	$0.16	$0.11	

[8]It is based on the actual cycle time of the current process obtained from time and motion study and the estimated cycle time with the robotic system. Note that if the actual cycle time data is unavailable, the engineering economist may have to seek help from industrial engineering staff, who will conduct a time study on the current process. This illustrates how additional tasks may creep into the engineering economic analysis.

[9]The new annual production could have been evaluated simply by multiplying the cycle time ratio with the current production as (28.8/26.8) × 780,000 ≈ 838,209.

We continue to consider the other implications of automation.

Quality Cost The automation is expected to improve process quality, resulting in less scrap. Under the current process the scrap is 3%, that is $0.03 \times 780{,}000 = 23{,}400$ parts per year. The new process is expected to reduce the scrap to 161 parts per week, that is $161 \times 52 = 8{,}370$ per year. The reduced scrap will result in an annual saving of $7,515.

$$\begin{aligned}\text{Parts saved annually through lower scraps} &= 23{,}400 - 8{,}370 \\ &= 15{,}030\end{aligned}$$

$$\begin{aligned}\text{Annual saving} &= \$0.50 \times 15{,}030 \\ &= \$7{,}515\end{aligned}$$

Maintenance Cost The robotic-system-based new process, being sophisticated, costs more to maintain than the current one. Its annual maintenance cost will require an additional sum of $2,500. Note that it is a cost that must be paid out of the benefit from the new process. In the data table, it is entered with a negative sign to signify loss.

Insurance Cost The robotic system will cost $100 less in insurance, which is a benefit.

Utilities Cost The cost of power (electricity) and compressed air for the robotic system will be $400 more than that for the current system. This data too is a cost of the new system, and therefore carries a negative sign.

On posting the data, the complete table for the new system is as follows. By summing up the various benefits (last column), the net annual benefit from the automation system is determined to be $64,255.

Item	Current Process	After Automation		Benefit
Value of part	$0.50	$0.50		
Annual production	780,000	838,240		$29,120
Labor cost	$121,680	$91,260		$30,420
Labor per part	$0.16	$0.11		
Quality of parts				$7,515
Maintenance				−$2,500
Insurance				$100
Utilities				−$400
			Total:	$64,255

The cost and benefit tables can now be consolidated.

Costs		Benefits	
Robot	$35,000	Annual production	$29,120
Installation	2,500	Labor cost	30,420
Training	1,500	Parts quality	7,515
End-of-arm tooling	3,500	Maintenance	−2,500
Maintenance equipment	250	Insurance	100
Guard	1,500	Utilities	−400
Total cost:	$44,250	Annual benefit:	$64,255

From the consolidated table,

$$\begin{aligned} \text{Payback period} &= \text{Total cost/Annual benefit} \\ &= \$44{,}250/(\$64{,}255 \text{ per year}) \\ &= 0.69 \text{ year} \\ &= 8.28 \text{ months} \end{aligned}$$

Note that the calculation for the payback period was simple once the two important data were known. Almost all the efforts of analyzing the automation project pertained to data collection. This is true of most engineering economics analyses in industry, where data gathering is usually more involved than the analysis itself. Throughout the text, we intentionally kept ourselves free of data-gathering efforts so that we could focus on the analysis aspects of the problem. This was done by providing the data in the problem statements. In the real world, data collection is a major part of the decision-making effort, as illustrated by this case study, and is usually time-consuming.

SUMMARY

Realistic economic analyses of engineering projects may involve several considerations, of which the important ones have been discussed in this chapter. Inflation can become pertinent in projects with long useful lives, especially if the inflation rate is high. Government projects have their own special features that must be considered in benefit–cost ratio analyses. Companies with limited capital may not be able to fund all the investment-worthy projects, so they ration the capital. The rationing is based usually on the ROR. The fixing of the MARR depends on the interest rates prevailing in the financial marketplace and the company's own business "health." Further difficulties in economic analyses arise from the uncertain nature of the data pertaining to the future. Statistical considerations enhance the accuracy and reliability of the decisions, but at the cost of analysis complexities. Economic analysis, especially the data collection, of real-world engineering projects is a lengthy process, as illustrated in this chapter through an automation project case study.

EXERCISES

Discussion Questions

15.1 Write a 200-word essay on inflation's effect on engineering economic analysis.

15.2 What is the inflation rate in your country? What has been its trend over the last ten years?

15.3 Write a 200-word essay on economic analysis of not-for-profit engineering projects.

15.4 Why do companies ration the capital for engineering investments, and how do they do it?

15.5 Write a 200-word essay on statistical considerations in engineering economics analysis.

Multiple-Choice Questions (Circle the *best* answer.)

15.6 Inflation is good for the
- a. lender.
- b. borrower.
- c. government.
- d. a or b, depending on the prevailing interest rate

15.7 A personal computer costs $1,850 today. If the inflation rate f is 6% per year, the computer's cost two years from now will be
- a. $1,875.34.
- b. $1,978.89.
- c. $2,009.66.
- d. $2,078.66.

15.8 The acronym CPI stands for
- a. current price information.
- b. current price index.
- c. consumer price index.
- d. consumer product information.

15.9 The notations μ and σ are associated with _____ distribution.
- a. exponential
- b. beta
- c. normal
- d. uniform

15.10 The notation σ is used to measure the _____ of a bell-shaped frequency distribution.
- a. central tendency
- b. dispersion
- c. height
- d. shape

Numerical Problems

15.11 The annual maintenance cost of a machine for the next 5 years is estimated to be $500. What is the present worth of the maintenance costs if inflation is (a) ignored, (b) expected to average 5% per year? Assume an annually compounded interest rate of 8% per year.

15.12 The maintenance cost of a machine is expected to be $500 this year. For the next five years it is likely to increase 5% per year due to machine wear and tear. What is the equivalent uniform annual maintenance cost if inflation is (a) ignored, (b) expected to average 5% per year? Assume an annually compounded interest rate of 7% per year.

15.13 If the inflation rate is 3% per year in Problem 15.12, what is the answer for part (b)?

15.14 Mary borrows $80,000 to buy a machining center. Beginning next month she will pay the lender $850 per month for the next 20 years. What ROR will the lender be enjoying if the annual inflation rate is 8%? (This is Problem 9.16 but with inflation.)

15.15 Two models of pollution control equipment are under consideration to meet the legal air emission standard. The basic model costs $8,500 and will last for 5 years; the deluxe model costs $12,000 and will last for 10 years. If the annual inflation rate is 6%, which model should be purchased? Assume the MARR = 12%. (*Hint*: Consider a 10-year analysis period, noting that the replacement cost of the basic model at the end of the fifth year will be higher due to inflation.)

15.16 Poonam earns $5,000 as an annual bonus and invests it for ten years at an annual interest rate of 12%, compounded yearly. If the annual inflation is projected to be 3% per year during the next five years and 5% thereafter, how much will she get as (a) actual dollars, (b) current dollars? (*Hint*: In (b) the purchasing power is maintained.)

15.17 The optimistic, most likely, and pessimistic costs of a new piece of equipment are $300, $320, and $350 respectively. What is the mean cost using weighting factors that approximate a β-distribution?

Bibliography

The focus in this introductory text has all along been on the basic principles of engineering economics. For readers interested in furthering their knowledge, textbooks dealing with advanced topics and/or in-depth analysis of engineering economics are listed here.

American Telephone and Telegraph (AT&T) Co. *Engineering Economy.* 3rd ed., McGraw-Hill, 1977.

Blank, Leland T., and Anthony J. Tarquin. *Engineering Economy*. 4th ed., McGraw-Hill, 1998.

Eschenbach, Ted G. *Engineering Economy*. Irwin, 1995.

Fleischer, Gerald A. *Introduction to Engineering Economy.* PWS Publishing Company, 1994.

Grant, Eugene L., Ireson, W. G., and R. S. Leavenworth. *Principles of Engineering Economy.* 8th ed., Wiley, 1990.

Newnan, Donald G., Lavelle, Jerome P., and Ted G. Eschenbach. *Engineering Economic Analysis.* 8th ed., Engineering Press, 2000.

Park, Chan S. *Contemporary Engineering Economics.* 2nd ed., Addison-Wesley, 1997.

Park, Chan S., and Gunter P. Sharp-Bette. *Advanced Engineering Economics.* Wiley, 1990.

Riggs, James L., Bedworth, David D., and Sabah U. Randhawa. *Engineering Economics.* 4th ed., McGraw-Hill, 1996.

Steiner, H. M. *Engineering Economic Principles.* McGraw-Hill, 1992.

Sepulveda, Jose A., Souder, William E., and Byron S. Gottfried. *Engineering Economics,* McGraw-Hill, 1st ed., 1984.

Smith, Gerald W. *Engineering Economy.* The Iowa State University Press, 1988.

Sullivan, William G., Bontadelli, James A., and Elin M. Wicks. *Engineering Economy.* 11th ed., Prentice-Hall, 2000.

Thuesen, Gerald J., and Wolter J. Fabrycky. *Engineering Economy.* 8th ed., Prentice-Hall, 1993.

White, John A., Case, Kenneth E., and David B. Pratt. *Principles of Engineering Economic Analysis.* 4th ed., Wiley, 1998.

Young, Donovan. *Modern Engineering Economy*, Wiley, 1993.

Index